ENERGY AND ECONOMIC THEORY

World Scientific Series on Environmental and Energy Economics and Policy

(ISSN: 2345-7503)

World Scientific Series on Environmental and Energy Economics and Policy – Vol. 9

ENERGY AND ECONOMIC THEORY

Ferdinand E. Banks

Uppsala University, Sweden

World Scientific

NEW JERSEY · LONDON · SINGAPORE · BEIJING · SHANGHAI · HONG KONG · TAIPEI · CHENNAI

Published by

World Scientific Publishing Co. Pte. Ltd.

5 Toh Tuck Link, Singapore 596224

USA office: 27 Warren Street, Suite 401-402, Hackensack, NJ 07601

UK office: 57 Shelton Street, Covent Garden, London WC2H 9HE

British Library Cataloguing-in-Publication Data
A catalogue record for this book is available from the British Library.

World Scientific Series on Environmental and Energy Economics and Policy — Vol. 9
ENERGY AND ECONOMIC THEORY

ISBN 978-981-4366-10-6

In-house Editor: Yvonne Tan

Typeset by Stallion Press
Email: enquiries@stallionpress.com

Printed in Singapore by B & Jo Enterprise Pte Ltd

For
Christian Otto, Claudia Soleil and Beatrice Gunilla

Preface

"Energy is the go of things," as James Clerk Maxwell pointed out many years ago, and as I repeated in the first line of my first energy economics textbook (2000). This book incorporates that notion, while surveying and summarizing some of the events and thinking in energy economics of recent years — events due, e.g., to wars and macroeconomic meltdowns, as well as thinking that changed when the oil price soared above \$100 per barrel. In addition, I provide a more thorough background for the study of topics like oil and nuclear energy than in my previous books, fill in some of the gaps in my previous textbooks, and correct the more egregious 'slips'/blunders in some of my publications.

Moreover, my intention this time is to provide a (limited) reference as well as a textbook. Frankly, I am amazed that the realities of the energy markets have not been generally absorbed in an optimal manner, despite some uncomfortable changes that have been taking place in things like the price and availability of oil, gas and electricity.

Teachers like myself must undoubtedly share some of the guilt here. The long-overdue increase in the importance of academic energy economics is at last coming about, and some of us are so gratified that we have forgotten that the subject still has some rough edges. For instance, all except the hopelessly naive of our students and colleagues must be aware that economics is not a science in the usual sense. It is like a science, which makes it a pleasure to study and teach, but there are probably many things that readers of this and other books want academic economics to tell them that they are not going to find out, no matter how hard they stare at the supply and demand curves and algebra that caused them such grief in Economics 101. The reason is provided by Friedrich Nietzsche, and it goes like this: "The future is as important for the present as the past."

Energy and Economic Theory

As bad luck would have it, correctly deciphering the future is an art that only a few lucky or talented people can master. Despite having studied the oil market for as long as I can remember, there have been periods in which I was almost as much a forecasting novice as those persons who thought that the oil price would fall below $10 a barrel in 2008, instead of escalating in perhaps the most dramatic manner in modern history. Three or four years ago some eminent journalists were prone to insist that oil price 'spikes' are no longer highly relevant for the explanation of economic downturns, and although at present we are experiencing what might be described as the mother of all macroeconomic dislocations, the role of the oil price in this catastrophe is not suitably acknowledged. Make no mistake: another extreme oil price rise could dampen an economic and financial market recovery or, when a full recovery arrives, restrain an economic upswing.

Most of this book should be accessible to anyone with a serious interest in energy economics, and if they are confronted with undesirable symbols, simply ignore them and move down a few paragraphs. *Moreover, the first three survey chapters (and also Chapter 8) should easily fit into introductory and intermediate courses in energy economics or microeconomics, and the last chapter is completely non-technical and written for everyone, regardless of their background!* I want to thank colleagues at Uppsala University for their help, particularly Tomas Guvå and Åke Qvarfort. Also, students in energy economics courses I taught in Stockholm, France and at the Asian Institute of Technology (Bangkok), and in addition the short course in environmental economics that I gave at Griffith University (Brisbane, Australia), which made it clear that I required a very different attitude when dealing with that subject than the one I displayed in my course in mathematical economics at the University of New South Wales (Sydney). I can also mention the persons supervising and contributing to those superb (and easily read) sites/forums EnergyPulse and 321 Energy, and also the brilliant online journal *Energy Politics*, edited by Jennifer Considine. Finally, I would like to thank my editor Yvonne Tan and World Scientific for publishing my energy economics and finance textbooks.

Contents

Chapter 1

The World of Energy 1: An Elementary Survey

"The production of energy is the moving force of world economic progress."

— *Vladimir Putin*

What I was primarily concerned with in my lectures in Bangkok, and in those I prepared for a short course in Spain, is an unambiguous description of oil and natural gas markets and their mechanics, nuclear energy, and some special topics that I first broached in my book, *Scarcity, Energy, and Economic Progress* (1977). Above all, I do not want to disappoint students of energy economics, because a few teachers are unaware of some of the most useful elements of the subject. As a result, even interested students could be in danger of experiencing unpleasant surprises in routine discussions with other students, future or potential employers, and genuine or bogus energy professionals. For instance, natural gas is a subject where we do not want this to happen, because for various reasons that resource may be more important than ever, particularly for countries with a large and growing natural gas deficit, but whose gas transportation and distribution network is intact. The UK seems to fit this description.

Before commencing a systematic examination of this topic, something should be made clear. According to Robert Bryce, managing editor of the *Energy Tribune Magazine*, coal dominated the energy picture in the 19th century, oil the 20th, and — in his opinion — natural gas will be the dominant "fuel" of the 21st century.

A statement like that gives the impression that gas is capable of out-shining coal and nuclear energy as, e.g., a source of base load (i.e., sustained and reliable) electricity throughout the entire 21st century, which I doubt. As pointed out by Len Gould in the forum *EnergyPulse* (www.energypulse.net), it is now possible to construct (in China) a 1,000 Megawatt (= 1,000 MW) nuclear facility in less than 5 years — from 'ground break' to grid power — and this reactor should have a 'life' of at least 60 years. For various reasons given in this book, that makes it a very attractive source of electricity.

Anne Lauvergeon, the former director of the French nuclear giant Areva, was also aware of the Chinese nuclear success, and regarded it as "worrying" — by which she meant worrying for the management and shareholders of her firm. Given that nuclear reactors do not release any carbon, and considering the likely supply of uranium and thorium, it is difficult to believe that nuclear-based power will be dominated by a fossil fuel (e.g., natural gas) whose global availability was once questioned by people like Alan Greenspan, the long-serving director of the U.S. Central Bank (i.e., the Federal Reserve System). On considering shale gas, Professor David Victor of the University of California states that "We don't know if it will be truly awesome or only theoretical in its impact."

As bad luck unfortunately has it, some of us know enough about energy issues to become unreservedly suspicious. Not only shale gas, but coal deserves some scrutiny. Coal is considered a near-toxic resource by a number of politicians and environmentalists, and so daily we hear about the strenuous efforts that will be made to replace it with renewables and/or natural gas (since natural gas has about 50% of the carbon dioxide (CO_2) emissions of coal). I have been informed about some of these goals for decades, and it is clear that in every part of the world there are politicians and civil servants who are serious about putting some sort of 'cap' on carbon emissions. Even so, I am afraid that a large fraction of these intentions might best be described as hypocrisy or bunkum. There is too much energy in coal for it to be dismissed.

For example, decision makers in China and India have not provided any proof that they are going to augment their high-minded rhetoric with tangible efforts to reduce their huge dependency on coal, and the scientific elite in Europe and North America are not making the efforts that they should make to provide us with the information we need to choose — or merely to think about — what might be called the optimal energy future, given the resources that are or will become available. There is also a great deal of what has been called "funny business" in Europe, where, as a result of the Fukushima tragedy, many politicians are "running scared" that they will be accused by voters and political rivals of supporting the wrong energy agenda. Danish and German politicians deserve mention here, because Denmark and Germany have the highest energy prices in Europe, and if they were unable to 'hook' into Swedish and Norwegian power supplies, those prices would go into orbit.

What was once a very long chapter has been divided into three (unequal) parts in this book. The reason is that readers of all backgrounds require a comprehensive, easily read, and for the most part non-technical survey of the world energy economy, and a chapter of over 150 pages would likely intimidate even advanced students and professionals. In addition, since this is a textbook as much as a reference book, questions and exercises are found at the end of these chapters, and they are easier to deal with if placed in three separate chapters rather than one chapter. Teachers should also be aware that the first three chapters are intended to form an elementary energy economics course, and to these three can be added Chapter 8, which contains no equations.

There are equations in the first three chapters; however, these are mostly located where they can be easily skipped if the reader does not feel like dealing with them right away. For example, a non-linear programming exercise has been placed in an Appendix to Chapter 1. However, in Chapter 3, there are some elementary quantitative materials that serious readers should examine until they understand them perfectly! These deal with units and terminology used in energy economics, and readers who want to impress friends and neighbors with their knowledge of this subject should absorb as much of them as they can. I have taught energy economics in many countries and universities, and in none of those did this matter of units and terminology cause my students any problems. Moreover, readers will feel better about themselves if they retain their equilibrium (or poise) when confronted with the special units and terminology of energy economics.

Like the remainder of the book, these three survey chapters have been structured to function as a supplement to my previous energy economics textbooks (2001, 2007), because many things have happened recently that were inconceivable when, e.g., the 2007 textbook was published. The main energy event was the oil price escalating to $147 a barrel (= $147/b) in 2008, which gave many observers the impression that it was going to go higher, and sooner rather than later. As pointed out in that book, and earlier, if the international macroeconomy was in an insecure state, and a spectacular oil price increase took place, it could contribute to a very bad macroeconomic scene. This should have been accepted as an indisputable fact, based on evidence from earlier oil price escalations.

What appears to have happened in 2008 is that the international macro-economy suffered a partial meltdown before the bad (oil price) news became worse. To use some terminology that became popular, this macroeconomic calamity 'destroyed' a great deal of the demand for oil (and gas), and thus restrained their prices. The New York oil price 'bottomed out' at $32/b early in 2009, and at about $33/b at London's International Petroleum Exchange. Conversely, when the troubles in Egypt began in 2011, oil prices moved above $90/b, and due to the situation in Libya some time later, the price in both the U.S. and UK moved above $100/b. Moreover, at the present time, it seems unlikely that they will ever fall below the middle eighties. Although the global economy has not fully recovered, some members of the Organization of the Petroleum Exporting Countries (OPEC) once expressed a belief that a fair price for oil is *at least* $75/b. If readers of this book do not learn anything else, they should attempt to understand that if OPEC wants $75/b for their oil, they are smart enough to obtain it if global economic conditions are not abysmal. The basic logic here was presented by Harry D. Saunders (1984) in a brilliant article.

Something else that must be considered in detail is the likelihood of a renaissance in nuclear energy. It is possible that some readers of this book may believe that nuclear energy is in serious trouble because of the events at Fukushima (Japan), but in Asia the opposite is blatantly true. Fortunately, the U.S. government is streamlining licensing procedures for new capacity, and providing many more loan guarantees. They know — although they may be afraid to shout it from the rooftops — that nuclear-based electricity is almost certainly optimal in the long run. A comprehensive introduction to the economics of nuclear energy is found in a long chapter later in this book, and while a few sections of that chapter are technical, it can and should be read by everyone desiring an insight into a high-priority topic.

Not much is said about environmental economics in this book; however, it should be emphasized that environmental economics is an issue that cannot be dismissed. Accordingly, I want to make it clear that our very grave energy problems cannot be solved without a massive input of renewables and/or 'alternative' energy resources. The key question is *what* renewables and alternatives? This may be the most important issue in all energy economics, and to get the correct answer governments and energy departments are going to need the help of people who read books such as this one, although this book is only an introduction. It is, however, an essential introduction, as is the book by David Goodstein (2004).

I am quite willing to argue that on grounds of *reliability* and *flexibility* a greater resort to nuclear energy may be necessary, particularly if macroeconomic and perhaps some climate issues are to be satisfactorily addressed. I regret to announce, however, that just now I have only a vague idea as to how much. I do know that we do not require a nuclear reactor on every street corner, but that a certain amount of nuclear energy might be essential in order to ensure an optimal performance from the very large investments in renewables and alternatives that will probably have to take place. The key thing here is to avoid forums and expensive 'talk-shops' and illogical ventures that shift attention from effective energy strategies to ineffectual economic 'flings' and posturing. Effective strategies are going to be more important than ever in a world in which the estimated population increase will be between 750,000 and a billion people every decade.

Furthermore, it might be true that things like energy, and perhaps the environment, should be given the status of public goods, or perhaps a better description is 'public-like goods', since they are just as important as parks or streetlights, or for that matter locally recruited armies that are sent to fight in strange wars thousands of miles away from their home country — strange wars that are also expensive wars, where *total* costs can reach billions of (U.S.) dollars a month. If we turn to the economic concept of *opportunity cost*, most of those billions would be more efficiently deployed if they were dedicated to financing suitable renewables and alternatives. If this happened, they would provide more service to local and other populations than soldiers and marines tramping through poppy fields and dusty streets, many years after their wars should have been declared over. This is the kind of miscalculation that turns friends into enemies, and enemies into fanatics!

In case readers have forgotten, opportunity costs are the goods and services that must be sacrificed in order to obtain other goods and services. A beautiful example was provided by U.S. President Dwight D. Eisenhower, Allied Commander-in-Chief in Western Europe during the Second World War (WWII), who stated that money unnecessarily spent on armaments reduces the standard of living of taxpayers and their families. It should also be clear that a well thought-out and successful attack on energy problems, commenced in the near rather than the distant future, could be of enormous help in solving climate enigmas. Consider, for example, the low level of greenhouse gas emissions from nuclear-intensive France and Sweden. More importantly, the 'profits' that should be realized from the optimal employment of nuclear energy could be used to help finance renewables and various alternatives. This situation may now prevail in China, because one often hears that the Chinese government favors a highly diversified energy portfolio in which there is adequate scope for renewables, alternatives, *and* a great deal of nuclear energy. Diversification intentions and realities aside, coal and nuclear energy are the key items in the present Chinese energy menu. Coal is discussed briefly in this book, but I am more concerned with imparting some positive information about nuclear energy, despite the recent tragedy at the Fukushima nuclear complex in Japan. An important nuclear physicist who apparently has a few reservations about my outlook on this subject is Michael Dittmar (2011), who works at CERN (in Geneva, Switzerland).

1. Another Energy Message for the 21st Century

> *"On every ship there is somebody who doesn't get the message."*
>
> — *U.S. Navy adage*

Gordon Gekko was the main character in the film *Wall Street*, and behind his sermons on the value of information, he did not exclude theft as a means for gaining enough of that 'commodity' to enhance his 'net worth'. But if we stick to legitimate endeavors, and the topic is energy, the value of correct information is immense, and this value is increasing every day. One reason is the uncertainty associated with the enormously costly investments required to guarantee an annual flow of energy to homes and businesses that could be valued at hundreds of billions of dollars. The European Energy Directorate has concluded that more than a trillion dollars will be required just to upgrade the European electricity network. Of course, the European Energy Directorate has made mistakes in the past, and this may also be one of them, but there is no point at the present time in believing that the future electricity network can be obtained on the cheap.

Perhaps just as importantly, genuinely good information helps us to avoid the tremendous amount of misinformation (and disinformation) in circulation. 2008 has come and gone, but in the many articles focusing on

Table 1.   Misleading or Incorrect Forecasts

Year	Prediction	Sources	U.S. output
1866	Synthetics available if oil production should end	U.S. Revenue Commission	3.5 mb/d
1885	Little or no chance for oil in California	U.S. Geological Survey	30.0 mb/d
1891	Little or no chance for oil in Kansas or Texas	U.S. Geological Survey	53.0 mb/d
1914	Total future production of no more than 5.7 billion barrels	U.S. Bureau of Mines	240 mb/d
1920	Domestic production almost at a peak	Director, U.S. Geological Survey	400 mb/d
1939	U.S. oil supplies will only last 13 years	Department of the Interior	1,300 mb/d
1949	End of U.S. oil supply almost in sight	U.S. Secretary of the Interior	1,900 mb/d

the dramatic energy and economic events taking place that year, there are some interesting and useful background materials that are highly relevant for every chapter in this book.

Let us see what some important persons and establishments have said about oil's availability. In Table 1, mb/d stands for million barrels (of oil) per day. Before reaching the last paragraph in this book, readers will have a more inclusive perception of the difficulties that can be encountered in the struggle to draw the right conclusions about the availability of many energy resources and the environment. What they will also notice while sharpening their ability to deal with these extremely important subjects, is that while most of the predictions in Table 1 are very wide of the mark, more useful appraisals are appearing of late. (Notice also that while 'm' signifies *million* in the table, for some resources 'M' is often used instead.) And let me offer an important piece of advice: do not base your energy opinions on casual remarks made in television interviews, or off-the-cuff remarks in local newspapers.

Pierre-René Bauquis (2003) has cited an *ex-ante* (or 'before the fact') estimate from 1998 about oil that was controversial from the first day it appeared, in that like some previous estimates it predicted that "global production of conventional oil will begin to decline sooner than most people think — probably within 10 years." *Ex-post* (or 'after the fact') this may turn out to be incorrect, but undeniably there has been a 'flattening' of the conventional global oil production that needs to be observed and discussed by everybody, and the same applies to OPEC. OPEC has become, in many respects, a much more rational organization where the economic interests of its members are concerned, as can be judged from the approximately one trillion dollars in export income received by that organization in 2011 and 2012. I would be surprised if they would settle for less in the future, which, everything considered, suggests only a moderate increase in the exports of OPEC oil.

Now we can move to the bottom line. What needs to be understood by readers of this book is the growing importance of energy studies of various sorts, the uncertainties about the supply of and demand for traditional energy resources, and above all the many bad things that could happen if decision-makers fail to do what is necessary to ensure an adequate and reliable supply of energy in the future. It might also be a good idea if these same readers pondered a future in which enough energy must be available by about, e.g., 2050 to allow a global population of 9 billion people — and perhaps a few more — to toil and frolic in at least a semblance of bourgeois comfort. Where population is concerned, readers might find it useful to turn immediately to Chapter 8 in this book, where there are some observations about population and associated matters, presented in a non-technical and down-to-earth manner. Here I can note that population issues are on their way into the economics mainstream, because it should be clear that this is a subject that deserves a great deal of contemplation — and just as important, contemplation by people who are interested in economics, regardless of their educational background.

In the course of his search for correct answers to use in energy controversies, the Chief Executive Officer (CEO) of the large oil company Chevron has insisted that even if renewable energy doubles or triples in the next few decades, fossil fuels (i.e., oil, natural gas, and coal) would continue to be the most important energy resources at mid-century. Mid-century is roughly 40 years from the publication date of this book, and predicting what the energy situation will be at that time is something that is not easy to do. I am not certain that the Chevron chief is more capable of doing this than I am, and I am ready to admit that I do not have much to say on that subject. I am prepared to claim, though, that nuclear energy will still be with us, and it might be much more highly valued in 2050 than it is today, although

Table 2. Percentage Consumption of
Energy Inputs

	1989 (%)	2008 (%)
Fossil fuels		
Coal	30.79	32.30
Natural Gas	29.40	32.10
Oil	23.50	14.20
Nuclear	8.00	11.50
Renewables		
Biomass	4.20	5.53
Hydro	4.00	3.53
Wind	0.03	0.71
Solar	0.08	0.13

the news from the Fukushima nuclear complex has led many observers to question whether it should be a part of their future.

Thus, it might be useful to commence our analysis by looking at some numbers instead of searching for conflicting theories and philosophies launched by persons with only a minimal background in conventional energy economics, but with considerable access to the media. Table 2 shows, in *percentages*, the situation for fossil fuels, nuclear energy, and renewables for the United States in 1989 and 2008. The figure for natural gas includes LNG (liquefied natural gas), which is natural gas that is liquefied, then loaded onto a specially designed ship, and finally turned back into gas and fed into pipelines at a very complex installation in a distant country, perhaps thousands of miles from the origin of the gas. The amount of LNG bought and sold is only about 7 percent of the total traded amount, but it is steadily increasing, as is the number of LNG 'tankers'. Unexpectedly, LNG might become a less significant item in the energy future of the U.S. because of the sudden increase in domestic reserves of gas that originated from shale deposits. In fact, talk has started to circulate that the U.S. will eventually become an exporter instead of an importer of natural gas. It is, however, much too early to accept this judgment as irrevocable. What can be accepted is that there are many very important questions that must be answered about natural gas, and large and well-financed establishments like the U.S. Department of Energy (USDOE) should do more to provide us with information that we desperately need.

These two columns of numbers each sum to 100 percent, and while we hear a great deal about the wonders of wind and solar power, the figures in Table 2 suggest that it is not wise to be unreasonably optimistic at the present time: there is still a great deal that must be accomplished before those two energy sources can achieve a sizable fraction of what their advocates claim is possible. Of course, absolutely no effort should be spared to see that everything possible is done, because it will be pointed out at various places in this book that renewables and alternatives have an important place in a healthy energy future. In the context of this table, the expression *primary energy* deserves mention. This is energy obtained from the direct burning of oil, natural gas, coal and biofuels, and also from electricity having a hydro or nuclear origin. Electricity obtained from the burning of, e.g., a fossil fuel (i.e., oil, natural gas and coal) is a secondary energy source.

In discussions about peak oil, there are still a number of researchers and concerned observers who say that it will never happen, and the leadership of several major energy companies do not seem reluctant to take this position. The peaking of oil production will be briefly examined in the next section, but although many of us are losing our interest in the 'peak issue', the insistence that peaking will not or cannot take place in a particular region, or more importantly, globally, is a wonderful example of the deliberate spreading of misinformation that was referred to previously. The directors of major oil firms know more about these matters than the rest of us, but some of them undoubtedly believe that it is in their interest and the interests of their stockholders to make claims about the oil future that are patently untrue. The same applies to OPEC. If we limit our attention to the corporate energy elite, it appears that only the director of Total (France) — Monsieur de Margarie — seems unconditionally prepared to deviate from this conduct where oil is concerned.

A similar argument may apply with respect to other resources. Panic-based though highly educated talk about the need to rush the development of renewables and alternatives has been plentiful for at least a decade, but according to the figures in Table 2, not much has been done about it in the U.S., and it should be appreciated that the same is true elsewhere, including Sweden. There is a great deal more talk about the environment and the need for radical changes in Sweden than in the U.S. or other countries, but at least up to now, engineering and economic realities have prevailed, despite a torrent of 'hype' and pseudo-science. It has been surmised that after a few reality checks, the rate of wind farm construction will decline in most of the world, although it may proceed on schedule in China. This may be because China is now the world's largest energy consumer, although at the present time, China is counting on coal (and to a certain extent, nuclear power) to produce most of the electricity needed to maintain its impressive macroeconomic growth.

As far as I can tell, in a great many universities, students of elementary economics and natural resource economics are still asked to learn the (Harold) Hotelling approach to price formation in the case of non-renewable (or exhaustible) resources. Oil has often served as an example, although many years ago Professor Hotelling's model was openly cursed by important members of OPEC because it led to expectations about the oil price that turned out to be completely false. In addition, a prominent teacher of energy economics once informed me that while Harold Hotelling's work on this topic was meaningless, it unfortunately was true that without it, he would not have had anything of an 'analytical nature' to present to his students.

That scholar was not completely correct, but in any event the main-stream Hotelling presentation is bothersome because it implies that the most important variable in the analysis of oil production is the rate of interest, r. Actually, in the case of oil (and gas), the most important consideration is the *deposit pressure*, which in a fairly straightforward manner determines the cost of extraction. As production takes place, deposit pressure decreases, and as this happens, the cost of obtaining additional units of, e.g., oil (i.e., the *marginal cost*) increases. Accordingly, we need an explicit and realistic reference to production cost in the mathematics associated with profit max-imization in a multi-period (i.e., intertemporal) setting for a non-renewable resource, as opposed to the feeble $(p_{t+1} - p_t)/p_t = r$, or $\Delta p/p = r$, which is the equation offered to thousands of students every year, where p_t is the present period price, and p_{t+1} is the price (or expected price) in the 'next' period.

An important point here though, as verified by some calculus, is that in all expositions, p is the market price of a unit of the resource (e.g., oil) *minus* the marginal cost of an additional unit of the resource, and is thus a *net* price. Also, while r is the rate of interest, it is the *real* rate of interest if prices are also real prices — i.e., adjusted for inflation. Otherwise, r is the market (or *nominal*) rate of interest. Δp, of course, is the *change* in the (net) price p, and so $\Delta p = (p_{t+1} - p_t)$, while $\Delta p/p = (p_{t+1} - p_t)/p_t$.

The above is the basic Hotelling result. Harold Hotelling was a brilliant economist, but his approach to exhaustible resources is only the beginning of a very complicated story. *A 'textbook' strategy for intertemporal profit maximization involves the equality of present values for every period in which production takes place, where a present value (PV) is a discounted future value!* For example, the present value of the amount F_t (from the beginning of period t) is $F_t/(1+r)^t$, and is 'recorded' at the beginning of the initial period, or period 1. This is the first step in a Hotelling-type explanation explaining why and how firms make adjustments that lead them toward *what they think* is profit maximization. They "think" it because generally future prices and costs are *estimated*, and so revisions of an n-period strategy might be necessary at any time, which could mean, e.g., altering the discount factor (i.e., r in this exposition), correcting estimated future prices and costs, changing the length of the planning horizon, etc.

We continue by taking a closer look at some details in a two-period model. Assume that oil can be extracted from a given stock S of reserves at a *constant* average (and thus constant marginal) cost, which is equal to $\hat{c}$ in both cases. Two periods will be taken as two years, because interest rates are generally defined for a year. Suppose that in the process of exhausting S, the present price (obtained from conventional supply-demand relationships) is p_t, while the (*expected*) price in the next period is p_{t+1}. If the decision is to extract S now, and invest (at an interest rate of r) the amount $[= (p_t - \hat{c})q_t = (p_t - \hat{c})S]$ obtained, or to wait and extract and sell S in period $t+1$, then to be *indifferent* between these two options — i.e., for one not to be better than the other — we must obviously have $S(p_t - \hat{c})(1 + r) = S(p_{t+1} - \hat{c})$. From this we get the Hotelling expression mentioned earlier, or $\Delta p/p = r$, where p is price minus marginal cost, or net price! Readers should confirm this! They can also confirm that in a two-period exercise, if one choice gives more profit than the other, then all production $(=S)$ will take place in that period.

In the above paragraph, you should focus on the specification *constant* average (and thus constant marginal) cost. (The total cost curve is thus a straight line going through the origin, *which you should immediately verify*.) In addition, the two costs are assumed to be the same in both periods. These simplifications greatly reduce the value of the results, although there are situations in which this is not important. A Norwegian economist once informed me that if the Norwegian Petroleum Directorate had been alert, the Hotelling paradigm suggested that the output of Norwegian oil should have been reduced when the price was close to $10/b, and thereby saved oil to be extracted when the price increased. Apparently, this kind of thinking carried no weight in elegant conference rooms where mainstream economic logic entered the discussion only as an afterthought. This might be one of the reasons why today some commentators believe that Norway's oil and gas future will be less favorable than its past.

Now let us extend the above discussion for the purpose of familiarizing readers with the terminology, rather than making heavy weather of the results. $p_1 - \hat{c}$ is the (net) price at the *beginning* of the first period of an n-period program, and assuming the interest (or discount) rate r and the marginal production cost $\hat{c}$ are *constant* over n periods, then for (*intertemporal*) profit maximization we must have an equality of (net) present values for prices in every period for which production takes place: production (q_i) in each of n periods must be determined from an initial stock S of a commodity in such a way as to satisfy Equation (1) for every period *in which production takes place*, given that $S = q_1 + q_2 + \cdots + q_n$, and $q_i \geq 0$ for $(i = 1, \ldots, n)$. Our *equilibrium* relationship — which readers should give some thought to now or later — is:

$$p_1 - \hat{c} = (p_2 - \hat{c})/(1 + r)$$
$$= (p_3 - \hat{c})/(1 + r)^2 = \cdots = (p_n - \hat{c})/(1 + r)^{n-1}. \qquad (1)$$

If this relationship is not obtained with a planned intertemporal production scheme, then optimal planning calls for rearranging (planned) output between periods until it does. The problem is that if we assume a constant average and marginal cost, production does not appear in this relationship, and so we must modify the above analysis in such a way that individual producers (or readers) will have something they can work with (e.g., output). Specifically, the cost relationship must be modified, and fortunately the modification moves in the direction of reality. Once again we derive an equation like (1), but with cost a visible function of output. (The derivation is in Chapter 3.) Then, if (1) does not hold, planned production for *each* period can be moved up or down, or remain the same, until an intertemporal equality takes place. The mathematical appendix to this chapter duplicates the above in a complex but orthodox fashion.

Planned production! What this means is that with the exception of p_1, prices (p) in Equation (1) are *predicted* or *ex-ante* prices, and the same might be true of costs. Of course there is no reason to believe that, *ex-post*, producers will always be able to look back, and with smiles on their faces congratulate themselves for being able to predict future prices and costs, and thus obtain optimal results where profits are concerned. It is likely that no mathematical presentation by me or you, or for that matter Albert Einstein or Isaac Newton, can provide a perfect or near-perfect calculation of the *optimal* (i.e., most profitable) future production from an oil or gas deposit that cannot be challenged by an engineer or manager who has worked with these items on a daily basis. The point is that for a firm (or an individual), the future almost always deserves serious consideration, and it might be a very bad mistake to exclude mathematics from your thinking.

Now for a slightly advanced exposition that some readers might prefer to avoid for the time being, but first, one thing should be understood: *if you have a serious problem with the mathematics, then bypass it and work to get an intuitive grasp of the issue!* I recommend an immediate visit to Chapter 8 for an intuitive presentation of many topics, and you should never hesitate to consult your teacher, or Google, and/or discuss these items with classmates and friends. Your goal where non-technical energy economics fundamentals are concerned should be *perfection*.

Once again we take a two-period model, but where average (and NOT marginal) costs are $c(q_t)$ and $c(q_{t+1})$, and so total costs (cq) ($= c(q)q$) are called $C(q_t)$ and $C(q_{t+1})$. *Marginal costs* [$= \partial C(t)/\partial q = C'(q)$] are thus $C'(q_t)$ and $C'(q_{t+1})$ for the two periods, and are greater than zero, *and like average costs are a function of the output (q) in each period*. Prices, costs and quantities are featured in an expression that, when operated on with some calculus, provides a *constrained* profit ($= V^*$) maximization. This two-period exercise begins with writing $V^* = [p_t - c(q_t)]q_t + [p_{t+1} - c(q_{t+1})]q_{t+1}/(1 + r) + \lambda[S - q_t - q_{t+1}]$, which has a vague similarity to previous analyses, although output stays in the picture as a result of its explicit inclusion in expressions for average, marginal and total costs. (Once again, as above, *if present values are unequal, we adjust production in some or all of the periods until they are equal!*) Moreover, $S = q_t + q_{t+1}$ has explicitly entered the expression to be maximized. (Note also that $c(q)$ and $C'(q)$ may not be equal!)

Noting that pq is revenue and cq is cost, elementary partial differentiation of V^* above (with respect to q_t and q_{t+1}) will give us the optimum condition for $n = 2$ periods. This is similar to Equation (1), or $[p_t - C'(q_t)] = [p_{t+1} - C'(q_{t+1})]/(1 + r)$, only now net present values are equalized, *and* we have explicit access to q_t and q_{t+1}. As for $[S - q_t - q_{t+1}]$ in the previous paragraph, S is again the total amount of, e.g., oil in the ground, which is exhausted in two periods, and the bracketed expression is called a *constraint*. As for λ, this is a *Lagrangian Multiplier*, which is defined as a change in profit (V) given *optimum* production behavior with a unit change in S (or $\partial V/\partial S$). The most important expression here is $[p_t - C'(q_t)] = [p_{t+1} - C'(q_{t+1})]/(1 + r)$. Write it down a couple of times!

Future profits are discounted (by, e.g., $(1 + r)$, $(1 + r)^2$, $(1 + r)^3 \ldots$) because they are future amounts, and (*ceteris paribus*) future money is considered less valuable than present money. In the mathematical appendix to this chapter, discounting plays a key role. [*Exercise: write the results in the previous paragraphs for 3 periods with $t = 1, 2, 3$ instead of t, $t + 1$, $t + 2$, making sure that quantities are visible in the expressions for costs, and, if possible, discuss what is meant by the equality of present values!*]

Some question might arise here about profits being considered at the beginning rather than the end of periods, which introduces an asymmetry into the discussion, but remember that these are *planned* or *estimated* profits, formulated by producers at the beginning of the first period for n periods. The goal of the analysis is to provide a theoretical framework for individual producers to consider and, if necessary, adjust present and future planned outputs over n periods. It should also be appreciated that this can be done without resorting to calculus and things like Lagrangian multipliers. I would also like to take this opportunity to inform interested parties that where economics is concerned, the study of history is as important as the study of mathematics.

2. Some Aspects of the Future Supply of Oil

The International Energy Agency (IEA) published a report in 2011 which said that a production peak has occurred in many of the largest oil fields in the world. With this easily verifiable information as a background, the chief economist of the IEA told a reporter from the (UK) *Independent* that a global peak was only about ten years away. Right or wrong, ten years was much too soon for Mr. Economist's superiors at the IEA, and so a politically acceptable 'alternative' was selected for journalists.

On the other hand, in a recent report (though very likely not the latest), the Energy Information Agency (EIA) of the United States Department of Energy (USDOE) asserted that oil demand will reach 118 mb/d in 2030. Given the considerable resources available to the IEA and EIA, the forecasts of these two establishments are always of some concern to energy professionals interested in energy scenarios, even though a few years ago both of these organizations concluded that in 2030, global oil output would be 121 mb/d. Although not publicized, the basic intention of these forecasts is to give the impression that a peaking of the global oil supply will not take place for a very long time, which will thrill certain forecasters and students of the oil market, certain bloggers, and certain eccentrics without any scientific insight into this topic. However, if the oil price can touch $147/b, as it did in 2008, and could conceivably go higher, an actual or likely peak is barely of secondary interest. The macroeconomic damage that might be caused by oil prices in the vicinity of $150/b could be devastating. *Please note though that conventional oil and various (oil-like) 'liquids' are often lumped together!*

There is something scientifically wrong with many energy forecasts: the ugly truth is that the less said about them the better, because they have a political dimension that makes them highly suspicious. Someone who has examined this issue in a brilliant article on oil and the IEA is Lionel Badal (2009). It can also be mentioned that in France, before the onset of the recent macroeconomic and financial market unpleasantness, one often saw 2015 — or even earlier — as the date when a global peaking of oil production was liable to take place. Rumor had it that this particular forecast originated in or close to the French government.

The position taken in this chapter is that the IEA and similar establishments sometimes confuse more than they clarify. Moreover, a brief glance at the past will tell us much of what we need to know about the oil future. The United States has 2.3% of the world's oil reserves, consumes about 22% of world oil production ($\approx$88 mb/d), imports more than 50% of the crude oil it uses, and its output in the 'lower 48' (states) peaked at the end of 1970 at a value of approximately 9.5 mb/d. Production then slid to 7.5 mb/d, but when the Prudhoe Bay field (in Alaska) came on line, total output (in the 50 states) went up. Unfortunately, the previous peak was never reached again, and eventually total U.S. production fell under 7.5 mb/d and (trendwise) continued to decline. Everyone should memorize this, along with the fact that the U.S. had the most impressive technology in the world when this was taking place, as well as access to the technologies of other countries, but even so could not reverse the decline until recently with the exploitation of shale oil.

Today, U.S. output of crude is just under 6 mb/d, although its oil (i.e., liquids) consumption is probably about 19–20 mb/d, where once again oil means an amalgamation of 'crude', natural gas liquids, biofuels, etc. (The U.S. has 4.5% of the world population, and consumes 22% of the world oil production. China consumes 8–9 mb/d.) The situation with oil is about the same as with the entire energy spectrum, in that large energy gains might be made possible in the U.S. (and everywhere) by the resolute application of existing technology. For instance, per-capita energy use in the European Union is only two-thirds of that in the U.S., and while it may be questionable that the U.S. should regard EU energy use as a role model, a comprehensive approach to energy saving might lead to large monetary benefits if optimal EU practices were adopted (or partially adopted). If true, these benefits could play an important role in financing the new energy system that the U.S. must absolutely acquire, and which would benefit importers of energy resources everywhere in the world.

Since the oil output of the U.S. has indisputably peaked, and given the superior technological, managerial and political competence of that country when the peaking took place, readers can feel comfortable arguing that a global peak is inevitable. If you find this dubious, examine production curves for the 300 largest oil fields in the world. Then, after noticing that output in a large majority of these fields has decreased (i.e., peaked), or is on a 'plateau' (which may be irregular), ask yourself how, in these circumstances, anyone can sincerely believe that a global peak will not take place. Note the word "sincerely". It means that there are people who know better than even energy professionals reading this book that a global peak will arrive, but have excellent reasons — of a career and financial nature — for claiming something else.

One of these people is the director of global oil and gas resources at an influential consulting firm, who has employed the picturesque word "garbage" to describe the work of peak-oil believers. If you encounter that gentleman some fine day, please inform him that the output of oil in the U.S. has peaked, and the same is true of oil in the UK and Norwegian North Sea. You can also mention that what was the second (or third) largest oil field in the world a few years ago — the Cantarell field in Mexico — is declining at a startling rate. Gregor Macdonald (2009b) has suggested that Mexico should forget about exporting oil, since a great deal of their output — which seems to be falling faster than ever — is being consumed domestically. Actually, the global output of *conventional* crude oil has definitely 'flattened', and may already have peaked, while claimed increases in output consist of increasing amounts of assorted liquids such as 'condensates' and hydrocarbons in the natural gas stream that are liquids or easily turned into liquids.

The Russian oil output is probably close to peaking. In any event, the director of one of the largest Russian firms says that his country will never produce more than 10 mb/d (and thus exports will be smaller). 10 is a nice round number (that may be a little too small), since at the present time Russian output is about 10.35 mb/d, which is the largest output in the world. According to the CEO of Total, the maximum for (conventional) world oil production will be 100 mb/d. He has also suggested in several articles in the most reliable energy publication in the world, *The Oil and Gas Journal*, that a large part of the output of the Middle East could peak before 2020. Readers should remember this when considering the strategy that is likely to be employed by governments in that region. And notice the word *strategy*. It means that those governments are not waiting for the kind of price signals discussed in the early chapters of your favorite economics textbook. They are active rather than passive when it comes to constructing production scenarios that will maximize profits.

If the above is not sufficient, consider the following. The *discovery* of what we call 'conventional oil' peaked in 1965. In 1982 the annual *consumption* of oil became larger than the annual discovery, and at the present time only 1 barrel of (conventional or near-conventional) oil is discovered for every 2.5 or 3 consumed. According to a recent British Petroleum (BP) *Statistical Review*, of 54 producing nations, only 14 still show increasing production. 30 are past peak output, while output *rates* are declining in 10 other countries.

As an aside and reminder, it should be noted that theoretically what we are dealing with here for each and every producer is intertemporal profit maximization in the presence of a constraint. The constraints are the (estimated) total amount of reserves available to be exploited, and perhaps the optimal rate of exploitation. Taking into account cost factors mentioned in this book results in a production curve which rises, reaches a peak, and either declines shortly after this peak or displays a 'plateau' that in some cases is very long, following which there is a rapid or gradual decline that in theory could be even longer. Often we see this output curve approximated with a bell-shaped construction, but according to some mathematics presented in the chapter on oil, attention should be paid to the use of a logistic function, which provides a bell-like frequency function. M. King Hubbert used this kind of function to estimate the output peak for the Lower 48 (states) of the United States. He said 1970 (or at the latest, 1971) would be the peak year, and late 1970 is what it turned out to be.

Occasionally claims have been forwarded by important observers that a global peaking of *conventional* oil has already taken place, perhaps several years ago, and the often-seen present output figures of perhaps 88–89 mb/d for global oil production includes, e.g., natural gas liquids and unconventional oils. This topic has been examined in brilliantly conceived articles by David Cohen, Marc Courtenay and Gregor Macdonald (which are noted in this chapter's bibliography). A dilemma that often surfaces when discussing oil appears to be identifying exactly what substances we are talking about, since increasing amounts of non-conventional or *near* non-conventional oils or liquids are possibly being consumed in many countries. Perhaps it would be a mistake to pay too much attention to this, however, because these 'new' items may not carry much weight when the present macroeconomic malaise is over, and the demand for oil begins to increase at or close to its 'traditional' rate. This averaged 1.76% from 1994 to 2006, with a possible high somewhat above 2.5% in 2003–2004.

Now we can turn to the main issue in this section. In a rather unique entry on his 'blog', one of the authors of the bestseller *Freakonomics* — Steven Levitt — made a few comments about his short stay in the United Arab

Emirates (UAE) state of Dubai. As most viewers of, e.g., CNN are aware, luxury has been the order of the day for many part- or full-time residents of that lucky community and its neighbors. However, in mulling over what is taking place in that part of the world, Professor Levitt failed to emphasize the key economic element behind the transformation of these 'states' into Middle Eastern versions of Monaco. That ingredient is *systematic diversification*, which means the emphasis put on the conservation rather than on the production and export of oil or gas (of which Dubai has less than other UAE states). Dubai has perhaps been encouraged in this undertaking by other Middle Eastern petroleum producers, especially those who want more refinery and petrochemical capacity in the region, which suggests welcoming an all-inclusive financial center to organize financial matters. The thing to take note of here is that both gross and net refining *margins* (i.e., profits) could be superb for Gulf refiners, and petrochemical profits should sweeten the outlook. (Gross margins are revenues from refinery products (e.g., gasoline and diesel) *minus* the cost of oil. Net margins add various other costs to the cost of oil.)

Unfortunately I know less about the goals of Dubai's government than a majority of the persons who have commented on and extended Professor Levitt's observations, but I do know that methodical OPEC behavior derived from mainstream development economics and elementary game theory appears to be superseding arbitrary or *ad hoc* responses to shifts in petroleum availability and demand. This sophistication could have a profound effect on the future export of oil, and might be the reason that several researchers have spoken of a "geopolitical peak". In other words, the decision makers in OPEC countries, or a sub-set of that grouping, have a meeting and decide that instead of slowly raising the production of their oil, as expected by many observers of the oil market, they will lower it if they find the price, or rate of exhaustion, unsatisfactory. To get some idea of what we are dealing with, it might be useful to sketch an argument that I presented to students in my course on oil and gas economics at the Asian Institute of Technology.

We can begin by considering arrangements in the key oil-exporting country, Saudi Arabia, in the early 1970s — just before nationalization — and then examine the situation after production facilities came under the complete control of Saudi officials.

The intention of foreign managers was to raise the production of crude oil (in phase with increasing demand) to a maximum of about 20 million barrels per day ($= 20\,\text{mb/d}$), and to keep it close to that level for as long as possible. Eventually, for reasons having to do with the increasing cost of production as depletion took place, output would be reduced. Note that the most important variable in this process — the thing that all readers should make every effort to understand — is cost: *ceteris paribus*, cost is a function of cumulative ($=$ past $+$ present) production. This is so because *deposit pressure* decreases as a result of cumulative production and something called 'natural depletion', and as a result it sooner or later becomes more expensive to increase output. (Natural depletion corresponds to what in capital theory is called *depreciation by evaporation*, and it will be briefly described later in this book with the help of some integral calculus, but you can think of it in terms of a machine or vehicle that gradually loses its efficiency because of 'wear-and-tear'. The Saudi firm Aramco says that natural depletion in its oil fields is on the order of 2 percent a year, as compared to what it believes to be a larger average for the rest of the world.)

Regardless of what you have or have not learned in various economics courses that you may or may not have taken, if you understand what has been discussed above, then you are likely to have an insight into how, after nationalization, the new owners of Saudi Arabian oil and gas properties decided to manage these assets. They decided to manage them the way that YOU would have managed them if YOU had been in their place, and desired to maximize national welfare! The first step consisted of implementing measures to obtain large profits for as long as possible, and in such a way that the *expected* discounted present value of these profits is maximized.

One of the things that distinguishes natural resource economics from conventional microeconomics is that extracting a commodity like oil from a well or field in the present period means that it will be unavailable in a later period (unless, trivially, it is stored). The foreign owners of Middle East oil recognized this, but they believed that when oil began showing signs of exhaustion in one locality, it would be possible to begin or expand operations elsewhere. Like many prominent economists, they took it for granted that the global output of oil in a given year could only be an insignificant fraction of the total stock in the earth's crust. But as a result of the sudden appearance of an oil price in 2008 that was almost $150/b, many prominent minds were forced to draw new conclusions. It now takes a very naïve researcher or decision maker to dismiss the misfortunes that depletion of an exhaustible resource like oil might eventually impose on consumers and producers alike, even if total remaining reserves are large.

In the case of producers, the downward pressure on oil prices that could result from the negative macroeconomic effects of previously high oil prices should be given some consideration. The profits of oil companies were lovely in the summer and early autumn of 2008, but falling demand due to stagnant incomes and increasing unemployment changed the prospects of many firms for a short time. As things now stand, however, the larger oil firms — and perhaps some of the others — are quite satisfied with the way that the oil price rebounded, as well as with predictions for its future.

As every serious researcher and/or student of oil economics should try to make clear to friends and neighbors, many of the countries and states in the Middle East possess the determination, wherewithal, and technical and managerial skills necessary to transform their oil and natural gas into the kind of (reproducible) physical assets that seemed as remote as the Milky Way to students of development economics when that subject became popular in Europe and North America. These kinds of assets were discussed at great length by perhaps the most brilliant development economist of the 20th century, Hollis Chenery, whose fondness for linear programming and input-output theory undoubtedly kept his work from being given the attention it deserved. They were also superbly reviewed by Professor A.A. Kubursi (1984) in an article that has turned out to be a preview of present-day economic development in the Middle East.

Saudi Arabia is not the only country that belongs in a discussion of this nature. Another is Qatar, which not only has oil, but together with Iran is the major natural gas producer in the Middle East. In addition, it is the largest gas-to-liquids (GTL) producer in the world. Last year, Qatar's Minister of Energy said that "Qatar is a country that has a duty to future generations," and in addition he clarified the difference between his country and a footloose energy company that could move to other locations after exhausting reserves in one of its host countries. This overt display of self-interest by a prosperous oil and gas owner has sometimes been labelled 'resource nationalism', and as Edward Morse (2005) pointed out, it could entail "much lower oil (or gas) supplies than would otherwise be available."

I do not call it resource nationalism, however; I call it *self-preservation*, which at one time was termed 'the first law of nature'.

Another observer, Neil King in the *Wall Street Journal* (December 12, 2007), also drew the logical conclusion that the Saudi "industrial drive will strain their oil export role". One certainly hopes that readers of this book do not have any doubts about the significance of the word "strain" for the price of oil.

As Mohamed Nagy Eltony points out in a recent issue of *Geopolitics of Energy* (2009), Kuwait's diversification away from oil has "largely been limited to the growth of the up-and-coming petrochemical industry". He seems to have some objections to the logic of this arrangement, although in terms of orthodox development economics it makes a lot of sense. For years, various publications have tried to convince governments and corporate executives that when gas-derived fuel and feedstocks for petrochemicals and plastics can be obtained in some Middle Eastern countries for 20 percent or less of the price paid elsewhere, then the output of those items in energy-rich countries will eventually expand much faster than elsewhere.

Regardless of what they actually do or do not do in the short run, or what directors and key politicians say or do not say, in the medium to long run many petroleum-rich exporters like Saudi Arabia will attempt to do everything possible to reduce their export of unprocessed oil. In line with orthodox development economics, "value will be added" by using much of the crude they lift as inputs in production activities for refinery products, just as many of those refinery products subsequently become inputs for petrochemicals. Moreover, with continued economic growth in oil-exporting countries, additional oil will be needed for domestic consumption and industrial activities that are directly (or indirectly) involved with petroleum.

As far as can be determined, at the present time Saudi Arabia could be consuming at least 2 million barrels of its own oil every day, and this amount can only increase as development in that country accelerates. Another interesting example is Tatarstan, which is (or was) in the Russian Federation. Their output of oil will be held constant, but much of it is destined for a new petrochemical installation. If a further comment on this arrangement is necessary, consider the following from a recent issue of the informative monthly publication *The Middle East*: "Wealth from oil and gas drives the expansion of an economy, increasing demand for energy that the (oil and gas) industry struggles to meet, while also satisfying the energy needs of much of the rest of the world. If it continues, this trend will see less oil exported and more kept back..." Simple arithmetic then suggests that the reduction in crude exports, in conjunction with the expansion in petrochemical exports, could greatly increase the wealth and influence of the countries that are at the producing end of the supply-demand chain.

Here, a paper by Professor Morris Adelman and his colleague Martin B. Zimmerman (1974) can be cited. These important energy economists perceived the writing on the wall long before many of their colleagues in academia and the business world believed that technicians and managers in oil-producing countries could construct and operate petrochemical facilities. They wrote: "... in the production of petrochemicals, most LDCs are at a severe and permanent disadvantage for lack of know-how, and the high opportunity cost of capital and feedstocks. Other countries, particularly OPEC members who do not face these obstacles, are expanding their petrochemical capacities. This too will drive prices down, lower the profitability of all plants built today, and force losses on many investors. Few can compete with lucky countries that get their feedstocks at a fraction of world prices, and initially are willing to earn low or negative rates of return."

Facing "low or negative rates of return" is not (and probably never was) an outcome that the new OPEC petrochemical producers anticipate experiencing: they not only will obtain their feedstocks at a low price, but their new plants are state-of-the-art in regard to cost and flexibility. Returning to Saudi Arabia, one of the indicators of their confidence is not just a rapid expansion in oil products (i.e., refinery outputs) and petrochemicals investment, but also other enterprises specifically designed to provide employment for an expanding population. Their plans include new 'economic' cities, power stations, smelters and facilities for processing and exporting large amounts of energy-intensive industrial products, and so on. (For instance, the Middle East has one of the fastest rates of growth in the world of electric power demand.) There is also a special industrial zone in Saudi Arabia that has received the *sobriquet* "Plastics Valley".

Before closing, it should be appreciated that one of the most celebrated economists of the 20th century, Professor Milton Friedman, predicted the downfall of OPEC and the collapse of the oil price. That was in the 1970s, shortly after those 'glamorous' days when the oil 'majors' (such as

ExxonMobil, Shell and British Petroleum) lost control of a large fraction of global oil reserves, and these properties came under the management of state-owned enterprises who were members of the OPEC 'family'.

Using incongruous arguments from Economics 101, Professor Friedman convinced many of his fans that he knew what he was talking about, but as things stand at the present time, we were extremely lucky not to confront a *sustained* oil price higher than $100/b during 2012, which might have had the effect of reintroducing or reinforcing a few macroeconomic discomforts. *Think about it: the oil price over $100/b and moving upward, and perhaps not stopping until it reaches the $200/b level predicted by people like Matthew Simmons (the late investment banker and former advisor to President Bush), and once predicted by the billionaire investor T. Boone Pickens, while Alexei Miller of Gazprom begins talking again about a price of $250/b!* Isn't this the kind of situation that starts oil importers wondering if the end of the world is approaching?

Hopefully, decision makers in oil-importing countries are not only aware of these trends and possibilities, but are fully conversant with their economic and political significance, and what they could eventually mean for the productivity of both large and small businesses, as well as lifestyles, in these nations. The recent macroeconomic meltdown came so suddenly that the part played by oil was initially unnoticed. However, while Michael Fitzsimmons and others may or may not be correct by claiming oil imports cost (or have cost) the U.S. almost a billion dollars a day, it is necessary to recognize that the increased cost of imported oil is only a part of the energy invoice. The oil price rise is at least partially responsible for an increase in the coal price (and initially natural gas), and it contributed to raising the price of energy-intensive goods and services, including agricultural products. There are also increased social costs caused by unexpected economic changes in the oil market. Some of the above issues and their extensions are examined in detail by Kevin Kane (2009).

As an example of what I am talking about, you can examine a plot of natural gas prices in the U.S. from the end of the 20th century until 2008. Naturally we perceive an instability, with occasional 'thin' spikes, but between 1999 and 2003 the fluctuations seemed to move around a price of just over $2 per million British thermal units (= $2/MBtu). In 2003–2004, however, just as OPEC's new strategy was put into practice, the gas price began a sustained climb — though still fluctuating — and on occasion spiked to $10/MBtu. This process lasted until 2008, when the price of oil crashed (although OPEC's strategy was so well thought-out that soon the oil price — though not the gas price — started to rise again). This should be remembered, because as soon as it becomes clear that the oil price could go into orbit again, the gas price might markedly increase.

A well-known academic in the United States once informed me that all will be sweet and lovely on the oil front because Saudi Arabia now has four large deposits in the initial phase of large-scale exploitation, which is a statement that is equivalent to perhaps the most bizarre fantasy put into circulation in the last few years. Instead, on one occasion, OPEC spokespersons declared that at least 35 of 150 drilling projects in their countries were delayed or cancelled, and word was circulated that projects involving 4 million barrels of oil a day (= 4 mb/d) had been delayed. (There was no reason to doubt this, because 4 mb/d was also the output that OPEC decided to 'cut back' when the oil price moved into a reverse mode late in 2008 or early in 2009.) In addition, investments that would add another 2 mb/d were supposedly cancelled. As I tell students and colleagues, "OPEC is now playing for keeps", and this arrangement deserves to be studied in detail, along with its macroeconomic implications.

Please do not make the mistake of paying too much attention to numbers that are included in the proclaimed intentions of oil producers. The bottom line here is that the oil- and gas-producing countries in the Middle East are going to produce as little oil and gas as possible, regardless of what they might or might not say. The decision makers in those countries may or may not have studied economics, but they know that more money is better than less money, and one of the ways to obtain more money is to ease up on the production of oil and gas. Just now OPEC may be exporting from 26 mb/d to 30 mb/d of oil, while (before the Libyan troubles) Middle Eastern shipments were approximately 17.5 mb/d, including those from non-OPEC members Oman and Yemen.

321 Energy is one of the most important sites on the Internet, and the oil price they give is the West Texas Intermediate (WTI) price, which reflects supply and demand conditions in the U.S. and Canadian oil markets. On Swedish television, however, we always get the Brent oil price, which is sometimes called the European price, but which — according to Elliot Gue (2011) — reflects global supply-demand conditions.

Normally the WTI price displays a premium to the Brent price, but there have been changes of late. In trading sessions in London, Brent once attained a premium over the WTI price of about \$30/b. The question arises as to why *arbitrage* is not taking place, with oil being bought at WTI prices and sold at Brent prices. As it happens, the issue here is physical and not financial arbitrage, and so it must be possible to physically deliver oil. If we are talking about North America that means *more* pipeline capacity is required, since in the U.S. there is not enough capacity to move oil between relevant market participants. This topic deserves more attention, and in addition we should have an *aggregate* oil price — e.g., a weighted average (or convex combination) — of WTI and Brent prices. The only recent thoughts on these topics that I have encountered are those of Dr. Sohbet Karbuz, Director of Hydrocarbons at the *Observatoire Mediterranéean de l'Energie (Paris)*, and a PhD student in Singapore, Wang Yan (2012).

In Gue's article he ties this shortage of (comprehensive) pipeline capacity to a reduction in local demands, but an increase in local supplies, as well as a need for additional inventories in a system where there is a lack of sufficient storage capacity. As you will find out in the chapter on oil (Chapter 4), this immediately suggests a decrease in the market price of oil. Of course, it needs to be pointed out that this suggestion follows from economic theory in a market in which the price is strictly formed by supply and demand and not by OPEC, which may no longer be the case.

I do not remember paying attention to a Brent price until a few years ago. That has changed because the oil demand has been stagnant or maybe falling in, e.g., Europe and North America, although rising in countries outside the OECD. Everyone who reads this book likely knows why: it is the soaring demand from China, and perhaps also India.

Suppose that more pipelines were constructed in North America in a situation where — for one reason or another — supply was rising while the demand for crude was falling. What would this mean for other transport assets? Given the outlook for oil supply and demand, more oil tankers would almost certainly be constructed. There are huge capital costs involved in constructing these ships — which usually have an effective 'life' of 20–25 years — but if people like Aristotle Onassis (who believed that he had a special advantage where commerce was concerned) were still in the tanker business, this kind of investment would almost certainly take place at the present time. Moreover, Onassis was a genius in deciding when and what tankers to 'lay up' (i.e., withdraw from his fleet), and for how long, and also what size tanker to sell and what size to buy.

This might be the place to say that, in general, I believe that readers in many countries have a right to be optimistic where the macroeconomic future is concerned, but perhaps not if there is a frequent return to oil import bills of the magnitude alluded to in various parts of this textbook. According to Ronald Cooke (2009), an oil shortage is inevitable, in which case the plain truth is that a weakening of the global macroeconomy is unavoidable. Let us hope that there are no ladies and gentlemen in responsible government positions in oil-importing countries who believe otherwise.

There will be mathematics in this book that many readers might choose to skip, though not Equation (2) below, or the explanation that goes with it. That equation plays an important role in this book! Here you should be aware that Nobel Laureates like Albert Einstein, and also Enrico Fermi (sometimes called 'The Pope' because he was never wrong), often argued that the use of mathematics should never exceed the essential. In my finance classes, Equation (2) was accorded the same distinction as Einstein's famous $E = mc^2$, where E is energy, m is mass, and c is the speed of light (which travels at a finite speed). Published by Albert Einstein in 1905, and initially ignored by all except the cognoscenti, Hollywood helped to explain the significance of this equation to the broad (movie-going) masses by having Paul Newman — in the role of General Leslie Groves (of the Manhattan Project) — state that if a nuclear bomb the size of a bath sponge was constructed and detonated, it could flatten a large part of Chicago. Equation (2) might inform someone borrowing money to purchase that part of Chicago after it had been rebuilt, how much (or P) they would have to pay the lender of the money at the end of, e.g., each year, for T years, given a rate of interest of r.

P is commonly referred to as the amortization cost (or charge) for a loan, and it takes into consideration the relevant interest charges. To begin, take P_0 as the asset price (where an asset is anything that has value). Put another way, P_0 can be considered the *investment cost* of an asset (such as a machine or a structure). Next we have P, which is one of the (equal) payments in a stream of payments over the period T, where this stream, e.g., repays the person or institution that 'lends' the P_0. P can also be thought of as the *amortized* capital cost per period, and is equal to the sum of the interest cost and the (*non-amortized*) capital cost for that period. This will be explained in full in Chapter 6. Finally, we have the discount rate r, which for pedagogical purposes is often taken as the market rate of interest. The equation is:

$$P = \left[\frac{r(1+r)^T}{(1+r)^T - 1} \right] P_0. \tag{2}$$

To see how this is used, we can construct a simple numerical example. If you buy an asset for $1,000 (the *investment* cost), and pay for it in two years, and the rate of interest is 10 percent ($= 10\% = 0.10$) per year, then at the end of the first and second years you pay the lender $P = \$576$, which can be referred to as the (*levelized*) *amortization* cost. Naturally you could have paid the $1,000 when you bought the asset, or with $r = 10\%$ you might be able to arrange to pay for the asset at the end of two years, in which case you would pay $P_0(1+r)^T = 1,000(1+0.1)^2 = \$1,210$. The reader can also think about the following. If the $1,000 were not borrowed, but paid for by the lender reaching into a wallet or purse and paying for the asset in cash, the calculation above is still valid. Now the cost per period (P) can be considered an opportunity cost: it is the cost of buying (and paying for) the asset instead of buying something else. A simple derivation of Equation (2) will be provided later in Chapter 3, and a more thorough — though still simple — explanation will be provided in the chapter on nuclear energy (Chapter 6). Readers who intend to buy property on Park Avenue in New York or Beverly Hills in L.A. should attempt to understand these matters, and they should consult Google to get the subtle distinction between investment and capital costs.

3. Deeper Thoughts than Usual about Nuclear Energy

> *"I'm a social scientist, Michael. That means I can't explain elec-tricity, or anything like that, but if you want to know about people, I'm your man."*
>
> — *J.B. Handelsman in Cartoonbank.com*
> (The New Yorker Collection, *1986*)

It might be asked why, regardless of whether you are a social scientist or a simple consumer of electricity, so many people seem willing to risk the economic futures of themselves and their families by falling in love with inferior economic and technological options. That question deserves an answer: it is because they have received so many untruths and misunderstandings about energy matters. In the long run, however, these untruths and misunderstandings will be discarded, which allows me to resort to some logic: *If voters continue to prefer more to less, then in the long run, nuclear energy will be in the picture!*

Perhaps the most straightforward reasoning in favor of nuclear-based electricity is in the non-technical article by Rhodes and Beller (2000). They say that "Because diversity and redundancy are important for safety and security, renewable energy sources ought to retain a place in the energy economy of the century to come." The meaning of this quotation is clear, especially if you add that we probably will never possess what is known in intermediate economic theory as the optimal amount of nuclear power, because the "energy economy" is unfortunately a mystery to most consumers, as well as to the people they vote for. Next, Rhodes and Beller (2000) explicitly state that "nuclear power should be central. . . . Nuclear power is environmentally safe, practical and affordable. It is not the problem — it is one of the solutions."

Central or not, this role for nuclear power is on its way to being realized in a number of countries. At the present time there are more than 450 nuclear reactors in operation in 30 countries, and they supply about 15 percent of the world's generating capacity (in, e.g., Megawatts). Their combined capacity is somewhere in the vicinity of 380,000 Megawatts (= 380 Gigawatts), where *Mega* indicates millions, and *Giga* billions. According to the energy commentator Richard (Rick) Mills, reactor fuel requirements are very high, but can be easily satisfied. Note the term *capacity* above. That is not the entire story: in terms of *energy* (in Megawatt-hours), the supply is probably more than 20 percent.

The U.S. has 104 reactors in operation, consuming about 51 million pounds of uranium per year. Although current U.S. production is only 4 million pounds per year, little is heard of this matter. This is because there is plenty of nuclear fuel in the world, thorium as well as uranium, and in addition it will eventually be possible to construct new types of reactors whose fuel inputs per unit of electricity produced are appreciably smaller. This is an extreme form of 'upgrading', which takes place continuously, and over the last few decades it has dramatically raised the *capacity factors* of nuclear equipment, or the average amount of time (in percent) that rated capacity is available over some given period. 90 percent is probably a good estimate.

China has 12 reactors in operation (November 2010), 24 under construction, and 100 in the planning stage. According to Anne Lauvergeon, the Chinese may be able to construct a 'Third Generation' (= Gen 3) reactor in 60 percent of the time required by her firm Areva to construct its new plant in Normandy, which she describes as "worrying". I am not worried, however, because it is quite unthinkable that other nuclear engineers will be unable to match the achievements of Chinese nuclear engineers.

What is more than worrying in macroeconomic terms is the intention of the Chinese government to possess the world's most technologically advanced reactor inventory, to be self-sufficient in reactor design and construction, and to optimize all other aspects of the nuclear fuel cycle. Put more directly, this is part of an enormous boosting of the competitiveness of the Chinese economy, which among other things means developing that economy in such a way that it cannot be challenged by countries with a 'less sophisticated' energy configuration — quantitatively or qualitatively.

For instance, the Chinese CPR-1000 (MW) reactor supposedly has a design life of 60 years, standard construction time will eventually be about 52 months, and the unit cost is supposed to be close to $1,500 per kilowatt (= $1,500/kw). Exports of this equipment were supposed to begin in 2013. The numbers presented above are goals or intentions, and if they are achieved — as may happen — competitors are going to find out later in this century that the Chinese 'workshop of the world' will be open for business around the clock, 365 days a year (with the possible exception of the Chinese New Year). According to Colette Lewiner, "when you speak about the nuclear renaissance, you are really speaking about Asia." And not just China; India plans to raise nuclear capacity from about 4,000 megawatts (= 4,000 Mw) today to perhaps 30,000 Mw by 2020.

Many readers of this and other books will not welcome facts of this nature. The construction of the Swedish nuclear sector and its later development was one of the most impressive engineering exploits of the 20th century; however, eventually a glib argument began to circulate that nuclear energy was just a "parenthesis" in world energy history, a "twilight" sector, and a recent Swedish prime minister called nuclear energy "obsolete". Just for the record, the Swedish nuclear sector — initially comprising 12 reactors, and supplying almost half of the Swedish electric power — was constructed in only 13 years, and once constructed it helped provide a dramatic boost in the productivity and profitability of both small businesses and large industrial enterprises, as well as raising the average aggregate income of persons living in Sweden. In the period before electric deregulation gained momentum, the *cost* of electricity generated in Swedish nuclear facilities was among the lowest in the world — and occasionally, the lowest. In addition, the Swedish electricity price was also relatively low, which was particularly true for the price that Swedish manufacturing firms paid for this indispensible input.

In the most nuclear-intensive country in the world — France — the intention from the initial investment in nuclear energy was to create a nuclear sector that would provide some of the lowest priced electricity in the world, and to use that electricity to make it possible for the country to optimize its macroeconomic performance. Unlike the situation in Sweden, the French decision makers initially made plans to stay at the forefront of nuclear development, and in addition, to continue to provide both the manufacturing and household sectors with reliable and comparatively inexpensive electricity. Later they also expressed the intention to maintain the relatively low level of carbon emissions that characterize nuclear-intensive France and Sweden. I have been informed, however, that some members of the French energy 'elite' are sympathetic to a changed outlook that has developed in Scandinavia during the last ten years, where the welfare of households and many firms has been endangered by an oddball digression into electric deregulation. See, for example, the short article by Wallace (2003), and also the discussion of electric deregulation (or restructuring) later in this book.

Please note the term "reliable", because it is often absent from the vocabularies of persons who would like to see further nuclear investment abandoned. Perhaps the simplest measure of reliability is the *capacity factor*. This compares the number of hours in a period that a facility operates at full capacity with the total number of hours in the period. The period is usually one year, and even including any 'down time' that is scheduled during a year for maintenance and refuelling, nuclear installations tend to have very high capacity factors. By way of contrast, in Germany, the capacity factor of wind turbines averages about 0.22 (= 22 percent), and there are places in the world where it is much lower.

By "they" and "decision makers" it is meant the French government. As Stephen Thomas, professor of energy studies at the University of Greenwich (UK), pointed out, "The two main French entities in nuclear power — Areva and EDF — originally were, and remain today, largely branches of the French government. They are directed as a matter of state policy and have benefited from extremely favourable government financing and credit assurances." I think it possible to add that few, if any, nuclear facilities have been constructed in a strictly 'free market' context, by which I mean without some measure of government support. Intelligent governments should gladly provide that support because it makes good economic sense. They also provide support for less profitable energy ventures, because where energy is concerned, the voters are often naive.

Accepting this reality reinforces my belief that nuclear power and — to a certain extent — many renewables and alternatives should unhesitatingly be treated as public or 'near-public' goods, available for the general public regardless of the opinions and prejudices of Wall Street, or the City of London or other financial havens, or various lobbyists and propagandists. It also needs to be emphasized that nuclear cost issues should be examined in greater detail by governments, and for a wider audience than is the case today, in order for voters and politicians to obtain an accurate picture of where the emphasis in electricity generation should be placed. For France, the basic comparison was between nuclear and coal, and given the various costs associated with importing and using coal, it was easy to conclude that nuclear was preferable — even if it were possible to ignore the environmental disadvantages associated with coal.

It might be possible to argue that this is not true for several other large energy-intensive countries, but the problem is more complicated than many observers think. For instance, there is this matter of the present and future supply of reactor fuel (= uranium + thorium), length of the reactor life, the lack of carbon emissions, the possibility of a radical improvement in reactor technology, and — something that should never be forgotten — expectations of a decline in the time required to construct reactors (which, as alluded to in the chapter on nuclear economics, is important for both the *investment* and the *capital cost*).

Before perusing these, I want to inform readers that despite propaganda about the U.S. overtaking Germany as the world's largest generator of wind energy, at the time this book is published, wind energy may not supply more than 1%–1.5% of U.S. energy requirements, and certain shortcomings of wind energy are raising nagging doubts about wind's future in many parts of Germany. Something causing considerably more than "nagging doubts" in Sweden is the increase in the *price* of electricity, which is the result of exporting *low-cost* (nuclear- and hydro-based) Swedish power to surrounding countries. According to a recent poll in Sweden, the price of electricity has surpassed the threat of unemployment as the main source of anxiety to Swedish adults.

One of the problems with carrying on a discussion of this topic has to do with the many estimates of the cost of a kilowatt of capacity of nuclear energy. There are too many. If we exclude China, they range from $1,500/kw, according to one director of a generating firm, to $9,000/kw by a gentleman who is convinced that he has most of the answers on the economics of nuclear energy, while he regards persons with different opinions as 'shills' of the nuclear industry, and their humble efforts braggadocio. Taking everything into consideration, the recommendation here is that $1,500–$2,500/kw should be used for the time being. It should also be noted that this might be labelled an 'overnight' cost, which refers to the capital cost of nuclear equipment. Readers are referred to Chapter 6 for further explanation, and also to Google. While you are in Google, you can also find out something about the *availability factor*.

To do some rudimentary thinking about this matter, consider the 3rd Generation (or 'Gen 3') reactor is being constructed (or perhaps has already been constructed) in Finland that has a capacity of 1,600 megawatts (= 1,600 Mw), which makes it the largest reactor in the world in terms of capacity. Initially the intention was to construct it in 5 years, and an early estimate of its investment cost was 5 billion dollars before it could be attached to the 'grid'. Now it seems that it will take 8+ years to construct this reactor, and it has been claimed that before grid power is attained, the final cost will be at least 8 billion dollars. (Roughly, a *grid* is a collection of electric power lines.)

This is nothing to be especially concerned about. There is a multinational workforce of several thousand construction workers and technicians working on the 'Olkiluoto 3' reactor, because in Finland (as elsewhere) years of anti-nuclear disinformation has resulted in a severe shortage of local persons qualified to work in the nuclear sector. This is a main cause of the large cost 'overrun' on that reactor. However, in Finland, the high quality of the educational system will eventually correct this absurdity, and perhaps sooner rather than later. In addition, the Finns were not so depressed by the project's delay and cost that they rejected nuclear energy. On the contrary, two more reactors are being considered, since like the Chinese, Finns understand the economic advantages that nuclear power could provide later in the century, especially if there should be a broad scarcity of fossil fuels.

A useful way to approach this issue is to consider the time from construction start to commercial operation of nuclear power plants in six important industrial countries. The figures that will be given below originate in the database of the International Atomic Energy Agency (IAEA), and are quoted in an important article by Roques, Nuttall, Newbery, de Neufville, and Connors (2006). I have questioned Fabien Roques — who wrote the chapter on nuclear energy in the latest IEA survey — and he told me that he finds them realistic.

They are quoted here employing the scheme [Country (Minimum Time, Maximum Time, Average Time)], where the times are of course construction times, and these are measured in years: China (4.5, 6.3, 5.1); France (4.9, 16.3, 7.1); Japan (3.3, 8.1, 4.7); Russia (2.1, 20.3, 6.8); UK (4.9, 23.5, 10.8); U.S. (3.4, 23.4, 9.2). In examining these, it should be clear that the average times must somehow be weighted in terms of capacity (i.e., power) or energy. For readers familiar with contemporary discussions of nuclear energy, the time required to construct nuclear reactors is a kind of proxy for the (*ceteris paribus*) cost of a reactor. In addition, it is not surprising to see that average construction times are much faster in China and Japan than elsewhere. When the governments of these two countries decide to ensure adequate energy for economic growth, they mean business. Let me also assure readers that Japan will be the *last* country to abandon nuclear energy, regardless of what you think or they say.

Worldwide, since 1991, the figures given by Roques *et al.* are (4.0, 8.0, 5.2). With an average construction time of 5.2 years, it might therefore be possible to argue that taking 8 or 9 years to construct a nuclear facility is an aberration, a first-of-a-kind cost, and it will not be too long before the average nuclear facility will be constructed in or slightly less than 5 years. The claim that I would like to make here though, and readers should remember, is that once the nuclear renaissance gets up steam, and it is possible to produce (and perhaps largely assemble) the components of nuclear plants in a manner similar to the way that large ships were constructed in the United States during the Second World War, rather than 'piecemeal' at the construction site (which was the case in Finland and perhaps elsewhere today), it should be possible to go from 'ground break' to 'grid power' for the average plant in well under 5 years.

Something that at least a few readers of this book might be interested in knowing, is that the (naval) Battle of Midway took place in 1942, and in the approximately three years following that major clash until the end of the war, the United States constructed 17 fleet (i.e., large) aircraft carriers, 10 medium carriers, and 86 escort carriers. In addition, crews and pilots were trained to efficiently and successfully utilize these assets, and hundreds of other warships were produced. Note that the building of ships and the training of crews began from what amounted to 'rock bottom', virtually 'zero', but even so, by the end of the war the U.S. Navy was as large as the sum of all the other navies in the world! Before the U.S. entered the war, nobody in their right mind would have claimed that the 'miracles' of modern technology and management skill that became commonplace in the U.S. during the war were possible.

It also needs to be stressed that the U.S. industrial capacity and education of the workforce in 1941 was far less robust than is the case today. A Japanese engineer with whom I worked at Camp Majestic (Japan), once said that the radio broadcast in which President Franklin D. Roosevelt called for tens of thousands of aircraft to be constructed was transmitted verbatim to himself and his colleagues, because it was considered an example of American pretentiousness. He then noted that late in the war, an American plane was brought down more or less intact, and examined by himself and his colleagues. As a result they understood that the war was lost. Achievements of this kind should make it clear to rational observers that eventually — though perhaps not this or next year — the construction of third generation reactors will become routine. When that takes place, another victory will be registered for modern technology, and the development of Gen 4 reactors might be speeded up. I was recently told that Gen 4 equipment will *never* be available, which is a claim that I dismissed, but I admit that its presence may lead to some complex economic and social issues.

Another interesting illustration was the U.S. armored force, which technologically is regarded by many students of the Second World War as qualitatively sub-optimal. The main U.S. battle tank, the Sherman, was produced in the thousands, even after it was discovered that it was an inferior piece of equipment. What is still not adequately understood is that it would have been possible to design and produce in large numbers a technologically superior tank. This might be classified as a gigantic failure to exploit existing technology, and the same kind of flaw is evident at the present time in the inability to greatly reduce the time required to construct nuclear plants. There is something strange here, because while the previous Energy Secretary of the U.S. is munificent in his praise of Japanese nuclear efforts, he is low-key when it comes to the future of nuclear energy in his own country, pronouncing himself "agnostic" on the subject.

According to Donald E. Carr in his brilliant book (1976), the Japanese were able to construct a nuclear plant in four years in the 1970s, which leads some of us to believe that they will not require more time when their decision makers and voters eventually comprehend what awaits their standard of living if they do not correctly ascertain the energy situation they are facing. It was also made clear to me many years ago that many Japanese decision makers are in no hurry to increase the size of the present nuclear inventory. What they want instead is to develop and introduce technologically superior breeder reactors, which would enable a much greater utilization of the energy in a reactor's fuel. Eventually this wish will come true, but it is not possible to declare whether it will be a happy occasion. It goes without saying that a more 'plutonium-intensive' economy presents certain security challenges that not every country may be capable of solving in an economically or socially acceptable manner.

Moreover, in thinking about the economic aspects of breeder introduction, it might be useful to know that world energy demand is forecast to increase by a very large amount by 2030. Although it is not certain, unless breeders are introduced, the price of reactor fuel (uranium and thorium) could reach insupportable heights.

It has also been argued that if at the present time it takes 5 years or more to construct a nuclear plant, then coal is a more economical resource for electricity generation than nuclear power. Many environmentalists who comment on this contention seem to prefer nuclear to coal, and since in the not too distant future it could take less than 5 years to construct a nuclear plant, any economical advantages possessed by coal over nuclear could disappear. But this does not mean that the coal industry will disappear, or that the use of coal will decline, especially in the short run, and probably not in the long run. When speaking of the global use of coal, it should be appreciated that some of the largest buyers of coal do not have the slightest intention of refraining from or reducing their consumption, nor is it likely — or perhaps possible — for them to 'clean' large quantities of that very dirty resource. The explanation here is simple: there is too much energy in the huge coal reserves scattered around the globe for it to be rejected.

If we take into consideration the environmental and health damages that might be caused by excessive and/or careless coal use — many of which are listed in the article by Rhodes and Beller (2000) — then the cost of coal might be much larger than that indicated by its market price. Various statistical estimates have been made of the amount (in dollars) that should be added to the market price of coal in order to obtain its 'social cost', but this research is excessively theoretical and is still in its preliminary stages.

Some observers believe that when the next (or 4th) generation of nuclear reactors appears, it probably will not make a great deal of difference if it does take 5 or 6 years to construct a nuclear facility. In theory, 4th Generation equipment is greatly superior to previous models; on the other hand, several nuclear physicists have suggested that it will be 20 years before a 4th Generation reactor makes an appearance. What those ladies and gentlemen often furtively mean is that they hope that this equipment will never be produced, because given things like population growth, they also suspect that if large numbers of these reactors are put into operation, their efficiency could raise the aggregate global monetary income by such an extent that there would be a very strong demand for other resources — for instance, non-fuel minerals and metals (that might be on the scarcity list), agricultural products, non-polluted air and water, etc. When (or if) this happens, we are very definitely in unknown territory politically, economically and socially, and might end up feeling that it is somewhere we do not want to be.

Readers will have to deal with this and similar issues, since though they are extremely important, they are much too complicated for a modest teacher of economics like myself! Readers will also have to deal with the controversies associated with nuclear waste, because it is difficult to understand why something that is hardly discussed in France is considered such a menace to public safety in many other countries. The French keep a large part of their nuclear waste stored where they have immediate access to it in case political or economic developments in France or somewhere else in the world make it essential to exploit the large amount of energy remaining in this '*dechet*', while the intention is to keep highly toxic waste stored deep underground, guarded by police and/or military personnel who have the qualifications and frame of mind for this important work, and where it also can be recovered in a very short time if — or when — new technology makes improved processing feasible, or for that matter this waste can be used in new types of reactors.

It also needs to be stressed that one of the reasons why American industry was able to bring about miracles during WWII was because, for the most part, pessimism and failure were not encouraged, which is not the case at the present time with both nuclear energy and the U.S. macroeconomy. Instead, in the U.S. and elsewhere, there are many beliefs about the energy future that do not make any engineering or economic sense at all. Where nuclear reactors are concerned, it should be clear that it will be much easier to produce new types of equipment and/or improve existing types than it was to design and produce the first reactors (in the 1950s), and there should also be significant progress in improving equipment and/or activities in the nuclear fuel cycle, including exploiting new sources of fuel. *Actual or likely technological progress is the main source of the superior flexibility of nuclear equipment!*

In Sweden the government once found it politically expedient to legally forbid research on nuclear energy, which was a decision that was regarded as inexplicable by the few persons who bothered to give it any serious thought. In fact, as compared to foreign physicists, many persons living in Sweden did not find out about it until long after this strange prohibition was introduced.

Most Swedes chose to ignore that eccentric decision because, like many things in life, perhaps subconsciously, it was understood that this was an irrational, spur-of-the-moment departure that would eventually be abandoned (which naturally happened). A country with hundreds of thousands of highly educated and/or intelligent persons in its population is unlikely to support a ban on research that is crucial for raising or maintaining their standard of living. As it happens, the only reason that can be given for this foolishness coming about in Sweden is the political process temporarily going off the rails, which is something that could happen anywhere, at any time.

In Sweden, as in some other countries, no political party can openly afford to offend anti-nuclear elements by aggressively disputing unsound adventures such as a ban on nuclear research, despite recognizing that behind this ban there might be a small group of unstable careerists and anti-nuclear extremists. What needs to be said here is that if there had been a referendum on the continuation of nuclear research in Sweden, then it is unlikely that this research would have been proscribed, although it is impossible to be absolutely certain. Regardless of how they personally felt in this matter, many media celebrities would have been only too happy to exploit the exposure they could have received by taking up the anti-nuclear cause.

But politicians and media celebrities are not the only villains here. In North America and Europe, only a small number of scientists, engineers and academics are willing to initiate or take part in a meaningful dialogue on energy with either the decision-makers or the general public. Furthermore, their tentative efforts are often amateurish or half-hearted. As a result, it is almost impossible to bring about a systematic attack on many of the major energy, environmental, social and macroeconomic problems of our time. Similarly, it appears that many Swedish civil servants are virtually terrified of displaying a pro-nuclear stance. Perhaps they know something that the rest of us do not, because it may still be true that sub-standard nuclear facilities exist fairly close to Sweden, and it cannot be excluded that in one of these installations somebody might press a button, or pull a lever, which results in a mishap that sours the Swedish voters on nuclear energy forever.

Two things remain in this elementary introduction: the location of uranium reserves, and a comment on the nuclear cost picture. The largest uranium reserves are in Australia (25%), Kazakhstan (17%), and Canada (9%). Canada is the largest exporter. 7% of uranium reserves are in both the U.S. and South Africa, 6% in Namibia and Brazil, 5% in Niger, and 4% in Russia. China has only 1%, and so belongs close to the bottom of the less-than-4% grouping, but it should be appreciated that thorium 232 can be converted (in an appropriate reactor) into a fissionable material, and may be at least as abundant in nature as uranium. Some researchers claim that it is much more abundant.

After the first oil price shock (in 1973), oil quickly lost ground in the electricity-generating sector in many countries. Thus, it was natural that nuclear energy and coal entered into competition to become the chief fuel selected for generating the electricity *base load* — which is the part of the load that is on the line for all, or a major portion, of a generating cycle (e.g., of a 24-hour day). The reason that coal and nuclear power loomed large on the generating scene can be inferred from the cost situation that existed until fairly recently, when the decline in natural gas prices (which could unexpectedly come to an end at any time), and especially a very large increase in the efficiency of gas-fuelled equipment, greatly increased the relative competitiveness of that energy medium. If nuclear and gas display the highest fixed and variable costs, respectively (see Table 3), we might have the following ranking (which will be explained at greater length later in this book), followed by an elaboration of the important concepts *base load* and *peak load*. These topics are also examined at some length in my previous energy economics textbooks.

The interpretation here is as follows. With $F_1 > F_2 > F_3$ and $V_3 > V_2 > V_1$, nuclear and coal facilities are relatively costly to build and equip and, as a result, it does not make economic sense to have them standing idle a large part of the time. This makes them prime candidates for carrying the *base load* — the load that is on the line all or almost all of the time. On the other hand, with comparatively inexpensive gas turbines that are easily

Table 3. Relative Costs for Nuclear, Coal and Gas (2005)

Fixed cost	Variable cost
Nuclear (F_1)	Gas (V_3)
Coal (F_2)	Coal (V_2)
Gas (F_3)	Nuclear (V_1)

switched on and off, but whose fuel costs have been relatively high, the ideal role was generating *peak loads*, which are loads that are only on the line a small percent of the time. Gas has suddenly become much less expensive in some places, and so this ranking might have to be altered. Hydro is also a relatively inexpensive source of electric power. Its main advantage is that it is optimal for carrying both the base and the peak load: its variable cost tends to be much lower than that of, e.g., nuclear, and it can be switched on (and off) almost as fast as gas. Norway generates most of its electricity with hydro, and electricity output *costs* in that country might often be the lowest in the world. ("Often", because this cost depends on rainfall and the amount of water stored.)

The combination of hydro and nuclear is why electricity costs — *costs* and not *prices* — in Sweden were among the lowest in the world. As it turned out though, Swedes were convinced by various experts that they had no reason to be satisfied with that arrangement, and so they accepted the deregulation of electricity. This allowed electricity producers in Sweden to adopt the same behavior as those people called "out-of-state criminals" by the governor of California when deregulation took place in that state. The maintenance of electricity-generating equipment was carried out at the least propitious time of the year, and as a result, electricity prices in the winter skyrocketed.

It is easy to find observers who say that a nuclear renaissance has not and should not begin, but the truth of the matter is that such a renaissance has already started. In the U.S., the Tennessee Valley Authority says that it is going to focus on providing clean base-load energy, and they are satisfied that nuclear should be a part of this program because it has a higher efficiency than any other energy source, and over the long run, a lower cost. One widely circulated prediction is that electricity demand will rise by 30 percent over the coming 25 years in the U.S., and it is clear that the environmental movement must not only recognize the unfolding of a new nuclear-based paradigm, but it would be best for all of us if they also worked with it for the purpose of obtaining the optimal energy portfolio, which includes many of their favorites as well as nuclear energy.

Robert Bryce (2010), in a short and easily read article, has gone beyond the usual terminology in order to define "power density" as the key imperative when comparing the performance of various energy sources. Power density refers to the energy flow that can be harnessed from a given unit of volume, area or mass. Put simply, it reduces to how much power can be associated with a certain amount of real estate. Nuclear energy comes out best here, while corn-based ethanol is worst. On this topic, former U.S. President Bill Clinton said that using too much corn for ethanol fuel could lead to higher food prices and riots in poor countries. It will almost certainly be possible, though, to develop a wide range of biofuels that do not impact on food prices.

As indicated in a recent EU 'poll' dealing with the next 20 years, wind and solar are considered the most desirable (or perhaps 'acceptable') energy sources of the future, while nuclear is at the bottom of the list. *Ceteris paribus* this is extremely discouraging to persons like myself, until we note that this excellent poll comprised only 26,000 citizens chosen at random throughout the European Union, and it may have happened that many of the polled thought that nuclear meant nuclear bombs, while solar had to do with the most desirable location of future vacations. "No, Ingrid," as they once said in the New York financial districts, "it's not just *mostly* about the money, it's *only* about the money!" — the money that can be made convincing many voters to buy and/or tolerate the presence of expensive and sub-optimal assets in their energy future because of their constitutional inability to separate dreams from reality.

4. Two (Nuclear) Hearts in Three-Quarter Time: Sweden and Italy

This is a continuation of the previous section, and despite the title it concerns more than Sweden and Italy; but if readers feel that they have read enough about nuclear energy, then they should continue to the next section, where a preliminary discussion of natural gas takes place. It can be mentioned, however, as demonstrated in a long and important discourse from the University of Sydney (Australia) by Bilek, Dey, Hardy, and Lenzen (2006), that the nuclear issue requires a great deal of patience in order for non-specialists to obtain the necessary fluency. As far as I can tell, many persons who could easily understand the logic and terminology of nuclear deployment and policy, are discouraged because they believe that this understanding requires a previous exposure to physics.

And fluency is exactly what is needed, because stumbling through a formal presentation (or an informal conversation) dealing with a controversial subject like nuclear energy is what it takes to bring frowns to influential faces, and to start influential heads shaking! Nuclear engineering is well-taught in a number of universities, but even most energy economists tend to be vague on the most important economic aspects of nuclear energy. Furthermore, as compared to the situation in courses on nuclear engineering, lectures on energy economics that are made by persons without an authoritative tone in their voices can degenerate into unproductive arguments.

In 1987 — following the Chernobyl reactor disaster — Italy voted to abandon nuclear energy, and a similar gesture was made by the good people of Sweden around the same time. According to the (UK) *Financial Times* (February 24, 2009), Italy's "relatively advanced nuclear capacity was (immediately) mothballed or dismantled." There were three nuclear power plants then in operation in Italy, and one of the reasons things moved so rapidly in that country was the belief by consumers — in their role as voters — that it was best if Italy made sure that they would not be exposed to a Chernobyl-type incident. In addition, Italy had access to a small amount of domestic oil, plus some hydropower, and it was clear that with a little effort (and luck), Italy might become the European leader in geothermal power. More importantly, there is a tremendous amount of natural gas in North Africa, and the construction of pipelines at the bottom of the Mediterranean Sea that terminate in Italy is a fairly simple engineering undertaking. Finally, it appeared that Italy's energy position could be improved by the large-scale exploitation of windpower. Italy has — or had — the sixth largest inventory (in megawatts) of wind turbines. Thus, it was comparatively simple for anti-nuclear enthusiasts to convince Italian voters that nuclear energy was a lost cause.

By way of contrast, Sweden was not in position to move so fast, although many Swedes thought that they were because the anti-nuclear movements in Sweden and neighboring Germany were the most aggressive in Europe. It was therefore extremely fortunate that Sweden specified 2010 as the date for concluding its nuclear engagement, because in the decade or so until electric deregulation was introduced, Swedish nuclear energy — together with hydro — occasionally provided Sweden with perhaps the lowest-*cost* electricity in the world. The *price* of electricity was also among the lowest in the world, especially for large industries. (In Sweden, households pay more for electricity than industries, although some neo-classical economists in Scandinavia and probably elsewhere have claimed that this is an inefficient arrangement. One reason for disagreeing with those ladies and gentlemen is because inexpensive electricity for industry can result in large benefits in the form of employment, and the subsequent tax payments of firms and employees support large welfare commitments having to do with items like health care and education.)

Something that should never be forgotten is that for geographical and industrial reasons, Sweden is one of the most energy-intensive countries in the world. As a result, a high energy consumption should be considered by the decision makers as a necessity rather than a luxury, and treated accordingly when contemplating the appropriate price policy for this invaluable good. Incidentally, it *is* considered a necessity by a few influential politicians and bureaucrats; however, it is not a good career move to become excessively interested in nuclear energy and its merits, and under no circumstances would it be politically or, depending on the milieu, even socially wise to suggest reprocessing nuclear waste. (When nuclear fuel rods are replaced because of the decline in their energy content, they are often referred to as waste; however, there is still usable uranium in them that can become a constituent of new fuel. These rods also contain a small amount of plutonium, which some day might be classified as a valuable fuel.)

It is fairly easy to detect that Italy is in a different economic situation from Sweden. Italy imports about 86% of its energy requirements, which makes it the most energy-dependent country of the G8 — i.e., the 'club' of rich (or 'important') countries. If new Italian nuclear facilities are approved, they would be constructed by the large Italian firm ENEL, with the cooperation of Electricité de France (EDF), who would probably play a key role in re-establishing an Italian nuclear sector. Of course, if there is a serious nuclear accident near that country, it would probably be a very long time before Italian politicians felt secure in ordering or just discussing a full-scale nuclear 'go ahead'. (I deduced this from an important 2009 workshop led by Professor Riccardo Basosi, of the Center for the Study of Complex Systems, University of Siena, Italy.)

But if a genuine nuclear renaissance were to begin and gain momentum in Europe and elsewhere, the notice of an increase in Italy's nuclear assets would hardly take up much space in European newspapers. Before the Fukushima incident, the German firm Siemens announced that at least 400 new reactors will be deployed globally by 2030, and many of these would soon be under construction. This might also be the time to clarify that when electric *energy* (in, e.g., Gwh) is an *output*, it is often specified by writing Gwh(e). This terminology will be clarified later in this book, but it should be understood now that on the *input* side of an activity providing 1,000 million watt-hours of electric energy — i.e., 1,000 Mwh(e) — on the output side to homes and factories, there might be 3,000 Mwh (thermal) required as an input from a fuel like, e.g., coal. This is another way of saying that when transforming a fuel into electricity, efficiencies can be very far from 100 percent, and this applies to all electricity-generating activities.

According to the World Nuclear Association, in April 2012, 53 reactors were under construction worldwide, and at least 20 were in China and Russia. A few years ago, Mr. Vladimir Spidla, who was the Czech prime minister at the time, visited Finland and the construction site for its new 1,600 Mw reactor, and he mentioned that while Finland and France were the only countries in Europe that displayed intentions to build an advanced (or 'Generation 3') nuclear facility, "soon almost all will join it."

They will join it if they prefer a higher standard of living to a lower standard! Italy's energy situation is a great deal better than many because of its access to the natural gas of North Africa (e.g., the expanding reserves of Algeria and Libya), but several things should be made clear where this topic is concerned. For example, when I first taught in Australia, the Maui gas field in New Zealand was thought to be inexhaustible; however, at the end of the present decade it may only be a memory. It is also interesting that the inaccurate calculation of Maui resources contributed to an unjustified optimism in early appraisals of New Zealand's intention to restructure its electricity market — an optimism that lured several academic stars and 'wannabes' into supporting deregulation exercises that should have immediately been identified as economically absurd.

Because of the ostensible "need" to reconstitute its nuclear safety authority, and to identify sites for the new installations, new Italian reactors are unlikely to appear during the present decade. Some of us do not think that this is important. The significant thing in Italy and elsewhere is that adequate psychological and financial preparations are made to accept this equipment before the price of oil and/or gas goes into orbit, which some observers now believe is inevitable. Put another way, nuclear energy is a *hedge* against runaway oil and gas (and perhaps coal) prices.

The nuclear reactors that might be constructed in European countries within the next decade or two would be 'European pressurized reactors' (EPR). They are often designated as 'third generation' (Gen 3) equipment such as the reactor in Finland, where the emphasis is on safety. (Safety with this equipment means that if control systems stop working, the reactor shuts down automatically, dissipates the heat produced by the reactions in its core, and stops both fuel and radioactive waste from escaping.) It might make economic sense though for some countries not to hurry nuclear investments, because there is no guarantee that things will go smoothly if too many foreigners are involved. There could always be problems of a cultural nature in a situation where, e.g., ENEL and EDF must coordinate their efforts in order to meet deadlines and keep costs under control when installing the equipment, while at the same time there is a large slice of the local population that is openly hostile to nuclear energy. Where hostility to nuclear energy is concerned, the recent earthquake and tsunami in Japan, and damage to reactors in that country, have elevated this hostility in almost every country in the world.

There have been some serious 'disagreements' of a cultural nature in Finland, where EDF and the Finnish power group TVO are constructing the 1,600 Mw(e) installation mentioned earlier. Anne Lauvergeon — the former CEO of Areva (France) — has blamed her Finnish collaborators for the delays and excessive costs experienced with this project, and without being especially familiar with the details of this accusation, I suspect that Madame Lauvergeon's displeasure is to some extent justified. As compared to nuclear-friendly France, there are a large number of persons in Finland who believe that renewables rather than nuclear energy should supply any additional electricity that Finland might require, and even if this delusion has not infected the air in the vicinity of the construction site, it has almost certainly resulted in the kind of frictions that are unavoidable in multicultural projects. This is an unpleasant fact of life.

It needs to be appreciated though that in a country like Finland, where the educational system seems to work as well or better than any in the world, it is not easy to confuse voters about something as important as energy — at least in the long run. The increased cost of the 'Gen 3' reactor in Finland will have to be absorbed by the French constructors of the reactor, and surprisingly, two more reactors are likely to be ordered by the Finns. There are probably a number of persons in Finland who would prefer another energy option, but in 30 or 40 years they will be congratulating each other on the foresight of their political masters where the provision of electricity is concerned.

With regard to the above comment of Prime Minister Spidla about the future of nuclear energy, it should be emphasized that the basic issue is the growing but not yet well-understood value of plentiful and reliable electricity in the life of persons in every country, in every walk of life, and especially individuals who are vulnerable to the economic discomforts that are likely to appear in a world characterized by a growing competition for resources, employment, space and various other amenities. In other words, a world in which the global population might eventually increase by almost a billion persons a decade, as is the present outlook. Naturally, it is impossible to go very deeply into this matter in the present book, although it is equally as important as any subject treated in your economics book, or physics book, or anything you are liable to read in the near or distant future. Moreover, it might be appropriate to believe that a nuclear facility taking between 4 and 5 years to construct, with a 'life' of *at least* 60 years, is an optimal piece of equipment for delivering large amounts of reliable and comparatively inexpensive electricity. In the chapter on nuclear energy, the matter of cost for nuclear-based electricity is examined in greater detail. Also, assuming that the reduction of emissions of carbon dioxide (CO_2) is also on the desired list, this is another 'plus' for nuclear energy.

The former Swedish prime minister was definitely not a friend of nuclear-based electricity, but it is very possible that the main source of his animosity was a strong desire to obtain enough votes from anti-nuclear voters to continue his presence at the head of the Swedish government, which fortunately did not happen. With all due respect, it is likely that in the confused and overworked mind of that gentleman, his gratuitous evaluation of the nuclear future was to some extent genuine, but unfortunately he was simply unable to deal with the complex architecture of energy economics, and he is not alone where this frailty is concerned. For instance, one of the recent prime ministers of Italy was deceived by the promise of plentiful and inexpensive natural gas due to trans-Mediterranean pipelines, but apparently the next prime minister did not agree, nor was it necessary for him to agree, because the price of gas may not have its present appeal to Italian consumers for many more decades, or perhaps even years. Although not certain, the price of natural gas is well below its equilibrium price, measured with respect to, e.g., the energy-equivalent price of oil. I should perhaps mention that the calculation of 'energy equivalents' is a fairly simple operation, and will be discussed later in this book.

Though unfortunately ignored by many energy economists, the optimal course of action for a country like Italy may be maintaining or increasing its consumption of natural gas during the period in which entry — or re-entry — onto the nuclear scene is being prepared. Despite the opinion of deeply concerned though amateur economists like the former Swedish prime minister, if Italian and Swedish voters do not have a guaranteed access to nuclear energy later in this century, they could find themselves without access to the kind of goods and services that they have no desire or intention to be without — although as yet most of them are not particularly anxious to acknowledge this state of affairs.

Many students of energy economics — and also energy professionals — are reluctant to recognize the tremendous importance of energy, and entertain a completely false concept of what the energy future will be like. Perhaps the basic problem is that those individuals are unable to adequately comprehend the mysteries of Economics 101, and in addition they may be members of a local 'anti-nuclear booster club'. Some of these ladies and gentlemen occupy important positions in the energy establishment or bureaucracies of their native lands because of their preference for fantasies, as well as an inflexible animosity toward mainstream logic.

Before continuing, it should be made clear that a main issue in this section is the illogical belief by some residents that the prosperity of countries like Sweden and Italy can be maintained if there is a comprehensive nuclear retreat, or perhaps even a reduction in the supply of inexpensive and reliable electricity. In addition, it would be pleasant if many readers of this book came to understand that in the long run, nuclear power is the least costly way to generate electricity. The energy content of proved uranium and thorium supplies is far greater than that of fossil fuels, even if many persons are not prepared to accept this easily provable fact. Among those deniers are many academics and energy professionals.

For some obscure reason, in 1978 all the major political parties in Sweden agreed that the growing controversy over the future of nuclear energy should be settled by a national referendum. The electorate was subsequently asked to choose between nuclear acceptance, or the more-or-less immediate closing of as many nuclear facilities as possible, or a gradual phase-out that was to be completed by 2010. Confronted by a whirlwind of neurotic fictions, the latter option was selected. Although still not fully understood by many Swedes, a key factor in that pseudo-scientific travesty was an assumption that the enviable and growing prosperity of Sweden at that time could be maintained even if the country's nuclear assets were liquidated. In other words, the choice between nuclear energy and 'something else' was reduced to a matter of taste, and to add insult to injury, the prevailing high energy intensity was pictured by many politicians as having little or nothing to do with a perpetuation of macroeconomic health, although in point of truth it was a key factor.

Note the observation "to be completed by 2010". This obviously was impossible, which even the anti-nuclear booster club accepted, and so the intention became to scrap nuclear assets when they reached the end of their productive life. This meant decades, and to show that they were serious — and also to recruit at least some of the members of that club to the Social Democratic Party — the reactors at Barsebäck (Malmö) were taken out of operation between 1999 and 2005. There have not been many discussions of the economics of that closure in seminars and publications, but environmentally, according to the parliament member Carl Hamilton (2009), if those two comparatively small reactors were still in operation, this part of the world would have avoided about 11.5 million tonnes/year of greenhouse gas emissions. Needless to say, it would also have meant less expensive electricity in Sweden, and perhaps in adjacent countries because of the cross-border electric transmission that takes place.

It is due to an intensified concern for the economic future that the irrational nuclear 'downsizing' in Sweden has apparently been halted. The key departure came earlier, and it involved upgrading the 10 remaining reactors so that they could produce approximately the same electric *energy* (in kilowatt-hours) as the original 12 reactors, where normally these 10 reactors provide at least 50% of the total electric *energy* generated in Sweden. It is the kind of flexibility explicit in these (and perhaps future) upgradings that should be a part of any scientific explanation for the low economic cost of nuclear energy.

The intention now should be to at least maintain the present output of energy, even if it means that new capacity must be constructed. The logic is straightforward, and cannot be altered by the resolute ignoring of mainstream economic history: *a high electric intensity for firms, combined with a high rate of industrial investment and the technological skill created by a modern educational system, will lead to a high productivity for large and small businesses. This in turn results in a steady increase in employment, real incomes, and the most important ingredients of social security (such as pensions and comprehensive health care).* Readers should examine the German intent to abandon nuclear energy, and then ponder the consequences of that abandonment on the German macroeconomy and also on the macroeconomies of surrounding countries.

Teachers of energy economics can employ the discussion in the previous paragraph to highlight one of my favorite arguments about nuclear energy, which is that if, *ex-post*, there has been an increase in employment, real incomes and welfare as a result of the expansion of the nuclear sector, then it may also be true that the 'subsidies' that a nuclear sector receives in order to function smoothly in the construction and early operating stages, are for the most part figments of the imagination. For instance, *taxpayers in Sweden (as a group) have gained rather than lost as a result of financing the introduction of nuclear energy into Sweden, and they have lost by passively accepting the reduction in Swedish nuclear capacity.* This financing can be described as a successful social investment.

To a considerable extent, the ill-founded assumption that Sweden without nuclear energy can be as prosperous as it has been with nuclear energy is now passé, which is why a majority of Swedish voters may no longer be hostile to it. With bad economic news rolling in from every corner of the world, and filling small as well as large newspapers in addition to small and large TV screens, the voters of this country may be losing their famous ability to tolerate economic nonsense. A few of them also know that Sweden has not suffered the economic torment of most other countries, and this is because of the availability of reasonably priced electricity for energy-intensive industrial activities.

Notice the expression "losing their famous ability". They have not completely lost it yet, because several years ago the Swedish government appointed an anti-nuclear gentleman with a PhD in physics to the highest position in the energy bureaucracy. His exact goals — other than to secure for himself a highly paid non-job in Brussels, or perhaps anywhere else between the southern tip of Sweden and the Capetown naval yard — were unclear, but it may have happened that he or one of his subordinates held a number of conversations with Ms. Mona Sahlin, who, before the Swedish electorate were given a comprehensive view of her talents, seemed to have an excellent chance to become the (Social Democratic) prime minister of Sweden after a recent election.

Ms. Sahlin has claimed that the production of renewable electricity — or 'green electricity' as it is called by environmentalists — has increased by 9 Terawatt-hours per year ($= 9$ Twh/year) in the period after the closing of the two reactors at Barsebäck (Malmö) took place. As is the case in every country on the face of the earth, this is another resort to a blatant untruth in order to conceal the kind of obtuseness that most textbooks in economics (and many on game theory) intimate cannot take place in a community inhabited by rational and intelligent human beings. Here it is possible to recall a testimonial by a talented American environmentalist conceding that even if Ms. Sahlin "isn't the (Albert) Einstein of energy, her aspirations and her ability to promote visionary goals are marvellous". Unfortunately, that gentleman missed the point. In the context of Swedish economics, she *is* the Einstein *and* Newton of energy, while the engineers, managers and scientists who shake their heads when they hear her and her foot soldiers sounding off about nuclear energy have been reduced to incapacitated bystanders.

The nuclear situation has played out exactly the way that many of us predicted that it would in our publications and lectures, with one exception. At a large international conference in Canberra (Australia) many years ago, after an American gentleman put in a good word for the breeder reactor, there was hardly a single person in the conference hall who thought that he was in his right mind, including the present author. Now, of course, it is only a matter of time before we have to deal with the presence of breeders, and if ladies and gentlemen with the mentality of those who contended that orthodox fission reactors can be completely replaced by wind and solar are involved in any way with the management of those breeders, their physical security and their outputs of plutonium, then we could be in very deep trouble.

There is also this matter mentioned earlier of the safety of nuclear reactors. The evidence seems to indicate that Swedish installations are as safe or safer than any in the world, and had it been possible to continue research in this country on several reactor types where the emphasis was on safety, they would be even safer; however, to the amazement of foreign scientists, nuclear research was prohibited in Sweden for several years. Therefore, a question must be asked as to the logic in scrapping nuclear facilities in Sweden when qualitatively inferior installations were in operation just across the Baltic. Of course, where the nuclear game is concerned, it is as pointless to ask sensible questions today as it was to ask sensible questions in Germany in April of 1945, when the American Air Force celebrated Adolf Hitler's birthday with a 1,000-plane raid on Berlin, at the same time that the Soviet army was closing in on that city.

A final point: While this book was being completed, the earthquake and tsunami in Japan took place. This slowed the pace of the nuclear revival, but not in countries that reject anti-nuclear fantasies that would depress their economic growth.

5. Natural Gas: In the Mood for Misunderstandings

Natural gas is a high-quality energy resource, and probably more valuable to consumers in the U.S. than indicated by its present low price, which averaged about $3.5/MBtu in 2013 (although approximately twice that in Europe, and three times that much in Asia). The comparatively low price seems to be due to large gas inventories and a fairly high exploitation of an 'old' resource, 'shale gas', whose availability has been drastically and unexpectedly increased because of technological advances. This situation tends to disturb some energy professionals, who believe that the gas-using public might draw wrong conclusions about the long-run availability of natural gas.

However, Dian Chu (2010) has provided a superb up-to-date analysis which indicates that a 'normalizing' of the gas market may happen in the fairly near future. Some observers also suggest that natural gas provides an ideal 'back-up' for, e.g., wind and solar power, due to the ease with which gas-burning equipment can be turned on and off. Moreover, natural gas has occasionally been called a superior 'bridging' fuel between present energy uncertainties and some sort of optimal energy future. About 15% of domestic U.S. gas production comes from offshore rigs, as compared to 30% of oil, but after the recent giant BP (British Petroleum) 'spill', the U.S. government may not be keen on granting new offshore leases. Moreover, where shale gas is concerned, the production method known as 'hydrofracking' has been accused of being a threat to groundwater, and various restrictions are constantly proposed by concerned observers. In fact, ecological issues seem to be leading to radical output restrictions in France.

It is not easy to solve the natural gas riddle, although I suspect that in a few years you will not have to be a weatherman to know in what direction the wind is blowing. At the present time, when I am asked about the future of natural gas, my reply always contains a mention of the purchase a year ago by ExxonMobil of the heavily natural gas leveraged XTO energy for 31 billion (U.S.) dollars. In *my* world ExxonMobil does not make 31-billion-dollar mistakes, and so I have no reason to believe that the present very low gas price will persist. Of course, it may be the case that the natural gas that ExxonMobil has gained control of contains a considerable amount of high-value natural gas liquids (i.e., high-value oil-like hydrocarbons produced along with gas in some deposits), and these liquids can be sold at well above the present low quoted price of gas, since (like expensive crude oil) those liquids are valuable for processing into valuable refined products (i.e., *oil* products).

Where bad energy news is concerned, the former Chairman of the U.S. Federal Reserve, Alan Greenspan, was at one time inclined to periodically warn his countrymen about the likelihood of spikes in the price of gas that could have very disagreeable macroeconomic consequences. He also once made the provocative statement that North American consumers will forever be confronted by a volatile and inefficient gas market unless a secure *and* unlimited access to the vast world reserves of gas could be obtained.

Exactly how securing "unlimited access" to natural gas would (or could) take place is uncertain, because the dilemmas imposed by uncertainty and information shortages — and in particular those due to the special nature of geology and the irrationality of some market actors — are enormous, and there is no reason to believe that Dr. Greenspan's former colleagues and friends in Washington (DC) would be more successful in dealing with gas-related issues today than they were in the past with, e.g., oil. Only a few years ago natural gas was considered expensive, though not as expensive as it would have become if large quantities of shale resources had not appeared.

Energy researchers should appreciate Alan Greenspan's concern. Natural gas may have satisfied about 25 percent of all U.S. energy requirements, and roughly 21 percent of the electric power generated in the U.S. is based on gas. It supplies heat to about 60 million households, and at least 40 percent of the primary energy for industries. (Remember that primary energy is energy obtained from the direct burning of oil, gas, coal and biofuels, as well as electricity originating from nuclear and hydro. Electricity originating from the first four of these resources is labeled a secondary energy source.) Dr. Greenspan seemed to think that the U.S. economy could someday become too dependent on gas imported from distant localities; however, today only 10 percent (or less) of the gas used in the U.S. is imported, with about 75 percent of that coming from Canada, and the rest in the form of LNG. In addition, unbridled optimism for the new shale gas emanates from the speakers' podium at almost every energy conference, and there is constant talk in circulation about the U.S. eventually becoming a major natural gas exporter.

Some observers claim that natural gas can restore the economic power of the U.S., particularly at the present time, when, according to the CEO of a large American natural gas producer, shale gas resources (i.e., reserves) plus other gas resources of the U.S. are twice as large as the oil reserves of Saudi Arabia. (This kind of cheerful reckoning will be touched on below and in Chapter 3.) One person who goes along with this is Michael Fitzsimmons (2010), who believes that the gas problem that Dr. Greenspan was once so concerned with is hardly worth mentioning, particularly when compared to what he sees as an enormous oil problem that natural gas is now in a position to help solve.

According to the calculations of Mr. Fitzsimmons, the U.S. is buying approximately one billion dollars a day of oil from abroad, and some of this oil is imported from countries that apparently do not have much sympathy for the cultural values of the U.S. ("One billion dollars" may be a miscalculation (i.e., too high) for the period dealt with by Fitzsimmons, even if imported oil products are included.) Most of this oil is needed to support the large vehicle ownership of the U.S., but technological developments have made it possible to operate vehicles on gas instead of oil. This is the case with, e.g., taxis in Hong Kong and several other places, and even in Sweden a few buses run on gas. These are 'fleet vehicles', where intelligent management involving things such as the obtaining and storage of fuel, and the price paid by (bus and taxi) users, can result in the profitable operation of gas-fueled vehicles belonging to a 'fleet'. (There is also a theory that vehicles using natural gas *and* electricity as 'fuel' are the future.)

But ownership of these vehicles by private persons might be much more expensive, which leads me to comment that the U.S. Department of Energy has the financial resources, and maybe the engineering and economics talent, to obtain the correct answers about the viability of natural gas-fueled vehicles in private ownership. Accordingly, natural gas as a large-scale replacement for oil may or may not make economic sense at the present time, although regardless of the actual situation, it is at least a mild dereliction of duty if the director of the USDOE does not have his foot soldiers ascertain the facts, and make them known to those of us further down the research food chain.

World gas production grew by 3.8% in 2008, which was almost a record increase, and had it not been for the onset of macroeconomic difficulties, a new record might have been set. Non-OECD consumption growth was larger in 2008 than OECD growth for the first time in history, and the record number of LNG tankers delivered to proud purchasers meant that the LNG fleet was expanded by almost 20%.

There will be a section in Chapter 3 that considers in some detail the units for important energy resources, but it might be useful to say a few things now about this matter. The price of natural gas is generally given in dollars per million British Thermal Units ($= \$/MBtu$), where Btu are a measure of the energy content (or heating value) of a resource. The price is also occasionally given in dollars per thousand cubic feet ($= \$/kcf = \$/kft^3$). Fortunately we know that 1,000 cubic feet of natural gas contains 1,035,000 Btu — which we approximate to 1,000,000 Btu — and so it becomes a simple matter to obtain the energy equivalency between oil and gas. Since the average barrel of oil contains 5,800,000 Btu of heat energy, a price of, e.g., \$3.5/MBtu corresponds to an oil-equivalent price of $3.5 \times 5.80 = \$20.3/b$. This price annoys *sellers* of natural gas, because they feel that the value of natural gas is unappreciated by the market. Similarly, an earlier price of \$10.00/MBtu came to an oil-equivalent price of $10 \times 5.8 = \$58/b$, which greatly infuriated many purchasers of natural gas when they had to pay it at a time when the market price of oil was far lower.

In the natural gas literature, cubic meters are sometimes used instead of cubic feet, and so that equivalency can also be given here. It is 1,000 cubic feet $= 28.3$ cubic meters, or 1 cubic meter $= 35.3 \, ft^3$. The consumption of natural gas in the European Union was 532 billion cubic meters a year in 2009, and as an exercise readers should convert this to cubic feet a year. The firm Gazprom (of Russia) has the largest production of natural gas in the world, or $1.55 \, bcm/d = 1.55 \, Gcm/d$ (where billion (B or b) can also be signified by G (or *giga*)). Its headquarters is in Moscow, its director is Alexei Miller, its revenues are well over 100 billion dollars a year, and the Russian state is its largest owner with 50.02 percent.

Russia is the largest (national) producer of natural gas in the world, as well as the largest supplier of gas to Europe (where it is followed by Norway and Algeria). It has been suggested by important natural gas experts like Jude Clementé that European dependence on Russia is too large, and from time to time I have encountered this conviction from other researchers of the Russian energy scene, but I am not sure whether this assessment is economically and politically valid. Natural gas plays a very important role in obtaining for the Russians the foreign exchange (i.e., money) that they require to keep their economy moving at a faster pace than was the situation before Messrs. Putin and Medvedev arrived in the Kremlin.

In case you are asked by a student or classmate, it might be useful to know that a Btu is the amount of heat required to increase the temperature of one pound of water by one degree Fahrenheit. Now let us look at a diagram that shows some elements of a typical natural gas pipeline, and for an important (but highly mathematical) analysis of this topic I can suggest Professor Ricardo Raineri (1997). He is particularly interested in what he calls "cruel battles" among competing corporations seeking higher profits.

In Figure 1, the pressure at the source of the gas is P_0, which is also taken as the pressure at the entrance to the first compressor (which is a piece of

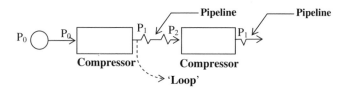

Figure 1.   Typical Section from a Natural Gas Pipeline

equipment that provides the energy to 'push' the gas through the pipeline). Naturally, the pressure at the place where the gas comes out of the ground might be higher than that at which it enters the first compressor, but there is no point in complicating this elementary discussion by going into that. In addition, determining the optimal distance between the compressors is an interesting but not particularly difficult engineering problem.

As shown in Figure 1, the compressor raises the pressure to P_1, but due to such things as friction losses in the pipeline, the pressure eventually falls to P_2, and the next compressor raises it again to P_1. As an example of this process, it might be useful to consider a pipeline from Australia's North West Shelf to Perth, the capital of Western Australia. This pipeline is 26 inches in diameter and runs 940 miles. There are five compressor stations through which the pipeline runs, as well as a large number of mainline valves. Something that needs to be pointed out here is that gas networks consist of three main segments — production, transportation, and distribution. Figure 1 shows the transportation segment, while the distribution segment (which features much smaller pipes) brings gas into your house or apartment, or perhaps the home of a neighbor. The engineering subjects thermodynamics and fluid mechanics deal with the flow of gas in pipelines, and there may be elementary books on those subjects.

The engineering associated with natural gas pipelines may appear comparatively straightforward, but there are many economic and political facets that require a great deal of *savoir-faire* on the part of engineers and managers. A key consideration is making sure that enough gas is available to keep the pipeline 'filled'. This often means filled for at least 20 years. Some readers have probably seen the expression 'stranded gas', and this usually refers to gas fields without enough reserves to economically justify transporting gas to distant markets, although an alternative might sometimes be to use this gas to generate electricity that can be used in the vicinity of the deposit, or maybe even turned into fuels in gas-to-liquids processes.

As for the 'loop' that is shown in Figure 1, this is a 'parallel' section of pipe that is sometimes added to the existing line in order to increase capacity, and it might be preceded by 'supercharging' relevant compressor stations. As discussed in the chapter on natural gas in this book, looping can sometimes be a profitable exercise. *Ceteris paribus*, considerable looping may take place in the future because of the increase in the production of natural gas. Estimates a year or two ago were that globally the reserves of natural gas could increase from approximately 3,000 billion cubic meters ($= 3,000 \times 35.3$ billion cubic feet) to 5,000 billion cubic meters ($= 5,000$ Bcm $\equiv 5,000$ Gcm), but the appearance of large amounts of shale gas in North America and elsewhere will probably cause these estimates to be raised. Note again that here, B (billion) is sometimes called G (or 'Giga'), just as million can be 'm' or 'M'.

Pipelines are not the only option for 'moving' gas. Another possibility begins with liquefaction, which involves collecting gas from a pipeline, and cooling it to -160 degrees ($= -160°$) Centigrade by a cryogenic cooling process. This reduces the volume of the gas by a factor of about 600, and from the liquefaction plant the liquefied natural gas (LNG) is transferred to a cryogenic carrier (i.e., ship) and transported to the consuming country, where it is turned back into gas and fed into a pipeline. It is sometimes claimed that the LNG trade is more flexible than piped natural gas because LNG carriers can be rerouted, while pipelines are fixed. Many observers believe that in the future there will be a growth in 'rerouting' and 'spot' trading, but the opinion here is that pipelines are still much more important now for trading gas than LNG, and should remain so over the foreseeable future.

One thing that should be understood is that the price of LNG in dollars/MBtu is almost always appreciably higher than for pipeline gas. In the last U.S. presidential election, the Republican vice-presidential candidate — Governor Sarah Palin of Alaska — said that she favored the construction of a pipeline that carried gas from her home state of Alaska to somewhere in the vicinity of Chicago, which was a suggestion that one of the directors of 'Big Gas' called politically wise but economically dumb. One of the reasons for this was that Mr. 'Big Gas' very likely wanted to transport that gas to Asia in LNG form and sell it there for the very high prices that have sometimes prevailed in that part of the world. The pipeline that Governor Palin wanted constructed has been discussed for several decades, and estimates were that it might take the better part of a decade to complete if it were started. The opinion here is that she was probably correct in so far as the politics *and* economics of this project are concerned if there is a very large amount of exploitable gas in Alaska, although it is clear that a considerable part of this gas might be siphoned off by *laterals* and consumed in Canada, or turned into LNG.

It might be worthwhile noting that as this is being written, Canadian gas exports to the United States show signs of decreasing, presumably because of the availability of shale gas in the eastern U.S., and new gas 'plays' (including pipelines) somewhere in the vicinity of the Rocky Mountains in the West. This might turn out to be an energy game changer, but intelligent decision makers with a long-term vision will not be distracted by the interim fortunes and misfortunes of exhaustible resources like oil, gas, and perhaps even coal. Individuals, firms, and 'plays' are destined to pass from the scene, but nations remain, and ideally their future energy requirements should not be underestimated by today's decision makers. Note the word "ideally". What it means is that in the real as opposed to the fantasy world, drastic energy mistakes are certain to be made.

There are constant claims in circulation about an increase in the discovery of natural gas, but LNG poses a problem. An LNG 'train' requires a very large amount of proved reserves (associated with a designated shipping point) to support 20 or more years of dependable shipments, which is often cited as the number of years necessary for a profitable operation, given expected gas prices and costs. What we have here are not just the costs of gas and pipelines, but also the large costs associated with processing and shipping the LNG to market. If there was a likelihood that the average price paid for LNG in Asia would remain at or close to that experienced a few years ago, then a purely economic calculation would indicate that the gas Governor Palin wanted to see in a pipeline terminating near Chicago should move West instead of South.

Several years ago Shell's CEO said that demand for LNG may double before the end of the next decade, and he seemed to think that since a considerable part of the world's gas reserves were still untapped, they would fit into the LNG picture. One of the reasons he might have said that is because Shell — a major oil producer for many decades — is ostensibly in the process of becoming as interested in natural gas (and particularly LNG) as it is in oil, or so some observers claim. But it unfortunately happens that a sizable fraction of those untapped reserves may not be in concentrations large enough to be profitably exploited as LNG, nor would their present owners be anxious to deepen their cooperation with Shell. It may, however, be profitable to turn this gas into motor fuels in the country of origin, which might simplify its transportation in large vessels. In addition, it has been suggested that as a result of recent technological advances, LNG could eventually replace oil as the leading fuel for ships of all categories.

Globally, the *growth* in the demand for gas may still exceed that of all energy media except renewables, and gas is still a very desirable input for electric power generation. (In both the U.S. and UK, a gigantic infusion of gas-based equipment was planned before some questions were raised about the supply-demand situation for gas.) A main reason for the popularity of gas was the advent of *combined cycle* gas-burning equipment with a very high efficiency. What happens in this activity is that in addition to the gas turbine, there is a secondary turbine producing steam from the waste gases/heat of the gas turbine. The kinetic energy in this steam is transformed into mechanical energy in the form of the turning flywheel of a generator, and as a result, produces more electricity. When compared with earlier equipment, the additional electricity produced for a given input of gas makes this a very satisfactory arrangement.

As often happens, there are very many wrong ideas in circulation about natural gas, the most pernicious of which — at least in Europe — have to do with the *restructuring* (i.e., deregulation/liberalization) of gas markets. Some questions need to be asked as to how these delusions came into existence, and one answer has to do with the inability of consumers and to a certain extent producers to judge the availability of gas. For instance, one of the arguments for deregulation turned on the ludicrous assumption that more 'competition' could compensate for the physical depletion of physical resources. On this topic I prefer the results published by Brau, Dononzo, Fiorio, and Florio (2010), which shows that gas restructuring in the EU has placed consumers at a disadvantage — something that I predicted years ago about gas *and* electricity. An earlier important analysis is presented by Professor Susanna Dorigoni and Sergio Portatadino (2009).

In addition, in some parts of the world, gas producers habitually expressed themselves in such a way as to give the impression that there was a huge amount of natural gas reserves (in one form or another) in politically secure locations that would eventually be available for exploitation if "regulators" and other so-called "meddlers" kept their distance. Similarly, many gas buyers were almost totally unaware of how supply and demand could develop in the long run, and instead continued to make plans for a future in which they would have access to all the gas that they would need, at prices that resembled those of the recent past. This might be a good place to note that in Brazil, some starry-eyed deregulators counted on gas-based electric power being cheaper than hydroelectricity and nuclear power. As they now admit, this supposition turned out to be incorrect.

In much of North America, despite propaganda to the contrary, and new shale gas 'plays', exploration and production have often yielded disappointing results, and expectations about, e.g., the Gulf of Mexico and gas imports into the U.S. by pipeline from Canada often have an air of unreality about them. In Europe, an interesting situation has unfolded in Finland. With copious potential gas supplies adjacent to Finland in Russia and Norway, the decision-makers in that country chose nuclear as the best option for additional power. They understood that given the likely future demand for gas in Europe, Asia and North America, in the long-run they might have found themselves relying on expensive imports from very distant sources — e.g., the Middle East and North Africa — and, in addition, dependence on Russia might increase to what they regard as an unhealthy level.

According to an estimate of the International Energy Agency (IEA) and other research establishments, fossil fuels will account for 80–90 percent of the world's primary energy mix in 2020. Global gas demand is expected to rise by 2.5–2.7 percent per year ($= 2.5$–$2.7\%/y$). In the U.S. one forecast is for a growth of at least 2 percent per year, taking into consideration the sudden and unexpected fall in the price of gas, and the appearance of shale gas. The big consuming area will almost certainly continue to be Asia, where it has been suggested that demand could increase by an average of 3.5 percent per year between 2001 and 2025. Some observers are even saying that the demand for natural gas from Asia will turn Australia into the Saudi Arabia of LNG.

The share of gas in world energy demand could greatly increase in the next decade or two. An earlier estimate had the average global gas production increasing by 2.75%/y until at least 2025, and gas overtaking coal in the global energy picture. This no longer sounds right, at least to me, and in considering the statistical exercises employed to obtain many of the estimates cited in this book, it might be useful to refer to the late and brilliant Danish mathematical economist Karl Vind, who once said that in economics, empirical work is not as important as theory. If the expression 'on the average' is inserted, then many of us will have no problem accepting that assertion. It then becomes: *on the average, in economics, empirical work is not as important as theory.* Of course, it is more important than the very mathematical theory often encountered in the learned economics journals, most of which are ignored or only read by students.

The average daily natural gas production in the U.S. increased from 51 billion cubic feet per day (=51 Gcf/d) in 2005 — when predictions of the future output of conventional natural gas began to greatly worry certain important people in that country — to about 66 Gcf/d in 2011, because of the contribution of shale gas. (The units in which energy is measured will receive a thorough treatment in Chapter 3 of this book.) There has recently been a great deal of attention accorded to *shale gas* extracted from tight, brittle shale formations, 2,000–10,000 feet deep in the earth, but shale gas wells apparently have exceptionally high *natural decline (natural depletion)* rates, and it may be true that many gas consumers in the U.S. are not being provided with the information they deserve in order to make optimal decisions about the kind of energy they should purchase. There also might be serious environmental problems associated with shale gas, due to the large amount of water and chemicals required for the 'hydraulic fracturing/fracking' that separates the gas from the rocks in which it is embedded and forces it to the surface.

But natural gas has always caused some environmentalists to raise embarrassing questions. The largest component of conventional natural gas is methane, which not only travels through pipes, but is also a part of the flammable ingredient that burns in kitchen stoves. No matter where it is, methane is like carbon dioxide (CO_2) in that it helps to block the atmospheric outlets through which our planet's daily output of heat escapes to outer space. "To me, methane is more scary than the better known greenhouse gas carbon dioxide," the biologist Patrick Zimmerman once said. Maybe so, but in various ways methane is directly related to the world's food supply, and given the growth in the human population, it is not something that can be easily tampered with.

And that means tampered with by the 259 investors from Asia, Africa, Australia, Europe, Latin America and the U.S., who, along with a large number of other high-value dudes and dames, made their presence felt at the latest climate 'gig' in Cancún (Mexico). Many of these concerned plutocrats are apparently like Barbara Krumsiek, chairman of the 'Financial Initiative' in the UN's environmental program, and CEO of the American investment firm Calvert Investments, whose evaluation of the present climate situation was apparently based on her interpretation of the *Stern Report* (on global warming). In any event, when it comes to making an acquaintance with (shale) natural gas, examine the work of Jude Clemente and J. David Hughes for opposing points of view, but start with the survey "Frack to the Future" by Michael Brooks in *New Scientist* (August 2013).

The natural gas industry in many countries seems to be working overtime to convince voters that natural gas is sufficiently 'clean' to allay all of their worries about global warming, and thus, in addition to generating electricity, it can be regarded as the ideal motor fuel. But this might be a ploy to divert attention from a *possible* overall superiority of electric vehicles, including plug-in hybrid electric vehicles (PHEVs), when or if the monetary 'premium' of these vehicles over non-hybrids declines. Premiums (i.e., extra purchasing costs) of several thousand dollars or more, in order to reduce the annual cost of operating a vehicle, might be difficult for many vehicle buyers to accept (although they might gain if they keep the vehicle for a long time). Moreover, 'coal-to-liquids' options have in theory some strong selling points, because existing gasoline (i.e., petrol) engines would not have to be replaced, as is the case with vehicles that are modified so that they can be fuelled by some variety of natural gas. Using more coal in any form, however, is something that is difficult for many persons to approve. There is also a theory making the rounds that the 'age of hydrogen' is closer than many people think.

In comparing the price and performance of electric vehicles (EVs) and PHEVs with similar (gasoline and diesel) cars and trucks powered by internal combustion engines, it seems to be the case that the conventional equipment will prevail for many years. But even so, the improvement of vehicle batteries is a crucial item if EVs and PHEVs are to become competitive. Given the uncertainties and fears attached to the present energy architecture, it is not just likely but *certain* that we will get the batteries we absolutely must have, although we may not get them as soon as we desire. There is some controversy here, but given what might happen to the price of conventional motor fuel, these batteries or large amounts of 'unconventional' fuels are absolutely necessary to keep our automobiles in the fast lane up to the skiing in Åre (Sweden) during this and future winters, or the skiing in the Midnight Sun late in the spring or early summer even farther north.

World natural gas prices declined precipitously following the onset of global macroeconomic troubles in 2008, but some observers believe that this is temporary. The IEA has often been mistaken in their forecasts of world gas prices, but it is difficult to completely reject their once-held belief that eventually a tightening of U.S. and Canadian gas supplies could take place, and this process could turn out to be very unpleasant for industrial buyers and households in North America. Of course, forecasts of the IEA sometimes leave a great deal to be desired. Their forecast for a wellhead price of $2.5/MBtu (in 1997 prices) for conventional U.S. gas in 2020 had an offbeat ring to it when it appeared earlier this century, and several years ago some of us suggested that unless the global macroeconomy greatly deteriorated — which unfortunately happened — sustained natural gas prices of perhaps $10/MBtu, or more, were not impossible.

The upshot of the above is that readers should get used to the likelihood that bargain basement natural gas could go out of style at any time. Although the (energy equivalent) price of natural gas has declined to well under that of oil, there are few competent observers who believe that this situation will persist when there is a macroeconomic recovery in the major industrial countries. In addition, some of us believe that a GAS-PEC is inevitable, and its members will be the most important gas exporters (with the possible exception of Australia and Norway). When I encounter objectors to this belief, I simply ask would you prefer a low to a high gas price if you were on the supply side of this market, although there are also intelligent arguments why a GAS-PEC might be difficult to form. An introductory analysis of this issue can be found in a short, unpublished working paper by Diego Villalobos Alberu.

That brings us to a closer look at restructuring. Here it is useful to emphasize that the same exaggerated claims made for electric deregulation have also been made for gas, though not so aggressively as a decade ago. It might be worth noting, though, that while the relevant laws of physics are similar for gas and electricity, their application may be dissimilar. It has been suggested that gas is simpler to model (and thus understand) than electricity, because Kirchoff's laws are not applicable, but geopolitical considerations greatly confuse the issue. For Europe, Russia and Algeria are two of the main gas suppliers (with the others being Norway and the Netherlands), and attempting to forecast their behavior is much more difficult than solving the kinds of equations that we amused ourselves with in undergraduate courses on electric circuit analysis.

The term "exaggerated" may also apply to the future of liquefied natural gas (LNG). In the U.S., the only place that LNG has been declared welcome is on the Gulf (of Mexico) Coast — although a friendly reception is no longer certain in, e.g., Louisiana. In the Northeast and on the West Coast, pipeline gas is preferred — although where this pipeline gas will originate is something that nobody seems to know. On the other hand, gas plays such an important role in the household as well as the industrial sector in the U.S., that if the price suddenly and unexpectedly escalated by a large amount, the NIMBY — not in my back yard — chorus that usually greets the announcement of a new or proposed LNG terminal might be considerably moderated.

A main shortcoming of the gas market debate has often been the presence of several academic economists without the necessary 'feel' for either the economics or the engineering aspects of the natural gas sector. This includes economists with a modicum of engineering training in their background. A question must therefore be raised as to how decision makers should treat the avalanche of misjudgments about this market in order to help prevent expensive, near-irreversible investments from taking place.

It could be argued that in this book it might be preferable to blithely skip over the deregulation issue raised above, because the exact situation with natural gas will soon be revealed by its increased price. Besides, unlike the electric deregulation travesty, gas deregulation was (and is) a blunder that would never be able to get up full steam. One reason for this was that in the U.S., and perhaps elsewhere, some important politicians and industry people, as well as a few genuine experts from the academic world and the U.S. Department of Energy, took issue with the more bizarre promises and objectives of natural gas deregulation (or restructuring). For instance, they pointed out that the natural gas market in the U.S. is *not* 'informationally' efficient. Roughly this means that gas prices at widely separate localities do *not* follow each other in a manner which makes it possible to conclude that — when *actual* as compared to *ideal* transportation costs are taken into consideration — these venues are in *one* market. Accordingly, the kind of ideal (or textbook) *arbitrage* cannot always take place which allows industrial consumers of natural gas that are faced with high prices in one region (or locality) to increase their profits by purchasing gas in another region at lower prices.

And scepticism is not just found in the U.S. A former CEO of British Gas contended that the "half-baked fracturing" of gas markets in order to bring about competition is essentially counterproductive. I can add that prospects for an 'efficient' global gas market that is brought about by greatly increased 'spot' (as compared to long-term) buying and selling are as much a delusion today as when first introduced.

Probably the most important observations on the ambitions of natural gas deregulators have been rendered by Professor David Teece (1990) of the University of California. According to him, market liberalization/deregulation/restructuring in the U.S. has already "jeopardized long-term supply security and created certain inefficiencies." He also notes that "While more flexible, a series of end-to-end, short-term contracts are not a substitute for vertical integration, since the incentives of the parties are different and usually contract terms are renegotiated at the time of contract renewable. That may sound attractive; however, there is no guarantee that contracting parties will be dealing with each other over the long term, and that specialized irreversible investments can be efficiently and competitively utilized."

For this reason I suggest that teachers should remind their students that large and complex gas systems operating in a climate of uncertainty are most efficiently operated on an integrated basis that emphasizes long-term contracting. This arrangement promotes optimally dimensioned installa-tions, and although it may not be mentioned in your economics textbook, if pipeline-compressor-processing systems which fully exploit increasing returns to scale in order to obtain minimum costs are to be readily financed, then — as the evidence seems to reveal — the kind of uncertainties asso-ciated with short- to medium-term market arrangements (that must be fre-quently adjusted or renegotiated) should be kept to a minimum. Failing to do so could cause a reduction in physical investment, and *ceteris paribus*, in the long run, unavoidably lead to higher rather than lower natural gas prices. Some spot transactions are useful though, and perhaps the most flexible and efficient gas market is one in which buyers and sellers establish a market structure in which transparency prevails, and liquidity is high enough so that spot transactions take place with ease when and/or if they are economically justified.

It is satisfying to note that many energy professionals are coming to their senses where topics in this chapter are concerned, and as icing on the cake, less tolerance is being shown to the ravings of flat-earth economists and their adherents when the discussion turns to future supplies of gas and oil. What is happening is that genuine experts have started paying closer attention to reality than to the kind of bizarre economics that became acceptable when some top executives at Enron assured a California governor that 'deregulated markets' would reduce electricity prices in his state by 20 percent. The bad thing about that situation was the failure of academic economists to warn the governor about the rashness of following Enron's advice.

A few years ago it appeared certain that domestic U.S. gas output had peaked, and according to the *BP Survey of World Energy*, U.S. (and North American) output was almost 'flat'; but that no longer sounds right because of the appearance of large amounts of, e.g., shale gas. Somewhat inexpli-cably, even before the gas price began to fall, the gas 'rig' count in that country also appeared to have peaked. (A rig is the drilling structure for gas or oil.) This suggests that more than a few important firms regarded North America as a hopeless case for further large-scale investments in the gas sector, even if the gas price had continued to rise. Furthermore, as in the U.S., increased drilling in Canada did not appear to be raising production by a substantial amount. The situation in both countries before shale gas boosted expectations could easily be summed up as follows: mature basins, smaller discoveries, and a high rate of *natural decline* from existing gas wells — which unavoidably translates into higher energy costs if the desire is to increase or even to maintain output (unless, of course, unconventional gas manages to turn the supply-demand picture around).

There was a time when there were any number of journalists, academics and assorted paid and unpaid propagandists inclined to inform everyone in their 'network' that the high oil and gas prices that began to appear were irrelevant from a macroeconomic and financial market point of view. Their arguments often claimed that today's economies are so sophisticated when it comes to energy saving and substitution, that even with oil prices around $100/b, and gas prices that might approach, e.g., one-half that level (in BTU terms), there could not be a threat to macroeconomic stability.

Since we may encounter this kind of implausible wisdom again some day, we can only hope that readers of this book make it their business to tune out at the first opportunity. In a recent conference of EU movers-and-shakers, it was proposed that the EU countries should formulate a joint strategy for dealing with their energy vulnerabilities, and while it is possible to sympathize with this goal to a certain extent, some questions must be raised as to how it conforms with the deregulation nonsense sponsored by the EU Energy Directorate, especially when there have been persons in that establishment who believe that 'peak oil' is *only* a theory, and whose ideas about electric or gas deregulation belong in cloud-cuckoo land.

Among other odd initiatives, the EU at least once called on Gazprom (Russia) to open its pipelines to 'third parties', by which they meant both independent producers and other gas-producing countries such as Kazakhstan. The then Russian president, Mr. Medvedev, greeted this outlandish proposal by calling it a "communist remnant", and made it clear that his country could easily avoid having to confront this kind of illogic by greatly increasing sales to China. Accordingly, it might be proper to suggest that all countries that now buy natural gas from Russia would be much better off if they ignored the precious intentions, advice, theories and prophecies of the EU Energy Directorate until that establishment takes the trouble to absorb at least the more rudimentary lessons of economic history and economic theory.

There are almost certainly influential persons in Europe and the U.S. who still believe that various deregulatory shortcomings can be remedied by greatly 'thickening' gas and electricity networks — i.e., thickening them with more pipes and wires. These persons certainly could be correct, although it must be clear to competent observers that spending enormous amounts of money in order to facilitate the smooth operation of spot (and perhaps also derivatives) markets *because of proposals to deregulate that should never have been offered or accepted,* is irrational and can be a drastic economic mistake.

Some reservations might also be in order about the use of the term *contestability,* that occasionally appears in sermons on natural gas deregulation, and which unexpectedly cropped up during one of my lectures in Hong Kong. This is a concept that for the most part is applicable to activities in which there are low sunk costs, and as a result, entry and departure from a contestable market is (in theory) almost costless. Street markets in Bangkok and Dakar might be an example here. As bad luck would have it though, there are very high sunk costs associated with natural gas (and electricity) networks, and so would-be 'players' who enter that particular world thinking that they will gain a reputation for analytical excellence should make sure that there are no gaps in their knowledge of Microeconomics 101. A few years ago in Hong Kong the issue was not gas but electricity, and although there was considerable talk about contestability, it dried up when the topic was carefully examined employing the kind of vocabulary and theory featured in elementary economics textbooks. Let us be explicit about this: contestability has little — or nothing — to offer the present discussion.

There have been suggestions that what mostly characterizes gas and electricity restructuring up to now is a reduction in economies of scale (due to sub-optimal investment strategies), increased prices, decreased reliability, excessive bonuses to top executives, and perhaps a threat to the security of supply. All or some of these flaws are visible in virtually every corner of the globe, and as yet show no sign of disappearing. The main reason for this questionable state of affairs is greed on the part of a few, and carelessness on the part of many. By carelessness I mean ignoring evidence that electric and natural gas deregulation are easily avoidable mistakes.

Occasionally somebody wants to know how the previous discussion fits into an empirical exercise. If empirical means econometric, then the following equation has been used by me to estimate short-run natural gas prices in the U.S., taking into account that the average natural gas price shows a tendency to be less unstable than the short-run price of oil. Taking I as inventories, C as consumption and p as price, we can write:

$$p_t = \alpha_0 + \alpha_1 \left(\frac{I_t}{C_t} - \frac{I_{t-1}}{C_{t-1}} \right) + \alpha_2 p_{t-1}. \tag{3}$$

This equation is the same type of equation used by Fisher, Cootner, and Baily (1972) for the estimation of the copper price, and is similar to the one I used for copper and zinc, and discussed in my books on copper and aluminum. It gave good results for copper and zinc, but only a fair result for natural gas; however, even so, it might be useful for gas if applied to different time periods and employing a different lag structure. The really important thing in this equation is the presence of inventories. Readers almost certainly did not see this sort of thing in Economics 101, but it should be stressed that in the past, inventories have had an important role to play in the determination of the oil price, and this is also true of gas. (In fact, it has a crucial role to play for oil, and readers should not avoid the discussion of inventories and short-run pricing in Chapter 4 of this book.)

As for doing some econometric work for the purpose of estimating a conventional demand elasticity — i.e., the percentage change in demand given a percentage change in price — an equation of this sort may not have much to offer. However, Professor Julie Urban (2006) of the University of Wisconsin has calculated a few short- and long-run elasticities in a paper that everyone interested in this topic should examine. More importantly, she states that the (average) productivity of gas wells in the U.S. peaked in 1971, and declined steadily up until 2004 — which was probably the last date for which she had data when preparing her paper. The productivity of gas wells in the U.S. is also discussed in my book on natural gas (1987).

That brings us to this business of indexing the price of gas to, e.g., oil or coal. The background to this is often the ability to substitute oil or coal for gas, although clearly this is not a short-run possibility. Indexing arrangements might then make sense for gas buyers with short maturity contracts. A typical indexing formula might be:

$$P_{gas} = 3.65 \left[\frac{average\ price\ of\ 5\ crudes}{27} \right]. \tag{4}$$

In the above expression, 27 was the average dollar price of a barrel of crude oil ($= \$27/b$) at the time this particular contract was signed, while $\$3.65/MBtu$ was the chosen base price for gas. The 'M' here stands for million. The 5 crudes are pre-designated, and usually determined by negotiation. As noted above, it happens that some contracts for gas are also indexed to coal or some other resource. There are many indexing formulas being applied, but the following is a little special:

$$P_{gas} = 16.8 \left[\frac{A}{400} + \frac{B}{1,350} + \frac{C}{600} \right]. \tag{5}$$

This indexing formula reflects 1975 price expectations in the agreement between the West Germans and Norwegians on the price of gas from the Ekofisk field. It gives the approximate price in deutschmarks/billion-calories. (Calories is a heating unit, and "approximate" suggests that *ad hoc* modifications/adjustments might be necessary.) Here, A is the price of heavy fuel oil to customers in Northern Germany, B is the price of gas oil, and C is the price of heavy fuel oil in Rotterdam. There have been suggestions that where indexing is concerned, coal should also put in an appearance in the indexing formulas. This is probably sensible, since gas and coal are competitors. I have also heard it said that indexing is wrong. I will not try to solve that riddle in this book, but it certainly makes sense for suppliers of natural gas in the U.S. at the present time.

Some radical changes seem to be taking place in the gas world, with, e.g., Russia's Gazprom contemplating a land-based pipeline to South Korea, and U.S. natural gas (in theory) going to Europe as part of an LNG venture — although only a few years ago LNG was a bad word in the U.S., when it was thought that large imports of that commodity would be required, and 'dangerous' installations would have to be constructed on American soil to receive and process these imports. Perhaps the most noteworthy oddity is news that the consumption of natural gas in the Middle East might outpace production, even though that part of the world is rich in gas. The shortcoming in this case is in the 'flow' as compared to the 'stock' of gas, resulting from the annoyance that much of the gas in the Middle East is 'associated gas', and cannot be obtained without increasing the production of oil — which is not on the program of many of the owners of this resource.

There is also a great deal of talk about the so-called 'Nabucco' pipeline in Europe, but according to leading students of that project like Professor Alberto Clo and Sohbet Karbuz, it has been either dumped or recast. I recently attended a large meeting at the Stockholm School of Economics which featured amateurish efforts to convince a large and attentive audience that European energy security is threatened by erratic behavior from Russian gas producers, or the Russian government, or some establishment in Russia. This is the kind of nonsense I heard a great deal about when serving in the U.S. Army in Asia and Europe.

One problem may be that the Russians will not accept the bizarre 'logic' appearing about ten years ago that was a part of pseudo-scientific proposals to restructure (deregulate) Europe's gas and electricity networks. Readers interested in this subject can turn to the work of Professor Susanna Dorigoni, who directs energy research at Bocconi University (Milan). (What might eventually happen is that the Russians get tired of these Cold War gestures and exertions, and abandon most of the European natural gas market in order to concentrate on Asia.)

Natural gas has often been thought of as the great competitor of coal, but lately coal prices have been so high that in the UK some utilities have turned to the use of wood pellets to generate electricity. The present UK prime minister, Mr. Cameron, has even gone so far as to endorse enough biomass from trees and vegetable matter to supply 6 million homes in his country (which sounds overly ambitious because of supply uncertainties). Biomass undoubtedly sounds good to environmentalists; however, it may be true that biomass produces about the same amount of greenhouse gas as coal.

6. Conclusions and Further Observations

I begin with further observations, by which I mean some up-to-date considerations that belong in this introductory chapter, where they are most likely to be noticed. When this book was being completed, there was a great deal of action in North Africa (Egypt, Tunisia and Libya), and there is talk of forthcoming tensions in the Gulf. In case you have not noticed, one cause of the trouble in that part of the world was pointed out by Mr. Malthus many years ago. It is population. It is the presence of large numbers of very intelligent people whose access to employment is limited. Similarly, the most interesting thing about the Libyan situation is the failure to understand that it is about oil. As this is being written, the price of West Texas Intermediate (WTI) crude oil was $103/b, while Brent crude was almost $20 more expensive. The falsehoods about going to war to protect Libyan citizens cannot conceal the real reason, which primarily has to do with restoring or increasing Libyan exports of high-quality crude.

Think about this. A potential output decline in Libya of 1.6 (or less) mb/d, which is about 1.85% of the global output of (conventional and non-conventional) oil, almost immediately led to an oil price rise of about 18.5%. This says something drastic about the power of oil! It also gives me the impression that a price rise of this nature is going to engender a great deal of fear in high places, and this fear might cause the governments of some oil-importing countries to do things that are best left undone. Moreover, many observers have not noticed that Libya exports high-quality (light sweet, low sulphur) crude, while most exports from, e.g., OPEC or Russia are sour crude with a high sulphur content, which costs more than Libyan oil to refine and process into oil products. Replacing Libyan oil with a lower quality liquid is possible, and perhaps in some importing countries simple, although it could greatly inconvenience a few buyers of oil and oil products in Europe, particularly Italy.

Given this situation, it might be time to start thinking about eventually replacing a large fraction of conventional oil. High hopes were once maintained for biofuel in the form of corn-based ethanol, but no less an authority than former U.S. President Bill Clinton recently said that using too much corn as an input for ethanol could lead to higher food prices and riots in poor countries. As it happens, a wide range of biofuels can be made available without having to rely on inputs like corn or similar food products, and ostensibly these other biofuels would not have a detrimental impact on food prices. Here it can be mentioned that biodiesel is apparently a more efficient fuel in most respects than ethanol, although an unnoticed problem is the large amount of water required to produce biofuels. This may turn out to be the cruellest dilemma of all.

Returning to oil, China alone added almost a million barrels a day to global demand in 2010, which has caused some of us to wonder just how the additional global demand that will appear in the near future can be satisfied.

On February 26, 2011, the West Texas Intermediate (WTI) oil price spiked to about $100/b, while the Brent oil price may have reached $120/b. Somewhat later the WTI price reached $112/b, though it soon declined. Very few of us thought that we would see prices of this nature so early during the year, though their presence was not accompanied by curses, which might have happened a few years ago. After what happened with the oil price in 2008, many observers feel that bad oil price news could reappear at any time, and in addition, the major importers of oil have lost all influence with the major exporters.

Hopefully, many readers now understand that there is a rhythm to energy economics, and as the famous musician Louis Armstrong once said: "If you don't have that swing, then you don't have a thing." Even worse, the future is going to tax our understanding of basic scientific considerations even more than the present. Anyone who does not believe this can examine the important contributions of Riccardo Basosi (2009) and Franco Ruzzenenti (2009), which provide a glimpse of the topics that will probably have to be treated in the energy economics of the future. My conclusion here is that we not only need more students and teachers of energy economics, but they should be prepared to deal with topics that are much more complex than those taken up in this book. For instance, I suspect that a crucial project will be outlining the exact role that electricity will be capable of playing in our lives. There is also this matter of the optimal mixture of renewables and alternatives, and somehow, unfortunately, someone is going to have to tell us a great deal more about shale oil and shale gas than we know today.

Getting the "swing" mentioned above is easy if you do not make the serious mistake of spending time reading 'off-target' literature. There are a large number of excellent articles, such as a survey by Professor Frank Clementé of Pennsylvania State University, which can be accessed via Google, a paper on shale oil by Elliot Gue (2010) and a brilliant paper on oil by Kenneth D. Worth (2011), both in *Seeking Alpha*, which are distinguished by a large number of superb comments. Also, basically non-technical papers on various energy subjects have been written by Michael Jefferson and Vlasios Voudouris, etc. These are listed in the Bibliography section at the end of this chapter. The best (completely non-technical) introductory book is by David Goodstein (2004).

Some of the best short articles and comments are found in a few forums and sites such as *EnergyPulse* and *321 Energy*, and also, if you enjoy arguing about (or discussing) energy issues, *Seeking Alpha*. I can also note some interesting comments on oil by Len Gould (2010) in *EnergyPulse*, who once said that at some point in the future, the voters in some countries might agree to go to war in order to obtain oil. According to Alan Greenspan in his book, *The Age of Turbulence*, the political movers-and-shakers in some oil-importing countries have already come to this conclusion. He cites the venture in Iraq, which Donald Trump — a one-time possible U.S. presidential candidate — has called a completely illogical undertaking. On the subject of illogical undertakings, physicist John Droz, Jr., discusses a few having to do with 'wind energy'. You can see his work on the Internet at *Energy Presentation Info.*

Exceptions to the dubious quality of some energy economics books are difficult to find, but I want to especially recommend the short and easily read book on oil by Dr. Mamdouh G. Salameh (2004), which is a 'must', and the same applies to a short paper by Tam Hunt (2011). I have also had the good fortune of lecturing at several outstanding academic energy establishments recently. I am thinking in particular of the Centré de Géopolitique de l'Energie et des Matiéres, at the University of Paris-Dauphine, whose director is Professor Jean-Marie Chevalier, and where the main reference is the book Professor Chevalier has edited entitled *The New Energy Crisis: Climate, Economics and Geopolitics* (2009). The chapter on oil in this book resulted from that talk, and from another talk I gave earlier at the Ecole Normale Superieure (Paris).

Now we go to what I call 'Key Concepts and Issues' and 'Questions for Discussion'. The Key Concepts and Issues form the basis for questions that, during or at the end of lectures, I ask class members to answer as rapidly and smoothly as possible. As for the Questions for Discussion, some do not require much discussion, but even so they are 'attacked' by teams of two or three students at the black or whiteboard, who are encouraged to use equations, diagrams and statistics. While this is taking place, I sit in the front row of the class and ask a few questions, and I encourage members of the class to ask their own questions as well, some of which lead to stimulating discussions.

Here I can mention that this practice does not go over too well at every university. I can recall one engineering school in which it did not go over well at all, and when I called on several students to do their duty at the whiteboard, they walked out. I will not mention what university this was, although I will mention that it and the educational system in the country in which it is located are very highly thought of by Swedish politicians, who mistakenly believe that educational standards in that country are higher than they are in Sweden. I of course ignored the departure of these ladies and gentlemen, because, as my Swedish students have often explained to me, being able to perform competently at black or whiteboards is an invaluable professional asset.

7. Mathematical Appendix: Easy Topics Made Difficult

The intention below is to complicate a simple topic, one that I first paid considerable attention to in my book, *Scarcity, Energy, and Economic Progress* (1977), and which begins with maximizing an intertemporal expression. This is the sum of discounted revenues between an initial period (t_1) and a final period (T), which leads us to write the expression for discounted profit given below in Equation (A1), which is:

$$V^* = \sum_{t=1}^{T} \frac{p_t q_t - c_t(q_t)q_t}{(1+r)^t} = \sum B_t(q_t) = \text{Maximum } (t = 1 - T).$$
(A1)

$\sum B_t(q_t)$ is the (abbreviated) *discounted value* of expressions such as $p_t q_t - c_t(q_t)q_t$ for periods 1 through T. The summed expression can be written as $B_1(q_1) + B_2(q_2) + \cdots + B_T(q_T) = B(q_1, q_2, \ldots, q_T)$. In non-linear programming we call this the objective function, and to complete the operation, add a constraint that takes into consideration the amount of, e.g., oil produced over the periods $t = 1$ to and including $t = T$. Specifically:

$$q_1 + q_2 + \cdots + q_T = \sum_{t=1}^{T} q_t \leq Q.$$
(A2)

We can write the relevant Lagrangian, with τ (or the Greek letter 'tau') as the multiplier:

$$V = B(q_1, q_2, \ldots, q_T) + \tau \left(Q - \sum_{t=1}^{T} q_t \right). \tag{A3}$$

Assuming that B is concave, the Kuhn–Tucker conditions for profit maximization can immediately be written as follows:

$$
\begin{aligned}
\frac{\partial V}{\partial q_1} &= \frac{\partial B}{\partial q_1} - \tau \le 0 &\qquad q_1 \frac{\partial V}{\partial q_1} &= 0 \\[2mm]
\frac{\partial V}{\partial q_2} &= \frac{\partial B}{\partial q_2} - \tau \le 0 &\qquad q_2 \frac{\partial V}{\partial q_2} &= 0 \\[2mm]
&\cdots\cdots\cdots & &\cdots\cdots\cdots \\[2mm]
\frac{\partial V}{\partial q_T} &= \frac{\partial B}{\partial q_T} - \tau \le 0 &\qquad q_T \frac{\partial V}{\partial q_T} &= 0 \\[2mm]
\frac{\partial V}{\partial \tau} &= Q - \sum_{t=1}^{T} q_t \ge 0 &\qquad \tau \frac{\partial V}{\partial \lambda} &= 0
\end{aligned}
\tag{A4}
$$

Here it should be appreciated that if $\partial B/\partial q_i > \tau$ for period i, then profit is increased by increasing q_i; whereas if $\partial B/\partial q_i < \tau$, profit is increased by decreasing q_i, unless it already equals zero. Moreover, and this is crucial, since we only have one value of τ, the discounted profit is the same for all periods. This might also be the place to mention that if $\sum_{t=1}^{T} q_t < Q$, then $\tau = 0$. The logic here turns on the fact that τ is the scarcity value (or shadow price) of Q, and if it is equal to zero, then being in possession of another unit of Q will not increase the profit (V^*) over the relevant time horizon $(1 - T)$, since all of the original amount of Q is not being utilized. The main thing to focus on here is that if λ is not equal for all periods, then 'change' is taking place: we are not at an (intertemporal) 'equilibrium'. This change consists of moving values of q between periods, or discounted marginal profits must be equal for an equilibrium.

The solution to our exercise is found in Equations (A4), but it is possible to avoid having to explicitly utilize them. Instead, I write expressions such as $\partial B/\partial q$ as $B'(q)$: for instance, $B'(q_t)/(1+r)^t$. In a pure Hotelling scheme, firms are price takers and average cost is equal to marginal cost, but while

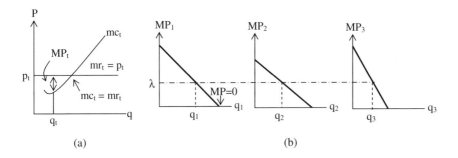

Figure A1.   Optimal Intertemporal Production

I might accept the first of these conditions, the second seems too unreal. Temporarily ignoring the discount factor, it is then possible to write $B(q_t) = p_t q_t - c(q_t)q_t$, and differentiating, we easily arrive at $B'(q_t) = p_t - c'(q_t)q_t - c(q_t)$. Looked at this way, p_t is marginal revenue, while the remainder of the expression is *augmented* marginal cost. The entire expression, *adding the discount factor*, can be called discounted marginal profit, and the conclusion is that discounted MPs should be equal across all periods. Graphically I will treat this proposition as shown in Figure A1(b), with $T = 3$, and for pedagogical reasons, linear MP curves are assumed for each period (which of course is very unlikely to be the case with output being a function of cost).

In Economics 101, marginal profit is zero, since marginal revenue equals marginal cost, but that will not work here because with Q given, we must explicitly take into consideration the fact that, e.g., a barrel of oil sold today will not be available when it might realize a higher profit in the future. Thus, in our typical Economics 101 diagram, or Figure A1(a), the equilibrium is not at $mc = mr$, but at q'', where some output that would normally be produced today might be saved to be produced later.

Several things should be noted very carefully here, and among the most important is that even if, for example, undiscounted $B_t = B_{t+1}$, they would not be equal when properly discounted: B_t would be larger than B_{t+1}, because *ceteris paribus*, 'early' money is more valuable than 'late' money. And now for a caveat: too many very talented economists have been involved with Hotelling's model to dismiss it completely. Perhaps, maybe, somehow it has something to offer, and readers should be on the alert for whatever that happens to be.

Key Concepts and Issues

Anne Lauvergeon	France's nuclear program
Barsebäck	Hotelling's theory (or model)
base load and peak load	IEA and EIA
Cantarell and Prudhoe Bay oil fields	indexing formulae
Finland and nuclear energy	Liquefied Natural Gas (LNG)
gas compressors and 'loops'	natural depletion
OPEC (and GAS-PEC)	NIMBY
diversification (out of oil)	peak oil
Dubai and other Gulf countries	spot (and long-term) trading

Questions for Discussion

1. What is your prediction of the oil, nuclear, and renewables future?
2. Hotelling's model does not appear to be very useful. Explain! Or do you think that it is useful to some extent? Discuss Hotelling's model in *words*!
3. Present and discuss the simple algebra of Hotelling's model. The deposit pressure is not a part of that model, but cost curves are. How do you feel about that? Suppose that you were teaching energy economics, and you wanted to present a numerical example of Hotelling's model to your students. Construct an example!
4. What is present value? Future value? How do you feel about the interest rate being a part of the discussion of the oil future? Can you tell your friends the difference between *interest rate* and *discount rate*?
5. Argue that 'peak oil' is a certainty, or argue that peak oil will never take place!
6. Suppose that you borrow $1,000 from a bank to buy a house, and you must pay back the loan in two years. According to Equation (2), how much do you pay at the end of the first and second years? (This is called *amortizing* a loan.) Now suppose you only made one payment on the loan, and that at the end of the second year. How much would you pay the bank?
7. Argue that 'peak oil' does not matter anymore!

8. Tell us something about the Cantarell oil field in Mexico and the Prudhoe field in Alaska. (Use Google to get some information on these oil fields.)
9. Things seem to have gone well for Dubai. Tell us why this might have happened.
10. Argue for an expansion of nuclear energy! Argue against an expansion!
11. The new 'Generation 3' nuclear plant in Finland took 8+ years to construct. What is special about a Gen 3 plant? Finland is going to order two more reactors, despite this first Gen 3 plant going about 3 billion dollars over budget. What might be their reasoning? That facility was constructed by Areva (of France), and they had to 'eat' the excessive cost. What do you think Ms. Lauvergeon told her board of directors about this situation?
12. Do you have any opinions about the future supply of gas to Europe by Russia? If you were in the OPEC council in Vienna, what strategy would you propose?
13. This chapter begins with a statement by Vladimir Putin, the current president of Russia. What do you think about his opinion of the importance of energy? (Was he right, wrong, or flying high in the friendly sky on vodka?)
14. China and India may purchase about 60% of the increased supply of conventional oil in the next two decades, most of which will be imported. Discuss!
15. Prices in Equation (1) are 'predicted' prices. Explain! What about quantities? Do you think that quantities are easier to predict than prices?

Bibliography

Adelman, Morris A. and Martin B. Zimmerman (1974). "Prices and profits in petrochemicals: an appraisal of investment by less developed countries." *Journal of Industrial Economics*, 22(4): 245–254.

Anouk, Honoré (2010). *European Natural Gas: Demand, Supply and Pricing.* Oxford: Oxford Institute for Energy Studies.

Badal, Lionel (2009). "What the IEA doesn't want you to know about peak oil." *Seeking Alpha* (September 6).

Baltscheffsky, S. (1997). "Världen samlas för att kyla klotet." *Svenska-Dagbladet.*

Banks, Ferdinand E. (2008). "The sum of many hopes and fears about the energy resources of the Middle East." Lecture given at the Ecole Normale Superieure (Paris), May 20, 2008.

———— (2007). *The Political Economy of World Energy: An Introductory Textbook.* Singapore and New York: World Scientific.

———— (2004). "A faith-based approach to global warming." *Energy and Environment*, 15(5): 837–852.

———— (2001). *Global Finance and Financial Markets.* London, New York and Singapore: World Scientific.

———— (2000). "The Kyoto negotiations on climate change: an economic perspective." *Energy Sources*, 22(6): 481–496.

———— (1987). *The Political Economy of Natural Gas.* London, Sydney and New York: Croom Helm.

———— (1985). *The Political Economy of Coal.* Boston: Lexington Books.

———— (1980). *The Political Economy of Oil.* Lexington and Toronto: D.C. Heath.

———— (1977). *Scarcity, Energy, and Economic Progress.* Lexington (Massachusetts) and Toronto: Lexington Books.

Basosi, Riccardo (2009). "Energy growth, efficiency and complexity." Conference Paper; Center for the Study of Complex Systems, University of Siena (Italy).

Bauquis, Pierre-René (2003). "Reappraisal of energy supply-demand in 2030 shows big role for fossil fuels, nuclear but not for non-nuclear renewables." *Oil and Gas Journal* (February 17).

Beyer, Jim (2007). "Comment on Banks." *EnergyPulse* (www.energypulse.net).

Bezat, Jean-Michel (2008). "Petrole: le pouvoir a changé de camp." *Le Monde* (May 20).

Bilek, Marcela, Christopher Dey, Clarence Hardy, and Manfred Lenzen (2006). *Life-cycle Energy Balance and Greenhouse Gas Emissions of Nuclear Energy in Australia.* The University of Sydney (November 3).

Brau, Rinaldo, Raffaele Dononzo, Carlo V. Fiorio, and Massimo Florio (2010). "EU gas industry reforms and consumers' prices." *The Energy Journal*, 31(4): 167–182.

Bryce, Robert (2010). "The real problem with renewables." *Forbes* (May 11).

Carr, Donald E. (1976). *Energy and the Earth Machine.* London: Abacus.

Chenery, Hollis and Paul Clarke (1962). *Inter-industry Economics.* New York: Wiley.

Chevalier, Jean-Marie (2009). *The New Energy Crisis: Climate, Economics and Geopolitics.* London: Palgrave MacMillan (with CGEMP, Paris-Dauphine).

Chew, Ken (2003). "The world's gas resources." *Petroleum Economist.*

Chu, Dian L. (2010). "Natural gas: better days ahead (in two years)." *Seeking Alpha* (November 22).

Cohen, David (2009). "Mr Market gets it wrong again." *321 Energy* (May 29).

Constanty, H. (1995). "Nucleaire: le grand trouble." *L'Expansion* (pp. 68–73).

Cook, Earl (1976). *Man, Energy, Society.* San Francisco: W.H. Freeman and Company.

Cooke, Ronald R. (2009). "The clean energy act is not going anywhere." *321 Energy* (July).

Courtenay, Marc (2009). "Peak oil: reality or a lie?" *Seeking Alpha* (August 27).

Crandall, Maureen (2005). "Realism on Caspian energy: over-hyped and under-risked." *IAEE Newsletter* (Second Quarter).

Cremer, Helmuth and Jean-Jacques Laffont (2002). "Competition in gas markets." *European Economic Review*, 46(4): 928–935.

Dittmar, Michael (2011). "The future of nuclear energy: facts and fiction." Conference Paper; The Institute of Particle Physics, Zurich, January 21, 2011.

Dorigoni, Susanna and Sergio Portatadino (2009). "When competition does not help the market." *Utilities Policy*, 17(3/4): 245–257.

Eltony, Mohamed Nagy (2009). "Oil dependence and economic development: the tale of Kuwait." *Geopolitics of Energy* (May).

Fisher, Franklin, Paul Cootner, and Martin Baily (1972). "An econometric model of the world copper industry." *Bell Journal of Economics and Management Science*, 3(2): 568–609.

Fitzsimmons, Michael (2010). "Foreign oil dependence: the root cause of America's economic pain." *Seeking Alpha* (November 28).

Goodstein, David (2004). *Out of Gas: The End of the Age of Oil.* New York: Norton.

Gould, Len (2010). "Comment on Banks ('Not so nice about oil')." *EnergyPulse.*

Greenspan, Alan (2008). *The Age of Turbulence.* London: Penguin Books.

Gue, Elliot (2011). "What does the growing spread between WTI and Brent crude oil mean for energy holdings?" *Seeking Alpha* (August 28).

_____ (2010). "The difference between oil shale and shale oil, and 6 ways to play it." *Seeking Alpha* (November).

Hamilton, Carl (2009). "SU's Karhuset, Oct. 30, 1959." Conference Paper.

Hamilton, James (2009). "Causes and consequences of the oil shock of 2007–2008." *Brookings Papers on Economic Activity* (Spring): 215–261.

Harlinger, Hildegard (1975). "Neue modelle für die zukunft der menshheit." IFO Institut für Wirtschaftsforschung (Munich).

Holmes, Bob and Nicola Jones (2003). "Brace yourself for the end of cheap oil." *New Scientist* (August 2).

Hunt, Tam (2011). "The peak oil catastrophe in waiting." *EnergyPulse* (March 10).

Höök, Mikael (2010). *Coal and Oil: The Dark Monarchs of Global Energy*. PhD Thesis, Uppsala: Global Energy Systems, Uppsala University.

Jakobsson, Kristofer, Bengt Söderbergh, Mikael Höök, and Kjell Aleklett (2009). "How reasonable are oil production scenarios from public agencies?" *Energy Policy*, 37(11): 4809–4818.

Jefferson, Michael and Vlasios Voudouris (2010). "Oil scenarios for long-term business planning: Royal Dutch Shell, 1960–2010."

Kane, Kevin P. (2009). *Oil Cross-Price Elasticity of Energy R&D Demand: A 12-Country Panel Analysis*. Thesis, Graduate School of International Studies, Seoul National University (South Korea).

Kubursi, A.A. (1984). "Industrialisation: a Ruhr without water." In *Prospects for the World Oil Industry*, edited by Tim Niblock and Richard Lawless. London and Sydney: Croom Helm.

Lorec, Phillipe and Fabrice Noilhan (2006). "La stratégie gasiére de las Russie et L'Union Européenne." *Góéconomie* (No. 38).

Macdonald, Gregor (2010). "A crude oil production forecast for 2015." *Seeking Alpha* (November 11).

———— (2009a). "Oil production: Brazil making the wisest choice of all." *Seeking Alpha* (August 18).

———— (2009b). "Should Mexico stop exporting oil?" *Seeking Alpha* (June 24).

Martin, J.-M. (1992). *Economie et Politique de L'energie*. Paris: Armand Colin.

Morse, Edward L. (2005). "Oil prices and new resource nationalism." *Geopolitics of Energy* (April).

Raineri, Ricardo (1997). "Gas pipeline transportation: competing within ex-ante increasing returns to scale and sunk costs." *Revista de Analisis Economico*.

Rhodes, Richard and Denis Beller (2000). "The need for nuclear power." *Foreign Affairs* (January–February).

Robelius, Fredrik (2007). *Giant Oil Fields — the Highway to Oil*. PhD Thesis, Uppsala: Global Energy Systems, Uppsala University.

Roques, Fabien, William J. Nuttall, David Newbery, Richard de Neufville, and Stephen Connors (2006). "Nuclear power: a hedge against uncertain gas and carbon prices." *The Energy Journal*, 27(4): 1–24.

Ruzzenenti, Franco (2009). "Spacial symmetry rupture and complexity change in energy systems." Conference Paper; Center for the Study of Complex Systems, University of Siena (Italy).

Salameh, Mamdouh G. (2004). *Over a Barrel*. Beirut: Joseph D. Raidy.

Saunders, Harry D. (1984). "Optimal oil producer behavior considering macro-feedbacks." *The Energy Journal*, 4(4): 1–35.

Stevens, Paul (2010). "The Shale Gas Revolution: Hype and Reality." London: A Chatham House Report (September).

Söderbergh, Bengt (2010). *Production from Giant Gas Fields in Norway and Russia, and Subsequent Implications for European Energy Security*. PhD Thesis, Uppsala: Global Energy Systems, Uppsala University.

Teece, David J. (1990). "Structure and organization in the natural gas industry." *The Energy Journal*, 11(3): 1–35.

Urban, Julie A. (2006). "New Age natural gas pricing." *The Journal of Energy and Development*, 31(1): 111–124.

Wallace, Charles P. (2003). "Power of the market." *Time* (March 3).

Watts, Price C. (2001). "The case against electricity deregulation." *The Electricity Journal*, 14(4): 19–24.

Worth, Kenneth D. (2011). "The myth of speculative demand for oil." *Seeking Alpha*.

Yan, Wang (2012). "The price disparity between Brent and WTI" (Stencil).

Chapter 2

The World of Energy 2: An Applied Survey

"Theory improves over time, given the energetic attempts to apply it and the repeated discoveries of serious shortcomings."

— *Professor Darwin C. Hall*

Repetition is perhaps the most important tactic in a strategy for successful teaching. As a result I would like to note once again that when I am confronted with overflowing optimism about the future oil price, I am inevitably reminded of Billy Joel's song "In a New York State of Mind", because wherever these oil optimists may wander, their thoughts are implacably focused on the next oil price meltdown, and some of them have described 'peak oil' as garbage.

I do not think, however, that I possess the 'distinction' or the 'chemistry' to always be taken seriously by certain oil experts, and so I commence this chapter by referring interested parties to the judgment of the star journalist Robert J. Samuelson in *Newsweek* (April 4, 2005). According to him, "Global demand is rising inexorably; global supply seems less expansive. Dependence on precarious Gulf oil will probably increase…" to which I add there is no "probably" about it. Samuelson continues with "The global economy remains hostage to uncertain or expensive fuel. Producing countries may become stronger, consuming countries weaker."

A suitable comment here is that "may" — as in "*may* become stronger" — should have been left out. Finally, Samuelson writes, "There may be more competition among consuming nations to secure long-term supply contracts." That can be extended to read 'more competition that in the long run might turn very ugly', which could mean almost anything, much of it extremely unpleasant, and leaving a bad taste in the mouths of all participants.

Having made that point, this chapter continues where the previous chapter left off. Once again the intention is to provide readers with an easily absorbed background to some crucial energy information and concepts, and in addition to strengthen their energy vocabularies. As I have found out from my adventures in academia and the blogosphere, there are too many bloggers and commentators and researchers who fail to connect because of gross deficiencies in their backgrounds characterized by vocabulary problems. As far as I am concerned, readers of energy economics literature need to be equipped to contribute to any discussion of energy issues, anywhere, at any time.

Not too long ago the International Energy Agency (IEA) warned that governments are not doing enough to prepare for the next spike in oil prices. Their latest prediction is that global demand will rise to 99 mb/d by 2035, which is a lower increase than the 105 mb/d they forecast in 2009, or the 121 mb/d (by 2030) that I discussed in my course on oil and gas economics at the Asian Institute of Technology (AIT) in 2007, and which all of my students eventually accepted as absurd.

I occasionally have doubts about the scientific offerings of the IEA, but I appreciate their pointing out that China is going to require a very large increase in energy consumption in order to attain their economic goals — a much larger increase than often realized, even if economic growth in that country is halved. Among their requirements, a large increase in natural gas imports is expected to take place — unless, possibly, their domestic shale gas reserves are as large as claimed by the Energy Information Agency (EIA) of the U.S. Department of Energy (USDOE).

On the supply side, the IEA pictures OPEC accounting for 50 percent of the world's oil supply in 2035, which may or may not be true; however, as far as I am concerned, this does not mean that OPEC will be producing more barrels of oil in 2035 than they are producing today. Instead, the global oil output may have peaked much earlier, and the output of OPEC oil in 2035 may be well below that predicted by the IEA and other forecasters. In my lectures at the AIT and more recently at the University of Paris, I assured everybody that to make IEA predictions come true for 2030, Saudi Arabia would have to produce 20 mb/d of oil, which they do not intend to do under any conceivable circumstances. My question now is how much of the 50 percent mentioned above will Saudi Arabia have to supply (i.e., export), and if it comes down to more than 8 or 9 mb/d, and perhaps less, then neither IEA predictions nor those of certain others are worth very much. Something that should never be forgotten here is that OPEC's strategy turns on producing as little oil as possible. *All* readers should attempt to explain why!

The IEA also predicted that 5.7 trillion dollars will be needed over 2010–2035 for investment in electricity generation. I am not going to question this, because the global population might have increased by many hundreds of millions by that time. However, the important thing here is that perhaps the IEA comprehends the tremendous and possibly growing importance of electricity. That organization also wants some subsidies phased out. I do not know exactly what they mean, but if they are talking about *all* energy subsidies, I consider it madness to follow or for that matter to pay attention to their advice.

A topic that needs more research is the shale gas 'boom' in the United States, and I would like to report that I am on top of that situation, but it is too early for me to provide the information that I hope to be able to provide at some point in the future. The CEO of a major producer of shale in the U.S. has said that that country has as much gas (energy-wise) in shale deposits as can be found in twice the oil in Saudi Arabia. This certainly sounds wonderful, because if true it means that gas can *really* and not just putatively provide a bridge to a wonderful energy future for the U.S., and this would help everybody. But I think that we will have to wait before ordering a boxcar or two of champagne. Moreover, I would really and truly like to hear what the U.S. Department of Energy has to say about the effect of this gas bonanza on the economy of the U.S., and to get their opinion on exactly how much gas is going to be available over the coming decade.

1. Three (Natural Gas) Kings: Russia, Iran and Qatar

> *"When the Qataris decide on something, they just go for it, they finance it, and they go big."*
>
> — *Wael Sawan (of Qatar Shell)*

Russia and Iran are countries that, in the middle of 2010, had the largest confirmed gas reserves in the world. Russia, which has the largest, is also an extremely important producer of this commodity, particularly where the European market is concerned. Qatar too is a lucky country when considering natural gas, and has become the leading producer in the world

　　　　　　　　　　Energy and Economic Theory

of liquefied natural gas (LNG), eclipsing such rivals as Indonesia, Algeria and Malaysia. Australia, however, has expressed a desire to take Qatar's place in the LNG league. There are many theories about the future actions of Russia where natural gas is concerned (almost too many in fact), but the two that might be the most attractive are that Russia would like to be the dominant natural gas power in the world, and in addition has the highest incentive to exercise market power — specifically, to initiate an upward movement in the price of gas by restricting production, and trying to convince other large producers that they should do the same.

A great deal of Russian gas comes from mature fields in West Siberia that, according to the IEA, are declining at a comparatively high rate. If true, this must be annoying for the Russian government. However, their annoyance is almost certainly eased by the likely presence on their territory (or offshore) of a large quantity of undeveloped resources, very likely including natural gas in one form or another, as well as the market that neighboring China can provide (at least until the Chinese develop their own reputed huge supplies of shale gas). Russia also has large hydroelectric possibilities, and has entered into an agreement with French interests to increase the exploitation of these assets.

Interestingly enough, in the oil-rich Middle East, only Qatar and Iran seem to have enough gas to play a preeminent role in local or international markets — or at least this is what certain observers think — and since Iran recently made arrangements to sell gas to the oil-rich United Arab Emirates (UAE), this may well be true. Qatar is also making some arrangements along that line. Dolphin Energy, a project initiated by the government of the city-state of Abu Dhabi (in the UAE), delivers Qatari gas in 364 kilometers of underwater pipelines down the Persian Gulf, with the supply of gas to Oman being a particular goal. The cost of what may be the first stage of that pipeline was about 3.5 billion dollars, and it is regarded as a project of the Gulf Cooperation Council (GCC).

The situation seems to be that some countries in that region want to use natural gas for electric power generation, while conserving oil as the main input for the production of petrochemicals and oil products (i.e., diesel, kerosene, fuel oil, various petrochemical inputs, etc.). This makes a great deal of economic sense at the present time, although some decision makers in that part of the world seem prepared to substitute uranium for natural gas in the production of electricity. According to recent information about the apparent plans of Qatar Petroleum (the national energy company), they intend to use approximately 12 billion cubic feet a day ($= 12\,Gft^3/d$) of natural gas in the production of LNG, and in addition they want to produce 300,000 barrels a day of liquids from gas-to-liquids (GTL) projects. If the latter is achieved, it will make Qatar Petroleum the largest GTL producer in the world.

Not everybody is happy at the present time about the plans of Qatar to produce and export more natural gas, because the price of gas is comparatively low in some markets, but still high in Asia. Exactly how Qatar plans to make its investments in this sector pay off is uncertain, but I doubt if they would be willing to behave in a manner that would cause the gas price to decline further. Dave Forest says that the answer for Qatar is in their production of natural gas by-products, and he names natural gas liquids and other high-value hydrocarbons as examples of by-products produced along with gas. Revenue from these by-products is apparently sufficient to make it profitable to continue producing gas at the present rate, even if the market price of natural gas remains fairly low. A problem here might be that a considerable amount of the gas of Iran and Qatar is associated gas, which means that it is produced together with oil (i.e., *joint products*). Today, Iranian gas production will almost certainly be focused on the super-giant offshore South Pars gas field. This gas structure extends into the Qatari sector of the southern Gulf, where it is known as the North Field. The combined structure is the single largest gas field in the world.

According to a natural gas model developed at Rice University (Texas), Qatar will be the largest exporter of gas from the Middle East until 2030, and seems to have plans to become a very large exporter to Asia. What happens after that is unclear, although if the model builders were asked, they would probably say that Iran will eventually have similar intentions, and perhaps some day they will be realized. At energy conferences a decade or so ago, many energy professionals and students of the gas markets felt that one of the most dramatic events in modern gas market history would feature the construction of large gas pipelines from Iran to Europe.

In order to make its LNG dreams come true, Qatar has started ordering QFlex and MFlex tankers, which are designed to take 210,000 and 260,000 tonnes of LNG, respectively. This should be compared with the 135,000–145,000 tonnes carried in the largest earlier tankers. As economic theory often suggests, moving up in size could mean a further exploitation of 'increasing returns to scale', which in turn will ensure that in the near future, natural gas originating in Qatar can compete with — or out-compete — any gas from any source in the world. This must sound good to some persons in Qatar, unless they conclude that selling more gas than other producers might involve accepting lower prices. (Exercise: At this point, readers can provide some insight into what it means to accept lower *unit* prices for an exhaustible resource!) This might be the time to say again that Qatar is more than an energy powerhouse, with 26 billion barrels of oil reserves, and 900 trillion cubic feet of natural gas reserves. They are also an investment powerhouse, and my favorite investment is their "Education City", which features facilities for the Qatar branch of three U.S. universities.

The best expression for Qatar might be 'a done deal'. Qatar is a member of the very-rich club that prefers doing business the right way instead of the wrong way, and so there must be some sensible reasoning behind their conduct. For instance, education plays a major role in preparing the country for that day in the future — although probably the distant future — when their gas resources are (or appear to be) in the final stages of depletion. Making the education sector as efficient as possible is the kind of action that every intelligent government facing an uncertain future should think about taking, and that includes countries like Sweden and the United States. Something of considerable interest is the extension of Kuwait's planning horizon by the present government. Earlier, a hundred years was mentioned, but why should they or any rich country, or for that matter any not-so-rich country, be satisfied with placing a limit on their ambitions? Incidentally, Qatar will host the 2022 World Cup (for football/soccer).

Qatar has indicated a strong willingness to do business with buyers in Europe and also North America. Qatar's attitude toward China is not particularly clear; however, since China ostensibly tripled its consumption of gas between 1997 and 2007, it seems likely that sooner or later China will become an important client of a large LNG producer like Qatar. Just recently China and Australia finalized a natural gas transaction involving 2.25 million tonnes of LNG per year for China, in return for which 41 billion (U.S.) dollars will go to Australia over a period of 20 years.

As compared to Russia and Qatar, and perhaps other resource-rich countries, Iran is a country that may someday be of maximum interest to curious development economists, in case any of those still exist. Everywhere in the world it is possible to encounter highly intelligent Iranians, and some of them have superb education, but even so the Iranian economy does not appear to be developing as fast as it should. This situation might very well change, because oil and gas prices may move in such a way as to greatly favor producers who make the right kind of investments in oil, gas, refining and petrochemicals.

On the same day that he received the Nobel Prize in economics, Professor Gunnar Myrdal said (in a loud voice) that he did not believe in Nobel prizes in economics. In the seminar he conducted at Stockholm University, and which I was lucky enough to attend, the reason he gave for this disbelief was that economists did not know how to study the most important topic in economics, which he believed to be development economics. Instead of concentrating on mainstream textbooks or articles in learned journals, he insisted that the only way to efficiently study development economics was to study in detail the dynamics of successful economies, which for every successful country meant the presence or evolution of efficient cultural patterns. It was clear that he considered the United States a very useful role model, although for pedagogical purposes he obviously thought the most useful was Sweden, which is a country where iron ore and forest products played to a certain extent the part that natural gas (perhaps in concert with oil) could play for Iran.

Although many economists have visited Sweden, only a few seem aware that in less than 60 years, Sweden advanced from Europe's Third or Fourth World to perhaps the richest country in Europe, and by 1976 it was a country with one of the highest per-capita incomes in the world. As noted above, the thing that set the Swedish economy in motion were exports of iron ore and forest products (and a sophisticated apparatus for training and utilizing engineers and technicians), and if Professor Myrdal were alive, I am sure he would propose that natural gas could possibly do the same thing for Iran. Moreover, when Swedish development began to accelerate, it became easier to borrow the money needed to construct an extensive electricity sector. That sector is probably the key element in the industrial development of most countries. For example, it played this part in the badly damaged post-war

Finland, and it might turn out that electricity is more important in Sweden and Finland in the future than it was in the past.

Professor Hollis Chenery — who obviously should have received a Nobel Prize for his work — was primarily interested in the application of quantitative methods to development issues, and his particular interest was linear programming, or activity analysis as it was sometimes called. (Activity analysis reduces to solving a linear program, but it makes use of an input-output type format.) His approach was also, in a manner of speaking, similar to what is known in game theory as 'backward induction'. He started out with a vision of how an economy would look when developed, and then systematically worked backwards to determine what this means in terms of production, exports and imports, and the exploitation of domestic resources including labor. The next step was to calculate the accounting prices (= shadow prices = opportunity costs) required to make this vision come true, given the non-human and human resources of the country. As a part of the solution, the relevant linear program would specify the optimal deployment of these resources to various sectors of an economy.

As in Economics 101, capital and various categories of labor (i.e., primary resources) played a key role in Professor Chenery's work. Natural resources were explicitly accounted for as inputs in production activities, or consumption activities, or in the export sector. Obviously, it could turn out that this 'vision' was unrealistic in terms of the assets that were available, in which case the vision would be modified and new calculations made. One country for which Chenery applied his techniques was post-war Italy, which may have been held against him by the Nobel committee. It might also be the case that in a sophisticated version of Chenery's basic model, there would be some indication of how labor should be educated as well as deployed. The government of Qatar may be thinking/planning in these terms.

The question then becomes, why did Chenery put so much emphasis on the 'variables' exports and imports? One answer here is that given the limitations in computing power when he developed his models, and also in individuals with the right attitudes and training to do the necessary computing, the 'message' was in the analytical method and not the realism of these models, nor any result that was or might be attained. With the computing power available today, it would be a different story, although, ironically,

'Planning' — with a capital P — is out of style. However, perhaps that does not make any difference. If Chenery-type logic was exploited in the manner that he outlined in some of his lectures, then the goal would be to steer the economy in such a way that a maximum growth and development effect would be obtained from a 'scientifically' precise utilization of resources such as natural gas. It seems likely that if Iran had been the object of his investigations, he would have concluded that the optimal strategy was constructing capital-intensive facilities for processing natural gas into higher value products, and doing so as soon as possible.

That brings us to Russia. There should be little doubt that if Professors Myrdal and Chenery were still alive and operating at maximum efficiency, they would find it possible to accept that Russia has as much or more to work with than virtually any country in the world. Not just energy resources, but enormous amounts of rich agricultural land, and an educational system that — theoretically — can be turned into the equal of any in the world.

The basic shortcoming seems to be that it was not until the arrival of President Putin that the Russian government began restructuring their economy in a facsimile of the way that Professor Yevsei Liberman suggested that it should be restructured many years earlier. Liberman's position was that emphasis should be placed on individual enterprise, profits and bonuses instead of political doctrine, which at the time was not a popular line to take with many of his countrymen. Thus, the 'inevitable' was postponed about 25 years, and except for a few researchers and journalists, Liberman's name seems to have been forgotten.

But there is still a Cold War shadow dogging Russia's footsteps. Perhaps it has to do with the fact that Russia ostensibly supplies one-third of Europe's energy, and they are getting closer to OPEC, which may account for as much as 30% of global oil production, of which most is still exported. It conceivably has something to do with what the former American ambassador to Sweden erroneously called "the invasion of Georgia", or even the so-called 'pestering' by the Russian government of foreign companies in some parts of Eastern Europe and elsewhere, which may or may not be happening. Or perhaps it is just bloody-mindedness on the part of various observers and would-be experts, because even the London *Economist* accepts that the Georgian president started a small-scale war involving Russia.

Almost 30 years ago a Soviet–Western Europe gas "deal" was announced, following which the U.S. president, Ronald Reagan, tried to convince his European allies that they should only purchase natural gas from politically acceptable regions. Even Argentina was mentioned as a possible supplier, although it was not clear why the Chief Executive or his experts concluded that that country possessed a large amount of exportable gas. Around the same time, General Alexander Haig, a former NATO commander, also assured the governments of Western Europe that it was a mistake to go to the Soviet Union for the gas they needed, since it could be found elsewhere. Haig was well aware that Argentina and a few others that were mentioned in or around the White House were unsuitable candidates for the export of gas to Europe, but he apparently decided that where this issue was concerned, it was best that he remained a team player, and if that involved saying things that scientifically and economically did not make the slightest bit of sense, then duty and career were more important than logic.

Professor Peter Odell, once a widely quoted energy expert and a professor of energy studies in Holland, informed Western European governments that all the gas they needed could be found in the North Sea, and when questions were raised about this grotesque belief, he accused firms operating in that locality of concealing information. Incidentally, this scholar once concluded that the oil supplies of the Gulf were superfluous, and that eventually the major importing countries could find all the oil they needed in places closer to home, like the Arctic Ocean. Professor Odell's thoughts unfortunately placed as No. 1 in a poll of articles in the *European Economic Review* — an Internet publication containing many contributions, a few good, although many are amateurish and completely without any value.

During that period, almost anything — any ridiculous assertion — was considered credible by a number of European academics and politicians whose primary goal in life was to rub elbows with Washington insiders. Eventually a compromise was reached allowing Russian gas exports to Western Europe under new contracts, but regrettably the pipes carrying this gas were under-dimensioned, due to a compromise devised by Mr. Reagan's experts. The ultimate meaning of all this was that energy buyers in Western Europe were given the privilege of paying hundreds of millions of dollars — or more — for their gas than they would have paid if those conduits had been of optimal size.

There are still doubts in some quarters about the suitability of the Russians as suppliers of energy. It would not be surprising to hear that the Swedes believe that the pipeline the Russians are now constructing at the bottom of the Baltic, from the vicinity of Saint Petersburg to a North German port, could someday be used to infiltrate Russian Special Forces soldiers (*Spetsnaz*) into the Stockholm archipelago, for the purpose of crashing those wonderful summer parties that take place in the Midnight Sun. What many students of energy markets fail to comprehend is that an increased purchase of Russian gas by Germany could reduce energy prices in, e.g., Sweden.

Although mostly overlooked, the petroleum major (Royal Dutch) Shell entered into a deal with Russia in 1996 that gave the Shell-controlled Sakhalin Energy Investment Corporation the right to recover all its costs plus a 17.5 percent rate of return on its operations on Sakhalin Island, which is north of Japan. Those costs apparently included pencils priced at $4 each, and salaries that were much higher than those for similar work anywhere else in the world. In other words, the intention of Shell was to make a fool of their host. When the Russians eventually turned the tables on Shell with a strong-arm version of that celebrated game 'Get the Guests', it was labelled harassment by some influential members of the financial press. It may be true that the problem on Sakhalin came about because it was Russia's first liquefied natural gas (LNG) project. Eventually Gazprom — the largest gas-producing firm in the world — may extend to very prosperous operations of this nature on Russian territory, although it has been suggested that at the present time Gazprom and other Russian firms may not be sufficiently mature to handle — in a socially optimum manner — the billions of dollars that will become available before Sakhalin's various natural resource riches are exhausted.

Australia's gas production is projected to grow at an average rate of 4 percent until about 2035, which will give them a leading role in what some observers predict will be a "golden age" of gas. There may also be an increased rate of growth of gas consumption in the Middle East, with much of this gas going into electric power stations. On the other hand, it was recently announced by the government of Qatar that they have the "flexibility" to deliver LNG to any country in the world where there is a market.

A large number of Americans apparently believe that the present century is going to be the "century of natural gas" for their country, and possibly for the world. I can understand why a financial 'mover-and-shaker' might come to that conclusion, although I have some difficulty understanding how his audience would be so naïve as to accept that prediction before the scientific evidence underlying it has been collected and analyzed by competent persons. Clearly, anyone who has done any serious thinking about natural gas must understand that it may be comparatively scarce after the middle of the present century, and definitely will be scarce toward the end of this century. If you do not believe this, try thinking about how gas and other resource markets will look around mid-century when there are two billion more consumers of energy resources on this planet than there are at the present time, and with the number continuing to increase.

This might also be the right place to notify readers who are planning to do any serious research in energy economics, that natural gas offers a great deal of scope, and there are still some questions about this subject that they can profitably examine.

2. Coming to Terms with Coal

It is considered fashionable in some quarters to ignore the coal market, but something happened in that market in 2010 that deserves our attention. The price of coal in the United States increased 31 percent (to $131/tonne), which was a price advance that was only matched by 'yellowcake' (i.e., processed *uranium* metal). This says something to me, because in my publications and lectures on coal, I could never have pictured the coal price attaining that level. *Ex-post*, though, I do understand why we have high prices for coal and uranium: the reasons turn on the fact that generators of electricity are not going to put their future into the hands of, e.g., renewables and alternatives — at least not yet, and not if they have a choice. Anyone who thinks that they would be so foolish should examine the U.S. Energy Information Administration's forecasts for the next 25 years. The growth in energy consumption is predicted to move as fast as ever, or even faster, although there is some doubt as to how this consumption is to be satisfied.

Although coal is still the fastest growing primary energy fuel, its consumption decline in the OECD in 2008 was the steepest since 1992 (and was undoubtedly caused by the macroeconomic difficulties that became severe in late 2008 and early 2009). The continued high growth rate of coal demand can largely be attributed to China and — to a lesser extent — India; despite a great deal of publicity about renewables and alternatives in China, about 70 percent of Chinese electricity is generated using coal. As this chapter is being written, some observers are claiming that the Chinese demand for coal will assure that high prices for that resource will be maintained.

Not too long ago, in the U.S., the directors of 10 firms producing electricity said that they will close at least 35 of their older coal-burning power plants. This may sound unfortunate to some observers, and fortunate to others, but since coal has been the fastest growing source of electricity in the world over the last decade or so (with the possible exception of solar energy, which probably has less than two percent of the global electricity market), and there are a large number of coal-fired electricity-generating plants in the U.S. alone, the main issue for these firms is how much investment they are going to make when the global macroeconomy improves. There is also some talk in circulation among executives in the coal industry about carbon capture and sequestration (CCS), but as far as I am concerned, this is mainly a play for the gallery.

If that is true, then the intention of the G-20 leaders at the 2010 summit in Seoul to stop subsidizing coal (and other fossil fuels) did not amount to much. China apparently intends to raise its consumption of coal in 2015 to 3.6 billion tons, and this should keep the world price of coal high enough so that subsidies to producers are unnecessary. Something worth mentioning here is that both the U.S. and Russia seem prepared to help China satisfy its demand for coal by increasing their exports of coal.

About half of the U.S. electric generation is coal-based, and the global reserve-production (R/q) ratio for coal is extremely large. U.S. production is approximately 2.8 million tonnes of coal a day. But when the total electricity output in the U.S. fell by 3.7 percent recently, which was the steepest decline in 72 years, it was clear that the demand for coal had to fall. There has also been a lot of talk about a rapid increase in the use of natural gas in the U.S. According to one 'expert', natural gas generation would supposedly represent more than 80 percent of additional electricity capacity in 2013, but I prefer to believe that this estimate was based on the present enthusiasm for shale gas, and some very intelligent people say that this enthusiasm is a mistake.

Although it is impossible to be certain, since we do not have reliable estimates of uncertain or 'hypothetical' reserves, coal is often designated as the most plentiful fossil fuel in the world (in terms of its energy content). In addition, until recently it was relatively inexpensive. It generates about 40 percent of the world's electricity, and there is no sign as yet that this amount is on the decline. How can it be, since the latest prediction is that global energy demand will increase by 40 to 50 percent over the next 25 or 30 years. Five or six years ago it was claimed that natural gas would soon be close to coal as a generator of electricity, but even if it moves closer, a possible rise in the price of gas may keep its consumption from reaching that level. The price of gas is very low at the present time (though not in Asia), but prior to the partial macroeconomic meltdown that began in 2008, it occasionally spiked to very high values. Perhaps the most widely noted selling point in favor of gas these days is that it is relatively 'cleaner' than oil and coal, and large reserves of 'shale gas' are being exploited in the U.S. by a new technology that can be applied elsewhere.

Naturally, the coal industry is going to great lengths to convince buyers and potential buyers that the cleaning of coal is a routine and relatively inexpensive operation. The technical aspects are of course interesting, and perhaps most interesting are the techniques for the gasification of coal in order to — ostensibly — reduce its environmental impact. Moreover, the resulting 'syngas' works well in combined-cycle electric power plants, and essentially the same technology that transforms conventional gas to liquids can convert coal into high-quality liquid fuels.

What happens here is that pressure and high heat are used on coal in order to obtain combustible gases, which are then processed into 'clean' (or relatively clean) transportation fuels. There are essentially five stages in this Integrated Gasification Combined Cycle (IGCC) activity, and these can be aggregated as follows. In the initial stage, coal reacts with oxygen and steam to form syngas, which is predominantly hydrogen. This syngas can be fed into a turbine whose revolving shaft generates electricity. To obtain the combined cycle effect, the residual gas can be burned to heat water, whose steam drives a separate turbine.

Don Hirshberg, an engineer and frequent contributor to the best energy 'site' (or forum), *EnergyPulse*, has pointed out that China produced 2.8 billion short tons of coal in 2007 — which it presumably consumed — while the world used 7.1 billion short tons. [A short ton is 2,000 pounds, as compared to a metric ton (or *tonne*) of 2,204.6 pounds ($\approx$ 2,205).] The plan now seems to be to increase Chinese coal production by about 30 percent in 6 years. Given these circumstances, Hirshberg considers it highly unlikely that the rest of the world can reduce emissions faster than China will be increasing them, and so the total *output* of emissions will increase (as will the *stock* of emissions in the upper atmosphere). Some of China's coal might be cleaned, but it would be naïve to think that this will happen to all or even most of it. It is also naïve to think that China, which still has greatly under-developed regions in the western part of the country, is prepared to reduce its rate of economic growth in order to brighten the sky over Sweden, to make the air cleaner over the ski resorts of Switzerland and France, or to enliven the debating get-togethers frequented by naïve environmentalists who are unable to understand how consumers with finite lifespans feel about sacrificing present consumption.

Of course, the thing to understand is that the coal issue broached above is mainly a game. There is too much energy in that resource for it to be abandoned, regardless of what anyone says or thinks, and regardless of whether it is cleaned or not cleaned. I have been told that there are high-ranking and would-be politicians in the U.S. who insist that if the opportunity presents itself, they will give the U.S. coal sector a hard way to go, but it so happens that in this world of ours, actions speak louder than words. Perhaps the most sensible approach here is to continue to think in terms of cleaning large amounts of coal and/or storing large amounts of carbon dioxide (CO_2), while — beginning immediately — only performing these operations to the extent that they make technological and economic sense. And probably just as importantly, doing everything possible to improve the relevant technology, while ridding the scene of bogus and illogical ideas that are constantly launched by self-appointed 'boffins' and 'gurus'.

There may be 600 coal-based plants for producing electric energy in the U.S. Despite claims occasionally presented in the media by various commentators and observers that the U.S. has too much electric-generating capacity, in the last five years 150 coal-based generating plants were at some point in the planning stage, although about 90 of these were cancelled or put on the delayed list. According to primary school arithmetic, this still leaves 60 additional plants capable of generating millions of megawatt-days of additional electric energy, and a great many tonnes of carbon dioxide (CO_2) emissions. More relevant, about 50% of U.S. electric capacity is coal-based, and according to energy insiders, there will be only a few percent less in 2030. This sounds pessimistic, but unfortunately, on the basis of economic theory, consumer psychology and some modest empiricism, it would not be easy to prove that this estimate is wrong.

Games are being played in other countries too. A great deal of effort has gone into convincing Italian voters that because of the wonderful amount of sun enjoyed by their country, solar energy will provide its brightest long-term prospects. This might be the case as long as the solar-based energy installations in the north of Italy have access to subsidies, because otherwise — on average — they might occasionally be as non-competitive in that sunny country as they are elsewhere. The question to ask here is that if solar-thermal power is relatively uncompetitive in Italy, then how accurate is the claim that it will take Germany by storm? The sad truth is that resources like solar and the even more aggressively hyped wind energy are unlikely to live up to their publicity, and so Italy and other countries will have to be more friendly to coal and/or nuclear energy. *Being more friendly does NOT mean abandoning the others, but keeping them in their place 'quantitatively', and this is true everywhere.* Where nuclear energy is concerned, since bringing large amounts of this kind of power on line cannot take place in the near future, the major Italian energy companies are investing in French nuclear facilities, so that electricity from fairly nearby French reactors can be imported into Italy.

In a recent article in *EnergyPulse*, Mark Gabriel (2008) points out that many observers have the wrong idea about the future of coal on the U.S. energy scene. Someone else who had a few words to say on that issue several years ago was the influential billionaire Mr. T. Boone Pickens, who claimed that the oil market situation did not look promising for vehicle owners in the U.S., and as a result they should encourage their political masters to pay more attention to turning large amounts of coal into motor fuel (CTL). (I got the impression that Mr. Pickens later dropped the coal gambit, and turned to natural gas, because although coal was typically one-half the price of gas over a long period, the demand for coal by China and India has meant that its price is increasing — or will increase — at a much faster rate.) The background to the earlier fondness for coal, as well as the observations by Mark Gabriel, is the copious quantities of high-quality coal available for exploitation in the U.S., and also the little-known fact that if the price of oil had not collapsed in the fall of 2008, the U.S. might eventually have been compelled to pay a breathtaking sum of money every year to import oil (and oil products). Moreover, when the price of crude oil was flirting with $147/b, the price of coal escalated to $80/tonne, and perhaps higher in some markets. It appears that the average price of coal may have briefly touched $130/t, and although in terms of its heat content it may not be more expensive than oil, this was certainly unexpected. When I wrote my coal book, I could not imagine the price of coal ever being much over $50/t.

One thing that is being proclaimed at the present time, though, is that the low price of gas (and earlier coal) has caused nuclear energy to become very unpopular where some investors are concerned. (The calculations that are relevant here are sketched later in this book, and it should be emphasized that they are perfectly straightforward.) My reaction to this situation is that it is perfectly understandable that private investors should behave in this manner, but since governments are supposed to take the long-term view, they should make sure that energy resources that will be more viable than exhaustible resources like gas and coal are favored where taxes and subsidies are concerned.

When the talk turns to abandoning coal, few observers and commentators realize how much of that resource is being used, nor how irreplaceable it is in the short run. According to the Energy Information Agency (EIA) of the U.S. Department of Energy, 40 percent of the global production of electricity is coal-based, as compared to 20 percent for gas, 5 percent for oil, and 33 percent for hydro, nuclear, biomass and wind. These figures, however, do not capture the dependence of, e.g., the U.S. and China on coal. Furthermore, coal's place on the global electricity scene is forecast by the EIA to greatly expand between now and 2030 (though in the U.S., coal may lose some ground to natural gas). As pointed out by Dave Forest, good coal-producing assets and properties are considered extremely valuable by Asian buyers, which in this case means China, Japan, India and — surprisingly — Thailand. The same is true for natural gas, although some questions remain as to whether the availability of natural gas has been correctly evaluated. Readers of this book should pay particular attention to this issue.

Regardless of how much coal there is or might be used, it is quite clear that many people do not want to burn coal. What they want is to solve the energy problems of the world with, e.g., wind and solar energy, perhaps natural gas, and 'scientifically' unproved departures like emissions trading. 'Clean coal' might be acceptable, but there are important questions concerning the cleaning of that resource that have not been satisfactorily posed or answered. As to be expected, those questions particularly concern the cost of attaining the advertised results, and where answers are available they often tend to be debatable. According to some basic figures in Victor and Rai (2009), it is easy to calculate that a 1,000-megawatt plant capable of reducing emissions by approximately 90% would cost between $3 billion and $8 billion, and hundreds of these plants would be necessary if climate warming goals are to be realized.

As reported by Professor James Hamilton, Professor Frank Wolak gave a talk at the economics roundtable of the University of California at San Diego in which he reviewed the outlook for wind. The impression given was that he was not very optimistic, although it would not have made any difference to some of us if he was ecstatic with enthusiasm: wind is not going to be able to provide the reliable electric energy that many concerned energy consumers want and have been led to expect, although we should avail ourselves of every watt-hour of wind-based energy that it makes economic and technical sense to obtain. For states like California and Florida, solar may have an important role to play; however, in the interest of realism, Professor Wolak suggested that greater attention should be paid to nuclear and clean coal. Whether this will happen is uncertain, particularly when a detailed (and non-technical) summary of the relative cost of the latter alternative becomes common knowledge.

CCS (carbon capture and storage) is receiving a great deal of attention in Germany, which is largely due to the efforts of Lars Josefsson, for many years the CEO of the Swedish utility Vattenfall, and who might still be an energy advisor to the German government. Vattenfall is of particular interest to many persons in Sweden, because that enterprise did not oppose the scam known as electric *deregulation* (or more correctly, *restructuring*), nor did it visibly protest the ridiculous decision of the Swedish government to consider a nuclear retreat. The reason of course is because these two economic distortions led to increasing the remuneration of the top executives of that firm by a large amount, and perhaps even better, enabled them to transfer a large part of their professional and social activities from Sweden to Germany. As for Mr. Josefsson's theories about clean coal, they belong in the same category as Mr. T. Boone Pickens' theories about a "wind energy corridor" from the Mexican border in the south to the Canadian border in the north, and which has as much economic justification today as the proposed colonizing of Mars or Pluto by zealous space pioneers.

The thing that readers should understand here is that the issue with CCS is not just the technology and the cost, because even if the cost were lower, it would take decades to replace a majority of existing installations. Josefsson supposedly aimed for CCS technology to be "fully commercialized by 2020", and claims that CCS is the way to 'carbon freedom' for Europe's electricity industry by 2050. He also wanted to reduce the CCS cost from 50 euros/tonne to 20, but in terms of orthodox economics, all this is half-baked bunkum or hypocrisy. Moreover, if this verdict fits what is proposed for Germany, it is equivalent to the same thing for the U.S. and the propositions for energy independence filling that country's blogosphere. See, e.g., some observations of Jeffrey Michel (Jeffrey.michel@gmx.net), who is an MIT engineering graduate and the leading commentator on this subject where Germany is concerned. He has referred to CCS as a thermodynamic travesty. I also strongly recommend the Internet site *321 Energy*.

To get some idea of what this might mean, it can be pointed out that the construction of a coal power station is the kind of project that engineers know a great deal about, and have refined considerably. The problems arrive when it comes to capturing the carbon dioxide as it is being produced (by the burning of the coal) and preparing it for storage. This is a very costly activity, and to this must be added another very large cost that is associated with pumping the carbon dioxide into a suitable storage place in the ground or into the ocean. According to some comments on *EnergyPulse* by James Carson, Jim Beyer and Richard Vesel, price-wise the cleaning of coal can be a very unattractive business. Vesel once remarked that at the 'comparatively' low capture and storage cost then in existence, the fuel cost of burning coal is doubled.

What Jeffrey Michel meant by a travesty probably refers to the loss of efficiency that takes place when carbon must be captured from the power station. Fiona Harvey (2006) has estimated this as reducing a modern coal-fired power plant of 45 percent efficiency to about 40 percent. Apparently more suitable innovations are on the way; however, these may not be worth considering, because these new processes and techniques will not appear in large numbers for a great many years, and when they do appear, most of the 'old' coal-burning installations will still be too 'young' to discard. Ms. Harvey does not bother to go into any of this, but the policy of her paper — *The Financial Times* (UK) — is to pretend that we live in the best of all possible worlds where cleaning the environment is concerned, and under no circumstances should the illusions of their readership be disturbed in environmental (and probably other) matters where disposable incomes might be threatened.

When the global macroeconomy gets back on the rails and incomes begin to rise all over the world, the price of energy — and particularly oil and gas — will likely move in such a way that the reluctance to use coal will gradually lose much of its urgency. Given the aversion to nuclear power by many voters, energy independence for the U.S. can only mean a greatly increased, and probably sub-optimal, consumption of coal — some clean, but for the most part dirty — and thus an outcome to avoid if other options are feasible, including a less adversarial attitude toward nuclear proponents. Reportedly, someone in or near the present U.S. government has spoken of deliberately bankrupting the U.S. coal industry in order to promote environmental sanity, but if this expression was actually (rather than reputedly) used, they probably meant that part of the coal industry whose mines are located in or near Beverly Hills or on Park Avenue.

In speaking of the global macroeconomy and its miseries, it perhaps should be pointed out that these do not apply to China to the extent that they did a few decades ago. That country seems to be moving along without a care in the world. The latest estimates of the growth of its Gross Domestic Product averaged between 8 and 9 percent for 2010, which happened to be a very bad year for the international economy. Here it might be appropriate to remember the words of President John Kennedy: "A rising tide raises all ships." By "rising tide" he meant "rising economic tide", which among other things raises the consumption of coal because of the relationship between energy use and national income. Few tides are rising as fast as that of China, where according to some observers, as much new coal-burning capacity is apparently added every two or three years as the total UK installed capacity, although the Chinese government preaches moderation in the consumption of coal.

On the academic front, there is talk of coal becoming what is known as a hot topic, which tells concerned students of energy economics that soon the TV audience might be spending more time pondering what coal has to offer their energy future than was the case a few years ago.

Professor David Rutledge of California Institute of Technology supports his estimate of (comparatively modest) coal reserves by saying that "curve fitting is a worthy competitor to a geological estimate". Rutledge is a talented engineer, but on the basis of even an elementary acquaintance with curve fitting, this contention is questionable. Professor Rutledge has apparently taken the same approach with coal statistics as M. King Hubbert took with oil statistics, when Dr. Hubbert astounded the scientific world with a correct prediction that the output of oil in the Lower 48 of the United States would peak in late 1970 or early 1971.

The fact of the matter is that the geology of an oil deposit is radically different from that of a coal deposit. Professor Rutledge did not identify the date of a coal peak, but apparently believes that the world will have produced 90 percent of its total coal reserves by 2069. On the other hand, Mikael Höök of the University of Uppsala and perhaps some of his colleagues predicted a coal peak by 2020, followed by a 30-year plateau. Like the other researchers in Global Energy Systems at the university, Mikael Höök is a very talented energy student and teacher, but he might need a short vacation from coal in order to spend more time in the wonderful student clubs of Uppsala University, especially on Friday and Saturday nights. A coal peak by 2020, along with a possible peaking of the global oil production about the same time or slightly earlier, is not the kind of situation that intelligent people would like to see in their crystal balls.

As mentioned, Mr. T. Boone Pickens at one time wanted more effort put into turning coal into liquids. The CTL process that he had in mind was the Fischer–Tropsch process employed by Germany during the Second World War, and which was improved by the Sasol Corporation of South Africa during the period in which oil exports to South Africa were embargoed. We heard a great deal about these programs; however, there is no reason to be overly impressed. The German output of 'synthetic oil' — or 'syncrude' as it is sometimes called — was barely enough to fuel the planes, tanks and trucks of a military establishment that was fighting a losing war on two fronts, and during the apartheid period a considerable amount of conventional oil was smuggled into South Africa via surrounding countries, although this was seldom mentioned by so-called concerned admirers of the 'Front States'. Instead, these observers preferred to believe that the remaining countries of Africa were willing to accept any sacrifice to bring the apartheid government to its knees, even though in reality this was very far from the truth.

It should perhaps be mentioned that globally, coal contains a very large amount of methane that potentially can be extracted as a gas. Although the subject is not discussed extensively at the present time, methane is one of the items that environmentalists like least of all, which means that its large-scale commercial exploitation will have to wait a while, but in a situation of energy scarcity we will undoubtedly find out exactly what this resource has to offer. Some students of energy matters feel that methane has a great deal to offer. For example, one of the most important commentators on the forum *EnergyPulse*, Jim Beyer, wants to see its use greatly expanded. He believes, and correctly, that this is a powerful energy source, but it may have certain disadvantages that cause a number of observers alarm.

One thing remains in this section, and that is to duplicate a calculation that makes a limited sense when applied to a resource such as coal, though not perhaps to oil and natural gas. To be specific, what I want to do now is to show how the effect of the growth rate g on the availability of coal influences the availability of coal, although it should be emphasized that the result obtained is only an approximation. But before getting started with this exercise, I would like to make a few uncomplicated comments on the important subject of discounting, which tends to come up a great deal when dealing with a subject like energy economics, where intertemporal relationships are crucial.

To keep matters simple, we consider an amount of money at time 'zero' equal to P. Then, with an interest rate of r, we calculate future value F. A key item in this exercise is the expression 'compounding', and as an exercise you can ask your friends and neighbors or computer what it means if you do not know, or try to figure it out yourself by putting some numbers in the expressions below:

1 year, 1 compounding/year	$F = P(1 + r)$
T years, 1 compounding/year	$F = P(1 + r)^T$
T years, n compoundings/year	$F = P(1 + r/n)^{nT}$
1 year, n compoundings/year	$F = P(1 + r/n)^n$

Something of interest in these matters is that e^{rT} can function as an approximation for $(1 + r)^T$ when r is small. Let us try this out for a one-year time horizon ($T = 1$), with two values of r [$r = 5\%$ ($= 0.05$) and $r = 50\%$ ($= 0.50$)] and with $P = 1,000$:

$$r = 5\% \quad F = 1,000(1 + 0.05) = 1,050 \quad F = 1,000e^{0.05} = 1015.3$$
$$r = 50\% \quad F = 1,000(1 + 0.50) = 1,500 \quad F = 1,000e^{0.50} = 1649.7$$

As we see, the approximation is good when r is small, but not when r is large.

Now we turn to the main order of business. Taking X_t as the consumption of coal in period t, we can write:

$$X_t = X_0 e^{gt}. \tag{1}$$

In this expression, g is the growth rate of X, while t is 'time'. The term X_0 is the consumption of the resource in the initial period. Cumulative resource use is then defined as the integral from the initial period to a terminal period T, whatever that happens to be:

$$X = \int_0^T X_t dt = \int_0^T X_0 e^{gt} dt \quad (\text{or}) \quad X = \frac{X_0}{g}(e^{gT} - 1). \tag{2}$$

If X^* is the total amount of the resource available at an initial period, we obtain the following relationship for the approximate time to exhaustion T_e:

$$T_e = \frac{1}{g} \ln\left(\frac{gX^*}{X_0} + 1\right). \tag{3}$$

It is also useful to observe the effect of changes in X^* on T_e. Differentiating, we get:

$$\frac{dT_e}{dX^*} = \frac{1}{gX^* + X_0}. \tag{4}$$

What we observe here is that substantial changes in X^* are not reflected in time to exhaustion (T_e). We can also compare the difference between the static time to exhaustion (where $g = 0$) and the dynamic time to exhaustion (where, e.g., the value of g is taken as 2.5%/year). The static value ($= X^*/X_0$) is approximately 260 years. Now, adjusting this for growth by employing Equation (3), we get $T_e = (1/0.025) \ln[(0.025 \times 260) + 1] = 80.5$ years for the 'dynamic' value, which is a sizable difference. Enough to make us wonder just how much comparatively inexpensive — and high quality — coal our great-grandchildren will actually have at their disposal.

The word "approximate" was used several times above. This is because although the *annual* rate of growth is, e.g., 2.5%, the compounding in the derivation is continuous, and so the *effective* growth rate turns out to be greater than 2.5%, and so g in the calculation is overstated. This means that the actual T_e would be greater than 80.5 years, and if increases in X^* took place, which is likely, there would be a further increase. Of course, T_e will not come anywhere near the static value. Readers who want to avoid the effect of continuous compounding can work with $(1 + r)^t$ instead of e^{rt}.

One person who believes that too much reliance might be placed on coal is Richard Heinberg of the Post Carbon Institute in California. In fact, he has said that "Energy policies relying on cheap coal have no future." That must certainly sound good to many environmentalists, and in a conference or seminar this may sound right, but not in the real world. In fact, his belief that peak coal may be only "years away" sounds a touch overstated, unless by years he means 20 or 30 or 40 or more. For better or worse, a large amount of coal is going to be used every year for a very long time for the simple reason that most people are more concerned with what happens today rather than in the misty future. This can be put another way: conventional or mainstream economic theory does not pay enough attention to the future. The global population has almost doubled in just 40 years, and so everybody needs to consider what the next 40 years will mean in terms of population and, by the same token, social and economic pressures.

3. An Invaluable Lesson on Electric Deregulation

> *"At the mercy of forces that show no mercy."*
>
> — *Former Governor Gray Davis of California*
> *(about deregulation)*

> *"Deregulation fosters turmoil in power markets!"*
>
> — *Headline in* New York Times *(July 15, 1998)*

The bottom line where this topic is concerned is undeniably simple: *electric deregulation has failed and is failing almost everywhere, and increasing numbers of observers are now prepared to admit that it cannot succeed in the real world, despite its occasional success in seminar rooms and conferences.* For instance, the majority of contributors to outstanding forums (such as *EnergyPulse* and *Seeking Alpha*) are now ready to accept that *as a result of deregulation, the average wholesale price of electricity in Southern California suddenly escalated to approximately ten times the average electricity price in 2008–2009.* What they do not know, however, is that in Sweden, where nuclear and hydro are the main generating assets, and as a result the *cost* of generating electricity should be — and occasionally has been — the lowest in the world, the *price* paid by households may recently have been the highest in Europe, and according to the latest forecast, it will continue to increase. We can thank deregulation for this fiasco.

Electric deregulation can perhaps be best described as an unsuccessful attempt to rescind the laws of mainstream economics. A justification for continuing the criticism of deregulation is the steadily increasing body of evidence at variance with fantasies about expected deregulation results, where by the latter I mean over-optimistic economic research that promised large amounts of reliable and inexpensive electricity if deregulation (i.e., *restructuring* or *liberalization*) were initiated and allowed to proceed without the meddling of politicians or bureaucrats. Electricity prices may indeed be lower today in some countries, but this is not because of deregulation. It is because of a fall in demand for electricity as a result of the partial macroe-conomic meltdown that has taken place. In spite of assorted tributes that are still occasionally being paid to deregulation, electricity prices in, e.g., Sweden have spiked to record levels.

In the United States — and perhaps elsewhere — it is not unthinkable that influential academics and others will once again rush to claim that successful deregulation ventures are possible, but this is a claim that will not be easy to foist on intelligent men and women whose memories still function. For example, according to Kimery C. Vories, an official in my former home state of Illinois, the "fruits of deregulation" in that state included price increases of 40–50 percent "all at once", lengthy electric outages, and increases in the salaries of the chief utility executives that ranged into the millions of dollars. He completed this catalogue of outrages by saying that "the role models of the corporate leaders of utilities in Illinois are the persons now serving jail time for their role in Enron's antics". Unfortunately, he failed to cite those who got away before the crash.

We continue with a little more background, beginning with the following quotations:

> *"What we've seen this winter is that the deregulated market works for (electric) companies, but not for consumers."*
>
> — *Hallgeir Langeland (Norwegian Parliament, 2003)*

> *"I have to expect to be a scapegoat (for deregulation failure), but people know they can't blame me for the policy."*
>
> — *Einar Steensnaes (Norwegian Energy Minister, 2003)*

Of course they can't blame you, Einar. The people to be blamed are the ladies and gentlemen who sent you letters questioning your intelligence when the 'spot' price for electricity in your country escalated from $4.62 per megawatt-hour to more than $140/Mwh during at least one brief period, but who earlier believed 'experts' on this topic, or for that matter concerned bloggers, who claimed that orthodox economic theory supported a freeing of (oligopolistic and in some countries monopolistic) electricity markets.

When I was a visiting professor and university fellow in Hong Kong, I gave a series of lectures on the shortcomings of electric deregulation, the last of which was presented at the Hong Kong Institution of Engineers during that marvellous period when the cracks in Enron's boisterous façade began to show. As Dr. Larry Chow, director of the Hong Kong Energy Research Centre, pointed out in the discussion following this lecture, engineers could only shake their heads when they heard economists talking about deregulation, but at that stage of the deregulation crusade this kind of mild physical gesture was far from enough to convince dedicated deregulation enthusiasts that they would be unable to succeed in their attempt to make the impossible possible.

As discovered later, Dr. Chow was not the only observer who came to suspect that a deception of monumental proportions was about to be perpetrated on taxpayers throughout the world. Joseph Somsel, a nuclear engineer and resident of California, has described the conclusions he reached whilst engaged in a business school project after leaving a (U.S.) West Coast utility (2007). This project focused on evaluating the impending California deregulation effort, and Somsel's research led him to conclude that (1) "Somebody would make a great deal of money since (2) there would be shortages causing prices to climb through the roof." In addition, (3) there would be excessive volatility. *Equally as significant, he concluded that the trading of electricity futures and options would not work!* In a sense he was wrong with the last observation, because they worked beautifully for the persons in the electricity exchanges — those lucky individuals in front of the computer screens who experienced little or no difficulty in brightening their lifestyles at the expense of consumers of 'physical' electricity.

As was inevitable, Somsel's astuteness was overlooked. It should be pointed out though that the initial acceptance by many retail electricity buyers (i.e., households) of a price regime in California that featured 'pegged' (i.e., fixed) prices can be largely attributed to the urging of the newly privatized utilities (or *distribution* companies), who believed various self-appointed savants when they said that competition among generators (i.e., wholesalers) would lead to these firms greatly reducing *wholesale* electricity prices. Thus, if retail prices were kept at about the same level (by, e.g., regulation), or were even pegged somewhat lower, utilities would still be in a position to enjoy lovely profits. Instead, when wholesale prices escalated, several utilities (i.e., distributors) found themselves billions of dollars in debt by virtue of having to buy electricity (from wholesalers) at prices above the (regulated) prices they were allowed to charge.

Most important of all, since deregulation was sold to the voters by promising that retail (distribution) prices would fall, allowing large increases in consumer prices for the noble purpose of 'clearing' the market was politically out of the question unless a large portion of the voting public was in one sense or another disenfranchised. This may have been one of the reasons why a California legislator, who was a prominent instigator in introducing deregulation, later joined Governor Davis in his efforts to protect California consumers from what the governor called "out-of-state criminals" — i.e., non-California generators (or wholesalers) who were in a magnificent position to utilize their (strong) oligopoly situations. Similarly, on the East Coast of the U.S., Senator Ernest Hollings brusquely abandoned the deregulation sinners who had seduced him into the ways of 'liberalization', and began to call himself a "born-again regulator".

Although many of us have studied this topic for years, and attempted to point out that electric deregulation was a gigantic mistake, few economists were able to predict the magnitude of the deregulation failure that actually took place in California (or, for that matter, in other regions). The normal generating (or 'wholesale') price of power in both these localities occasionally spiked to grotesque levels, where in Sydney (Australia) the cause of the trouble was the belief espoused by a Nobel Prize winner in economics — and accepted by less distinguished scholars — that deregulation would lead to more 'trading', and ostensibly the greater the amount of trading, the better for all concerned.

'Gaming' (or strategic behavior that usually features illegally (or in some cases legally) manipulating supply), together with the shortage of local generating capacity, and the absence of rain for hydroelectric installations, explained most of the bad news in California. However, it should never be forgotten that while deregulation prohibited the large California utilities (i.e., 'distributors' or 'retailers') from signing long-term contracts that might have provided some protection from these excessive prices, the large utilities were not allowed to pass those prices to, e.g., households and small businesses. Why was this? It was because consumer (retail) prices could have escalated by as much as 200%, and as Governor Gray Davis made clear, the Californian economy might have been shocked into recession.

Pacific Gas and Electric Company, perhaps the largest utility, moved into bankruptcy — although this was after providing occupants of its executive suite with millions of dollars in the form of a bonus 'kitty' to divide. In the course of establishing itself as a kind of intermediary for purchasing electric power from wholesalers, and then dispatching this power to utilities/distributors, the Californian government spent somewhere in the neighborhood of 4 billion dollars, with most of the electricity purchased apparently originating from the very same generating firms that Governor Davis described as out-of-state criminals. Terminology of this undiplomatic nature unexpectedly came into general use when it was surmised that at least some wholesalers intentionally reduced their supply of electricity in order to increase its market-clearing price. The flimsy excuses they used to justify their behavior are similar to those recently employed by certain electricity wholesalers in Europe. However, the ultimate blame for this state of affairs mostly belongs to voters and so-called economics commentators and journalists, who, had they only been awake and sober, could hardly have failed to be aware of the deception that had taken place in California.

A suggestion originating in academia was that generators should somehow be broken up into smaller units in order to become *price takers* instead of *price makers*. A well-known and very talented American professor of economics prescribed this solution for the UK. Unfortunately he must have been suffering from jet lag at the time, because if there is a single industry on the face of the earth in which returns to scale are important, it is electricity generation. Of course, it might happen that he was naïve instead of ignorant, in which case a remark from Graham Greene's novel *The Quiet American* deserves citing: "Such naiveté is a form of madness."

My work on electric deregulation began more than a decade ago, and the agony has continued, and in some cases escalated. As already noted, in examining the history of electric deregulation across the world, it appears that failure is the usual outcome, and often sooner rather than later. In the United States, electric deregulation has crashed in almost *every* state in which it has been attempted, although recently I heard that the situation has improved in Texas. This may be true, but when discussing the Southern part of the U.S., the regulated and vertically integrated firm Southern Power — with about 40,000 Megawatts (MW) of generating capacity, serving more than 4 million customers in 4 southern states — is constantly cited as an unequivocal success.

An important energy observer, Thomas Tanton (at that time with the Institute for Energy Research, Houston), was unhappy with a comparison of this nature that I made in some lecture or paper, pointing out the different conditions that prevail in Texas as compared to the districts in which Southern Power operates. In principle, students of energy economics should be sufficiently generous to appreciate Mr. Tanton's comments, only it happens that deregulation's record is so incredibly poor just about EVERYWHERE that, while Texas is indeed a very special state, it is not so special in the context of American states and regions that various economic shortcomings could be attributed to uncommon qualities and circumstances.

In the matter of transmission lines, Governor Davis of California made the following observation: "In a deregulated environment, investment chases the highest returns, and the highest return is not in upgrading transmission." By way of educating American voters on this dilemma, a former energy minister in the U.S. — Mr. Bill Richardson — once declared the transmission system in the U.S. suitable for a Third World country, but not for the only remaining superpower. Had 'wannabe' presidential candidate Richardson made it clear that, prior to embarking upon the deregulation fling, the U.S. transmission network was generally regarded as sufficient to provide almost all Americans with reliable and economical power, he might have deserved to be judged a legitimate contender for the most important political post in the world; but for observers who have taken a deep interest in the mysteries of electric power transmission, it may have sounded peculiar to hear that the wires above their heads in Chicago or San Francisco were in the same qualitative and quantitative category as those on the rim of the Kalahari or in the Chaco.

Academic economists make a point of misunderstanding this situation, and so perhaps a short clarification is in order. Deregulation can bring about a rapid shift of the loads on *grids* (i.e., the network of wires that carry electricity). Instead of in a vertically integrated system with (approximately) known final demands, there can (in theory) be a shifting amount of buying and selling *between* (as compared to within) networks and regions. As David Buchan — then of the *Financial Times* — pointed out (April 11, 2002), this was why Enron — the failed U.S. energy trader — was such a fervent advocate of deregulation: the more trading, the more income Enron would enjoy! What Buchan did not mention was that in Europe, Enron wanted comprehensive 'liberalization' regardless of the cost, because an orgy of trading might have added enough billions of dollars to corporate profits to compensate for the embarrassing business shortcomings that were later revealed when the directors of that enterprise found themselves expressing their hopes and dreams in a court of law instead of in a special edition of *Fortune*, *Forbes*, or that compendium of London wine bar gossip, the *Economist*.

As for salaries, a short time after deregulation in the UK, the first page in the business section of a London newspaper displayed a gallery of power-company executives whom deregulation had catapulted into affluence. This observation might also apply to the major Swedish company Vattenfall, although admittedly executives of that firm were underpaid before deregulation. After deregulation, its managing director shifted his focus to Germany, where he busied himself with promising that his firm's intention was to bolster the electricity supply in all of Northern Europe, and to make it 'green' as well. It should be stressed, however, that some industrial sources in Germany have noticed Vattenfall's interest in highly polluting coal, which several observers feel is why Vattenfall did not protest more aggressively the decision by the Swedish government to consider a nuclear retreat.

My previous energy economics textbook (2007) is filled with unpleasant facts about electric (and natural gas) deregulation, but as they once said in the U.S. Navy, "on every ship there is someone who does not get the message." Getting the message seems to have become more difficult than ever, because in the case of Sweden it is hard to imagine anyone who, when in the grip of sobriety, is capable of voluntarily describing the Swedish electricity experiment as anything other than a grotesque mistake that has increased the financial burdens on households and small businesses, and even some large businesses.

The failure of electric deregulation in Sweden and elsewhere should be a grim reminder to Swedish voters and non-voters, or for that matter anybody living in Europe, that a tolerance of greed and unawareness has found its way into the Swedish lifestyle, and it might make life miserable for many purchasers of electricity in this country in the near future. Professor Susanna Dorigoni, director of research at Bocconi University in Milan, Italy, and her colleagues and associates have pointed out that the same judgment applies to natural gas in Italy, which comes as no surprise to me, because in my textbooks and lectures I have often claimed that the desire for a comprehensive deregulation of natural gas pipelines in Europe is a serious mistake. Moreover, when I lectured in Hong Kong, I maintained that if a successful comprehensive electric deregulation could not be achieved in a country as rich, technically advanced, and market-oriented as the U.S., then it could not be realized anywhere on the face of the earth, at least in the short to medium run, unless enormous costs were assumed in order to revise and/or expand the existing pattern of wires and — when relevant — pipelines.

There have been important persons who doubt this judgment. For instance, another former U.S. Energy Minister, Ms. Hazel O'Leary, once pronounced deregulation efforts in Pakistan the best in the world, but before this evaluation received widespread approval, some market actors in that country began to shoot at each other with real guns loaded with real bullets, and so a new assessment became essential. This resulted in the Pakistan deregulation experiment being downgraded and reclassified as a lamentable botch that was distinguished only by an exceptionally heavy slice of corruption.

What do we mean by failed or failing when considering electric deregulation? Undoubtedly, the most important reason for deregulation coming into fashion was the promise by various advocates and their paid or unpaid propagandists of lower prices and higher or unchanged reliability, all of which would be served up against a background of increased 'choice' for households and businesses. Choice was inevitably mentioned in academic circles, because it is a standard of value in academic circles that if there is 'genuine' choice on the consumer and producer side of the electricity market, or any market for that matter, it would lead to the kind of optimality described in almost every chapter of your favorite economics textbook.

The word "efficiency" was also frequently introduced into discussions of the paradise-on-earth that was supposed to result from electric (and also natural gas) deregulation, although hardly anyone understood its scientific meaning. In classroom situations, regardless of the level, I elected to present a clarification that begins with some materials from introductory economics, which underscore that *excess profits* are profits that are only available on a temporary basis in markets with free entry and full transparency. Accordingly, an efficient electricity market denotes one in which if there is free entrance, free access by all actors to existing technology, and where consumers have the right to change supplier when and as often as they choose, then firms will supposedly be unable to realize excess (or 'supernormal') profits for other than a short time, and consumers with conventional preferences will not have to experience the outlandish charges for electricity that have become commonplace in many countries.

The extension of this to the present topic is straightforward: according to orthodox or mainstream economics, finance, engineering and common sense, there *would* not — nor *could* not — be anything resembling lower prices in a situation where unregulated monopolies or strong oligopolies are allowed to practice what the great American songwriter Irving Berlin called "doing what comes naturally". This is essentially because of the prodigious advantages enjoyed by, e.g., monopolies due to increasing returns to scale (= decreasing unit costs), and in addition high investment costs for actual or potential competitors. For instance, residents of Sweden have been permitted to enjoy freedom of choice in the matter of their electricity supplier, but what difference does it make when it has become impossible to avoid price increases that occasionally were considerably more than twice the consumer-price inflation rate, in some months a great deal more, with the likelihood of even higher prices because of a geographical expansion of the market resulting in a much higher demand that is immediately translated into higher prices because of the presence of an electricity 'exchange'.

This might be the right place to quote U.S. Senator Byron Dorgan: "I've had a belly full of being restructured and deregulated, only to find out that everybody else gets rich and the rest of the people lose their shirts!" (*Financial Times*, April 22, 2003). Exactly how a statement of this nature was received by his colleagues is difficult to say, although I was tempted to write to Senator Dorgan and inform him that one of the most important researchers at the very conservative Cato Institute (in Washington, D.C.) was also prepared to admit that electric deregulation had failed, and moreover its shortcomings were incurable.

"Now before you ask whether I am still asleep or dreaming or had something extra in my coffee this morning," the chairman of the independent Electricity System Operator of Ontario (Canada) remarked several years ago, "let me qualify this by noting that I have not given a timetable to arrive at this destination", where "this destination" included a "reliable, efficient, effective, transparent, accountable, credible and competent" supply of deregulated electricity. That is putting it mildly, because on the date when the contents of Madame Chairman's morning coffee came into question, Ontario had less generating capacity than it possessed a decade earlier, and according to the president of the Association of Major Power Consumers of Ontario, a bungled deregulation agenda had resulted in that province losing an invaluable competitive advantage. If you go through some 'back presentations' on *EnergyPulse*, the comments of Len Gould and Bob Amorosi, both engineers, have taken up in some detail travesties of this nature in Canada, although they are no different from those evident in most other countries and regions.

The time has now arrived for a slightly more technical recapitulation. Suppose that you were a reputable member of the deregulation booster club, and you were given the opportunity to clarify for the television audience why electricity deregulation is even more beautiful than love's young dream. You might start by dramatically insisting that every economics textbook in the world spells out in detail the advantages of a wider and more thorough competition (or what on the electricity scene has come to be called *liberalization*). What you would not say is that this storyline appears in the first part of these books, while in later chapters — which students and teachers often do not bother to read — there might be a detailed explanation of why things could go wrong in the case of industries like electricity and gas. In non-technical language, this reduces to the following:

(1) The unambiguous inability to establish the kind of (theoretically) ideal competitive arrangements found in the first part of your favorite textbook. As already noted, this is due for the most part to increasing returns to scale (i.e., decreasing unit costs) up to a certain output. The way this potentially embarrassing topic was originally handled by deregulators was simply to state that economic — as opposed to technical — increasing returns to scale do not exist. How did they get this brilliant result? The answer is that they assumed that there would be a fall in the rate of growth in demand for electricity and gas, and so investments that were intended to take advantage of technical returns to

scale would take so long to pay off that, in terms of discounted profits, they did not make economic sense! This was perhaps the most unsound reasoning of all.

(2) Lack of investment in new capacity. As it unfortunately happens, deregulation has provided firms with an increased opportunity to engage in *strategic behavior* (as it is called in game theory), or gaming the system, as it is often termed in common parlance. Accordingly, these firms utilize the well-known fact that the smaller the total capacity, the higher the price, and depending upon costs, the greater the profits and bonuses. As pointed out at one time by many deregulation enthusiasts, and correctly, for deregulation to not only be successful but commendable, adequate facilities should be available for *hedging* (i.e., insuring against) the price risk that accompanies deregulation. Despite what these fine people believed, and perhaps still believe, efficient or credible facilities of this nature are *not* available, and they are unlikely to appear. This dilemma is constantly referred to in the business press.

(3) Because of its importance, it should be stressed that electricity price risk/uncertainty cannot be hedged on conventional derivatives (e.g., futures and options) markets of the kind that function excellently for various commodities and financial assets. As we found out in California — and many other localities — the electricity market is radically different. In fact, there is a simple way to approach this difference. Professor Lennart Berg teaches financial economics at Uppsala University, and he has about 100 finance books of all types in his room. Every new book that is published on this subject comes to him. I have examined at least half of these books, and estimate that there are probably less than a total of five pages on electricity and natural gas derivatives in all his books, or for that matter in any collection of finance books between the ski slopes of northern Sweden and those of Chile or Argentina. Five pages out of thousands or tens of thousands of pages. What none of them bother to point out is that the most important derivatives exchange in the world, NYMEX in New York, delisted its electricity contracts several years ago, along with at least one of its gas contracts. They may, of course, have reinstated these in one form or another, because the memories of many people who lost money on those contracts are too short for their own good.

In the June 18, 2002 issue of the *Financial Times*, there were two mentions of violence in connection with the deregulation of electricity. The countries involved were the Dominican Republic and Peru. The second of these was of particular interest to me, because some years earlier, in Lima (Peru), I delivered a keynote address at a conference that was attended by a large number of energy economists and engineers from the Caribbean and South America, and for various reasons I chose to remind participants of the kind of advice that the Brazilian government was forced to give citizens of that country when a demented deregulation concept, in conjunction with a rainfall deficiency, resulted in catastrophic electricity shortages: Brazilians were encouraged to pray to Saint Peter who, according to Brazil's frantic deregulation authorities, is in charge of the rain department. Something else that happened in Brazil was an important executive in the electric sector expressing his approval of regulation.

Shortly after the difficulties in Brazil surfaced, one of the best known poster boys for electricity deregulation insisted that New Zealand gave no consideration at all to the experience of other countries in its own deregulation efforts. What he overlooked was that there was hardly any to examine when that country initiated this "goofy" experiment, as someone in Canada called it. Instead, the deregulators in that fair land focused their attention on the large supply of domestic natural gas, whose price — by one means or another — was kept below the scarcity/free-market level in order to ensure the blessings of deregulation. As things often happen, that large supply of domestic natural gas has become small, and as a result the failure of New Zealand's deregulation 'experiment' became visible for everyone to observe and interpret in any way they chose.

The managing director of one of the largest firms in Sweden, called SCA, once said that (*ceteris paribus*) his firm would not be making a planned, very large investment because of the high price of electricity in electric-deregulated Sweden. Somewhat earlier, the directors of other large industries stated that they will form a syndicate in order to purchase electricity from countries in Eastern Europe.

This is extremely interesting, whether they persist with their intentions or not, because what deregulation initially did by raising electricity prices was to threaten to partially eliminate the traditional and highly favorable comparative advantage that Sweden has enjoyed in some of its major export activities over the past 40 years, and which to a considerable extent accounts for the prosperity of the country. This kind of discomfort has also been noted in Canada, particularly in Ontario, and could be experienced everywhere. If deregulation is once again allowed to bloom, or perhaps just to continue in its present form, *and is sustainable*, many electricity-intensive firms (and other types of firms) in this country will not just reduce their investing, but could eventually move everything movable out of the country, in which case Sweden could find itself having to deal with the kind of unemployment and social problems that were unthinkable when I first arrived in Scandinavia.

Readers who want more of the above should examine Casazza (2001) and Watts (2001). If you desire a taste of the other side of this argument, turn to the very important forum *EnergyPulse* (www.energypulse.net) and particularly the many comments attached to articles on electricity regulation and deregulation, as well as, e.g., global warming and nuclear energy. I am thinking in particular of the consultant Jose Antonio Vanderhorst-Silverio (a current or former resident of the Dominican Republic). He admits that deregulation has been a disaster, but he also sincerely believes that somewhere out there in the great world of economic theory is a magic formula for making everything right.

My comment here reduces to the following: In the real world, no playing of games on the consumer side of the electricity market can possibly offset the upward pressure on prices resulting from conventional profit-maximizing behavior on the supply side of that market in the presence of deregulation. (This same comment applies to the global market for crude oil.) Put another way, given the technological configuration of the electricity sector (and the presence of electricity exchanges), deregulation inevitably leads to much higher and more volatile electricity prices.

A final point needs to be brought out here, which is that the real scandal of deregulation is the failure of a great majority of economics teachers and students to join in the opposition to electric deregulation, even when their electricity bills zoomed up. The reason is that the introductory courses in economics fail to spend enough time on the materials in the latter chapters of textbooks. These teachers and students live in classrooms and daydreams featuring perfectly rational buyers and sellers, operating in a world where they have all the information they need to make wise decisions.

4. Emissions Trading

"The theory is based on unscientific assumptions that are hindering the implementation of viable economic solutions for global warming and other menacing environmental problems."

— *Robert Nadeau (2008)*

The best approach to the interior mechanics of emissions trading begins with going into any library, and examining everything you can find on this subject, particularly in economics textbooks. If you do that, then you will make a painful discovery.

The difference between economics and what are often called 'hard' sciences is that the latter deals with the behavior of nature, while too many practitioners of the former are obsessed with manipulating and theorizing about the behavior of models, although many of those models have no relation to any state of nature that exists on this planet, or is likely between now and doomsday.

Under these circumstances, it is easy to understand why emissions trading came to be considered an excellent way to get rid of excessive carbon dioxide (CO_2). In thinking about this project, a researcher might choose to employ what the game theorist Ken Binmore calls a "toy market", by which he means a market that is devoid of inconveniences like risk (or uncertainty), monopoly, irrationality, dishonesty and anything else that prevents a few simple equations from being put on a blackboard for the delight of drowsy or, for that matter, upbeat teachers and students. Despite a statement in *Newsweek* that in order to suppress excessive CO_2, the United States needs a cap-and-trade system of the kind providing 'wonderful' results in Europe, the ugly truth is that the European arrangement is without any economic or environmental value except to the brokers and 'intermediaries' who expect to get rich by playing games with 'emissions credits'. Cap-and-trade is a cynical deception, which Richard Karn — managing editor of the *Energy Trends Report* and a committed environmentalist — calls a carnival sideshow whose aim is to financially abuse the world.

One of the best short accounts of the flaws of emissions trading originates with Emma Johansson (2003), who points out that "...a utility that runs fossil fuel-fired plants will be exposed to an additional price risk that affects risk and return. As a result, the classic spark spread (i.e., the price difference between the price of electricity and the price of the fuel used to generate the electricity) will have an additional component." What Ms. Johansson forgot to add was that according to mainstream economic theory, this increase in risk will almost certainly reduce the investment in physical capital, and therefore many of the new electricity-generating plants that are essential for maintaining our standard of living later in this century might not be constructed. After reading the Johansson paper, interested parties can scrutinize a short article by Uwe Maassen (2003) in the same publication, which examines some further tribulations that can come about if an unreasonable confidence is placed in 'carbon trading'.

In a 2004 issue of the *Newsletter of the International Association for Energy Economics (IAEE)*, Mr. Erling Mork accused this author of failing to understand the magnificent service that he feels is being rendered to electricity consumers and producers by the presence on this earth of the Nordic Electricity Exchange (Nordpool). The truth is that too few persons in the academic world are fully cognizant of the makeup and functioning of that establishment, in that, among other things, they are unaware that without establishments like Nordpool and the electric deregulation 'scam' of which they are a part, their electricity bills would be considerably lower. Moreover, organizations like Nordpool are now putting themselves into position to substantially boost their earnings by taking part in carbon-trading schemes.

Fortunately, Mr. Mork opened his critique by admitting that Nordpool has suffered liquidity problems, which was a signal for the sophisticated among his readers to relax and forget about searching his unhappy communication for any imperfections in logic, because it is a well-known fact in theoretical economics — dating back at least to the work of Leon Walras in the 19th century — that if an auction market has inadequate liquidity, then it is best for both its present and potential customers if it closes its doors (or cancels a particular contract). As it happens though, electricity exchanges have every intention to continue their elegant 'hustle', because with a little luck and some help from paid and unpaid publicists, the buying and selling of marketable emissions contracts might help turn any financial troubles that, e.g., Nordpool might be experiencing into opportunities for a few neophyte 'masters of the universe' to bank a few million dollars. By way of contrast, like the various components of the electric deregulation trickery, there is no guarantee that 'carbon trading' is capable of providing appreciable benefits for producers or consumers or anybody else not actively engaged in eulogizing and hawking emission-based assets to a gullible audience.

As to be expected, the head of the New York Mercantile Exchange (NYMEX) claims that emissions trading will "evolve to be big business", but at the same time he has made it clear that NYMEX will not swing into really serious action until, in his words, he is able to determine which of the existing contracts "developed gravity". By "gravity" he could conceivably mean liquidity, although it is not certain that a bizarre expression like "developed gravity" has a background in standard English. It should be made clear though, that in these perilous times, when the financial sector seems to be under attack by relentless bloggers and amateur economists, those of us who have made a serious career of studying and teaching finance should have no difficulty confessing our confidence in NYMEX and conventional exchanges, at least for the time being.

According to Andrei Marcu of the International Emissions Trading Association, "Europe is now clearly committed to action on climate change, whatever happens to the Kyoto treaty." I am sure that he is sincere in this belief, because his salary (and bonuses) will depend on the trading successes of carbon permits, and not on the fate of the Kyoto Protocol nor a reduction in the stock of physical carbon in the atmosphere. For him and his collaborators, cash comes first, with carbon in its various forms somewhere to the rear.

Another heavyweight player in this burlesque, Professor Michael Grubb of London's Imperial College as well as the ludicrously named 'Carbon Trust', informed *The Economist* (UK) that "Kyoto was designed for the rich countries to miss their domestic targets. That's why we included international emissions trading" (April 3, 2004). The identity of the "we" to whom he was referring was not clarified; however, for the purpose of the present contribution, it could apply to everyone with expectations of a first-class ticket on what they hope will become a carbon-trading gravy train, which means at least a battalion of 'wannabes', observers, analysts, publicists and the like.

To give some idea of Professor Grubb's knowledge of energy economics, about a decade ago that gentleman published an article on wind power in *New Science*, in which he clarified for the unlearned that the most common mistake when considering wind power is to think of it as useful for peak power — which is usually required on an irregular basis between 10 and 20 percent of the time. However, according to him, the ideal role for wind power involves providing the *base load* (or load that is on the line all of the time). It is difficult to imagine anyone who is so dense as to believe that wind is 'right' for peak loads, because when the peak appears, there may not be any wind. As for the base load, wind is hardly useful unless there is a 'back-up'. In theory, a 'smart grid' might function so that wind turbines are always ready to produce when there is wind, but when their output 'dips', as is inevitable, energy from the back-up is immediately switched in.

Smart grids are occasionally discussed in local newspapers in Sweden and elsewhere, but after reading accounts of them you might decide that the less you know about this innovation the better it is for your peace of mind, because at the present time they seem to be something intended for dilettantes and busybodies. Yes, the grid should be 'smartened' if possible, but not under the supervision of, or because of the approval of, ill-informed academics and journalists, pandering to an unenlightened audience.

On several occasions the previous Swedish Minister of Industry announced that Swedish firms that are heavy users of energy will have their energy taxes reduced if they take steps to become more energy-efficient. *Ceteris paribus* this makes sense, because the minister and her advisors might have been thinking in terms of a technology-based approach to CO_2 suppression that could serve as an alternative to emissions trading, or, for that matter, a substitute for badly thought-out political initiatives that would force energy-intensive firms to move out of the country. But it is likely that the Swedish government could best assist in carbon suppression by reopening the nuclear facilities in the south of Sweden that were prematurely closed, or even better, constructing several large nuclear facilities that make full use of the latest technology. If this were done, then any commitments Sweden has accepted for reducing CO_2 could be efficiently satisfied without having to utilize the services of retrograde organizations of the Nordpool variety, because nuclear reactors do not emit CO_2.

Discussing this sort of tradeoff in Sweden is exasperating, because what we have in this country (Sweden) is a variant of the situation in Berlin when the Soviet Army entered the suburbs of that exciting city just before the end of the Second World War, and Josef Goebbels stood on a balcony and gave a thrilling speech to several thousand over-age combatants before he sent them to the front. Here, in the land of the Midnight Sun, politicians and certain high-level civil servants diligently practice a servile conformity to international pressures and crank concepts, in order to be rewarded with a quantum of recognition when they strut or slouch through the corridors of the European Union's headquarters in Brussels. If this means bad news for Swedish taxpayers, then the discomfort of those ladies and gentlemen is one of the penalties that must be accepted for choosing to live in Scandinavia instead of Pago-Pago.

Some persons may be dissatisfied with the terminology in this section, as well as with the manner in which certain issues have been approached — for example, comparing emissions trading to the widely acknowledged fiasco of electricity deregulation. The key thing here, however, is that the directors of many energy-intensive companies in Sweden — despite their traditional preference for market-based solutions — have for the last few years informed friends, neighbors and colleagues in this country and else-where that emissions trading and electric deregulation are two of the worst ideas ever hatched, and might cause serious harm to consumers as well as the Swedish industrial sector.

Unfortunately this author can remember failing to convince one of his former mathematical economics students that he should accept the above assertions at face value. This student wanted some 'algebra', but since none was near at hand, I gave that gentleman a lesson in primary school arith-metic. A few years ago the Swedish government planned to issue one group of companies emission permits for 250,000 tonnes of CO_2 per year. The emissions from these companies were reportedly 450,000 tonnes the pre-vious year, and so primary school arithmetic indicated that if those enter-prises wanted to maintain the output of the previous year, then they would have to go into a 'market' (or what the prominent New Zealand economist Owen McShane called a "pseudo market") and purchase — at an unknown price — emission permits that gave them the right to emit about 200,000 tonnes of CO_2.

Purchase at an unknown price! Isn't this the kind of dilemma that Emma Johansson was talking about, and which may be one of the reasons why even Jerry Taylor — Senior Fellow and environmental researcher at the conser-vative Cato Institute (in Washington, D.C.) — has expressed a preference for carbon taxes over cap-and-trade foolishness? Some of us have a preference for carbon taxes because, among other things, it may be possible to design an income-neutral system in which the tax revenues can be returned (in some simple or complicated way) to the aggregate of enterprises and/or individuals paying these taxes. This seems reasonable, because the goal of this tax should not be to punish consumers or firms, but to *very gradually* increase the cost of 'producing' certain kinds of harmful emissions, and thus encourage the adoption of new technology. Specifically, new technology that will be a valuable component of the future energy economy, and thus adopters of this technology would have their taxes reduced. As a result, in theory at least, it might be possible to talk of an income-neutral system.

Although not called a tax, the emissions trading referred to in this section is equivalent to a tax, and as we know from elementary and intermediate microeconomics textbooks, the tax will be divided between the producers and the persons who purchase energy from them — in amounts dependent upon the *elasticities* of supply and demand. Of course, consumers who use the services of energy firms could avoid to some extent increased electricity prices by the simple expedient of spending at least a part of their winters on Devil's Island, or, if that option is too expensive or uncomfortable, they could try wearing thick fur underwear and fur-lined baseball caps indoors; but despite the proven ability of the Swedish electorate to accept all sorts of nonsense from their political masters, it is likely that they will soon find themselves wondering if they really and truly want emissions trading complicating their lives.

Before readers turn to the elegant employees of CNN or Fox News for some help to obtain answers to climate riddles, it can be suggested once more that emissions trading (or carbon trading) or cap-and-trading is one of the most absurd departures in modern economics, and is exceeded only by the scandalous resort to electric deregulation. Both of these endeavors 'work', if that is the correct expression, only in the early chapters of elementary economics textbooks that view supply and demand through the rosy prism of full competition. In case readers have forgotten what that means, it might help to cite the criteria given by O'Sullivan and Sheffrin (2008) in their introductory textbook: *informed buyers and sellers, perfect competition, and no spillover benefits and costs.*

What about uncertainty? Well, sharp and alert consumers and producers could in theory be beautifully informed about present economic phenomena, but they might have a problem estimating the course of events in the coming years. If this is true, then it might be a good idea to refer to Nietzsche's warning that "the future is as important for the present as the past." Since the murky future that characterizes emissions trading might contain ugly surprises that can distress all of us, perhaps we should give some extra thought to the way we organize the present!

Readers who might want a more extensive acquaintance with this topic should turn to the publications of the Carbon Tax Center, which can be located via Google, as well as the article by David Victor and Danny Cullenward (2007) entitled "Making Carbon Markets Work". Also of interest is a paper by Ruth Greenspan Bell (2006).

And finally, it might be appropriate to mention — without naming — a 'newsletter' in the great city of New York, whose personnel seem to have a strong professional belief in the efficacy of emissions trading, and undoubtedly believe that if it is adopted on a large scale, it will make their lives sweet and rich. Unfortunately, the increasing dishonesty in energy matters seems to have contaminated much of the formal as well as the informal energy literature, and as a result those readers who teach or will someday teach this subject might find it increasingly difficult to reach their students, and teach them the few extensions to Economics 101 that are absolutely necessary if they in turn are to provide friends and neighbors with a suitable helping of valuable information.

5. Energy and Some Macroeconomics

In an article in *Seeking Alpha*, Michael Fitzsimmons (2010) stated that the source of the "economic pain" in the United States can be traced to sending almost a billion dollars a day abroad to pay for oil. This figure was probably too large when first proposed, and is invalid at the present time because of the increased availability of oil (and gas) in the U.S. There were many comments published about that article, and it was not always possible for some of us to agree that oil should be replaced by natural gas on the U.S. energy scene. (Natural gas is not easily compressed and stored, and as a result is not an ideal transportation fuel in its regular form. Converting it first to a liquid (like diesel) might be preferable.)

Let us start the discussion with a taste of what happens in a major oil-consuming country when the oil price suddenly and unexpectedly rises. Among other things, the inflation experienced in the 1970s was the most severe peacetime inflation in U.S. history (up to that date), and according to many observers, the cause of that inflation was OPEC coming of age, and in 1973 — and with the war in the Middle East — unexpectedly taking actions that caused the price of oil to rise from approximately $3.50/b to approximately $13/b, and to do so in a very short time.

The October War in the Middle East gave OPEC the opportunity to influence the price of oil that they had dreamed of since their formation many years earlier, and in 1979–1980 the Iranian Revolution provided another opportunity (or excuse) to push the oil price up. This time it increased to over $35/b, and once again the global economy was treated to inflationary pressures (of a cost-push nature) that were felt almost everywhere. In looking more closely at these two episodes, we note that the rise in the oil price subsequently increased (in one way or another) the cost of producing most goods and services in the major oil-consuming countries, because energy-intensive activities and processes that made economic sense when the price of oil was $3.50/b became sub-optimal when the price reached, e.g., $13/b, and just as importantly, created expectations that where oil prices were concerned, more bad things were going to happen, and probably sooner rather than later. Faced with higher costs due to the increase in energy prices, many firms in oil-importing countries had no choice but to increase their prices.

That increase in the macroeconomic price level lowered the *real wage* (i.e., the nominal wage divided by some index of the price level), and to get employees to accept this situation, an increase in unemployment appeared to have taken place. This in turn reduced output (or the rate of growth of output), and so we might claim that the rise in the oil price impacted *negatively* on output, though, unlike the situation in your Economics 101 textbook, prices rose. What we have here (during 1973–1975) is *stagflation*: a combination of stagnant output and a rise in the aggregate price level. In addition, in 1974–1975 a theory began to circulate that if the price of oil continued to rise in the manner that it had earlier — as various OPEC personalities and 'wannabes' often threatened — then war was likely; a war that would begin with an attempt to occupy the Gulf oil fields — or perhaps just some of them — by paratroopers and marines from the larger oil-importing countries. Palestinian President Yasser Arafat once discussed this possibility in an interview, and he seemed to think that the final act in this kind of drama would involve a giant bonfire from one end of the Gulf to the other. The significance of this kind of bonfire on the oil price was obvious to just about everybody.

Investments (in machines and structures) were also affected by the rise in oil prices, and a number of firms cancelled some investment projects, or shifted to less energy-intensive commitments. In addition, on the whole, the average capital intensity in some countries declined slightly, which (*ceteris paribus*) switched these economies onto a lower growth track. Even if capital intensities remained the same for many firms, demand for their output often declined, and sometimes drastically (thanks to the *multiplier* you studied in Macroeconomics 101). Moreover, as an aside, if the demand for the output of, e.g., the U.S. industry involves increasing exports of goods and services to the oil exporters, it could be argued that exporting machinery, vehicles, reactors, aircraft, etc., in order to obtain an input (oil) that even at today's price costs about seven or eight times as much as it did when the new century arrived, is not an enviable arrangement.

In some countries, an amalgam of all of the above resulted in a lower growth rate for the Gross Domestic Product. This was indeed bad news, because it may be true that the enormous debt burdens that plague many countries can only be lessened if economic growth can be maintained at or close to the 'traditional' levels! Clearly, the oil-importing nations of the world found themselves facing a monumental challenge.

In my opinion, this challenge was surmounted until recently (i.e., 2004–2005), even though the cost-push effects of oil price spikes from 1973 to 2004–2005 may have caused millions of persons in the oil-importing countries to become unemployed, or have to accept for shorter or longer periods employment that was incommensurate with their training or preferences, etc. But as we all know, everything is relative in this world of ours, and despite various inconveniences, those years may eventually be judged as golden years for a fairly large fraction of the world's employees, by which

I mean golden as compared to what existed earlier, and perhaps what they might undergo in an energy-deficient future.

In considering Michael Fitzsimmons' thesis about economic pain, it should be clear to everyone reading this book that an oil price increase functions as a tax. It is a tax on the users of oil and oil products which, in many cases, they cannot avoid or choose not to avoid. As suggested above, the major oil price spikes of the 20th century caused considerable discomfort, while the oil price run-up that began in 2004 and continued until the end of 2008 transferred hundreds of billions of dollars from oil consumers/importers to oil exporters. Paying that tax caused residents in some oil-importing countries to reduce a number of expenditures, some of which were crucial in that they had to do with housing, medical care, etc. As recently as 2011, the bad news apparently continued.

Moreover, countries like the U.S. are often in dire need of the goods exported in order to pay for something (e.g., oil) that to a considerable extent will be "burned up in the air", to use an expression coined by the last Shah of Iran. Those goods, as well as the expertise used to produce them, are going to be needed by countries like the U.S. in order to help offset what might be a declining productivity with respect to foreign competitors, beginning with China.

6. A Conclusion

Readers of this book have already hopefully learned a few things about the great world of energy markets. When dealing with oil, the history of the oil supply in various parts of the world is of key importance, and it should be learned *perfectly*. Learning these materials perfectly is the best way to deal with persons attempting to sell the wrong ideas, bogus histories, and half-baked predictions. Oil outputs have peaked in some of the great producing areas of the world, and in the coming decade, might peak in others. It is also necessary to comprehend that — at the present time — replacements for oil are not being made available at an optimal rate. In the United States there are people who believe that natural gas should fuel vehicles instead of gasoline, and they point to a few successful experiments with, e.g., buses and taxis in some large cities. 'Experiments' do not always suffice, however, and besides, there are many observers who believe that electricity is the best vehicle 'fuel'. Some readers of this book should take a careful look at this issue, and eventually devote some time and gusto to making their findings known to political and economic decision-makers. Hopefully, their ideas and work will help their governments to grasp and deal with an extremely important issue.

Coal plays an important role in the global energy economy and will continue to do so, and in conjunction with this topic, readers should focus on the following. Cap-and-trade is the method sometimes favored in the U.S. for combating CO_2 emissions, largely because it has a capitalist aura, and some articulate and/or influential people believe that it has succeeded in Europe, where it is called the European Union's Emissions Trading Scheme (or EUETS). In actual fact, the EUETS has failed or is failing, and is predicted to fail in Europe by almost everyone who does not have something to gain in financial or career terms by declaring otherwise. The essential thing that needs to be recognized and often repeated is that many of the economists who initially proposed this scheme have now declared it hopeless.

For example, the *Wall Street Journal* (August 13, 2009) reported that the 'inventor' of this concept, Professor Thomas Crocker, has denied its utility, and in California an important energy and environmental executive openly referred to cap-and-trade as a scam. (See Jon Hilsenrath (2009).) Similarly, despite their seminal work on the topic, the late Professor John Dales and Professor W. David Montgomery said that it was useless for real-world applications. By way of contrast, a U.S. government confidant and expert, Joseph Aldy, once rejected the judgments of the above scholars, and insisted that cap-and-trade arrangements display a flexibility that makes them a useful policy device. But what else could he say if he wanted to continue strutting his stuff in various White House meetings and social occasions.

In the U.S. Department of Energy, and also the U.S. Congress, educated men and women are trying to make the impossible possible by accepting various ill-considered energy departures. Without going into detail, I would like to claim that it would be better if they attempted to assimilate the rhythms and logic of mainstream economic theory, instead of latching onto crank concepts proposed by itinerant delegates at international conferences.

There is no praise in this book for electric or natural gas deregulation. Instead, I focused on the situation in California, where everything that could go wrong went wrong, and the same is true for other countries. Some ingenious falsehoods about deregulation were put into circulation in Sweden, and kept in circulation even after the failure of electric deregulation became known in every corner of the kingdom. What happened was that important economists and politicians became partisans of the deregulation farce, while non-believers were systematically denied access to the 'media', and their research denigrated. One economist who has not been misled by deregulation fantasies is Professor Susanna Dorigoni, research director at IEFE, Bocconi University (in Milan). See her 2009 paper with Sergio Portatadino, "Natural gas distribution in Italy: when competition doesn't help the market."

Strange things are also happening on the climate scene. Many scholars who originally pronounced emissions trading as the ideal approach to improving the environment have now confessed that they were mistaken. They see it as an interesting theoretical exercise, though without any genuine scientific efficacy. At the same time, it is obvious that emissions trading can bring prosperity to some of the people who are on the right side — i.e., the 'selling' side — of this nefarious practice, or "scam" as one of my students called it. This might be a good place to remember Meyer Lansky, the most brilliant member of 'The Outfit', as organized crime in the U.S. was sometimes called. According to Mr. Lansky, in enterprises in which speculation (i.e., gambling) plays a major role, the winners are those who organize the speculation, while sooner or later almost everybody else will be a loser. We see this in Scandinavia, where Nordpool is steadily grinding out profits for its founders and executives, but on average is not providing any real comfort for most participants in this infamous scheme, but a great deal of discomfort for electricity consumers, although on ideological grounds some of them refuse to acknowledge the source of this malaise.

Organizations like Nordpool employ a legal and uncomplicated scheme to enrich the few at the expense of the many. Nuclear and hydro facilities should provide Sweden with the least costly electricity in Europe and maybe the world, but deregulation, the introduction of Nordpool, and the untruths associated with those two items have changed that prospect; today, the price of electricity to many households in Sweden and elsewhere is a source of great worry. The research that needs to be done here has to do with the reasons why intelligent and competent (and of course influential) economists who were not the recipients of bribes accepted this malicious scheme, and when they discovered its inadequacy, did not denounce its originators.

One of the things that might be extra important in future energy economics textbooks is hydrogen: according to rumors, its time is coming. Enthusiasm for this resource goes up and down, but remembering the success of the Apollo flights, where hydrogen was a rocket fuel, and its use in the so-called Skylab, it cannot be ignored, which is something that I have heard from several researchers. I see no reason at all why its advantages should not outweigh its disadvantages, by which I do not mean now, but soon.

Key Concepts and Issues

Byron Dorgan (U.S. Senator)	Hazel O'Leary (U.S. Energy Minister)
CCS	IGCC
Chinese coal consumption	Dr. M. King Hubbert
CTL process, and 'syncrude'	New Zealand natural gas
distributors/retailers	Nordpool
electric deregulation	NYMEX
Gray Davis	Qatar's energy plans
grids	Russian gas reserves

Questions for Discussion

1. What is the situation with China's energy demands?
2. What is the Gulf Cooperation Council? (Check with Google.)
3. Who was Professor Hollis Chenery?
4. What are Russian gas ambitions in Europe and elsewhere?
5. Who is T. Boone Pickens? What are some of his beliefs and intentions?
6. Who are Jeffrey Michel and Fiona Harvey? What have they said or claimed?
7. Who said "A rising tide raises all ships", and what did he mean?
8. Who are Ernest Hollings, 'Madame Chairman' (in this chapter), Michael Grubb, and Gunnar Myrdal? Say something about them!
9. What are QFlex and MFlex tankers?

10. Most of the Middle East oil-producing countries are rich, and have made plans to stay that way. Discuss!
11. Almost 30 years ago a Soviet–Western Europe gas 'deal' was arranged. Discuss!
12. What is it that Professor Ken Binmore calls a "toy market"? Why would somebody use the word "toy" to describe a market?

Bibliography

Adelman, Morris A. and Martin B. Zimmerman (1974). "Prices and profits in petro-chemicals: an appraisal of investment by less developed countries." *Journal of Industrial Economics*, 22(4): 245–254.

Amundsen, Eirik S. and Lars Bergman (2005). "Why has the Nordic Electricity Market worked so well?" Department of Economics, University of Bergen (Norway).

Arrow, Kenneth (1951). *Social Choice and Individual Values*. New York: John Wiley.

Baldwin, Tony (2004). "Electricity market: price volatility no flaw." *International Association for Energy Economics Newsletter*, 2nd Quarter.

Banks, Ferdinand E. (2008). "The sum of many hopes and fears about the energy resources of the Middle East." Lecture given at the Ecole Normale Superieure (Paris), May 20, 2008.

_____ (2007). *The Political Economy of World Energy: An Introductory Textbook*. Singapore and New York: World Scientific.

_____ (2004). "A faith-based approach to global warming." *Energy and Environment*, 15(5): 837–852.

_____ (2001). *Global Finance and Financial Markets*. London, New York and Singapore: World Scientific.

_____ (2000). "The Kyoto negotiations on climate change: an economic perspective." *Energy Sources*, 22(6): 481–496.

_____ (1996). "Economics of electricity deregulation and privatization: an introductory survey." *Energy — The International Journal*, 21(4): 249–261.

_____ (1987). *The Political Economy of Natural Gas*. London and Sydney: Croom Helm.

_____ (1980). *The Political Economy of Oil*. Lexington and Toronto: D.C. Heath.

Bell, Ruth Greenspan (2006). "The Kyoto Placebo." *Issues in Science and Technology*, 22(2).

Bilek, Marcela, Christopher Dey, Clarence Hardy, and Manfred Lenzen (2006). *Life-cycle Energy Balance and Greenhouse Gas Emissions of Nuclear Energy in Australia*. The University of Sydney (November 3).

Brau, Rinaldo, Raffaele Dononzo, Carlo V. Fiorio, and Massimo Florio (2010). "EU gas industry reforms and consumers' prices." *The Energy Journal*, 31(4): 167–182.

Carr, Donald E. (1976). *Energy and the Earth Machine*. London: Abacus.

Casazza, Jack A. (2001). "Pick your poison." *Public Utilities Fortnightly* (March 1).

Chenery, Hollis and Paul Clarke (1962). *Inter-industry Economics*. New York: Wiley.

Constanty, H. (1995). "Nucleaire: le grand trouble." *L'Expansion* (pp. 68–73).

Cook, Earl (1976). *Man, Energy, Society*. San Francisco: W.H. Freeman and Company.

Cooke, Ronald R. (2009). "The clean energy act is not going anywhere." *321 Energy* (July).

Dales, H.H. (1968). *Pollution, Property and Prices*. Toronto: University of Toronto Press.

Dorigoni, Susanna and Sergio Portatadino (2009). "When competition does not help the market." *Utilities Policy*, 17(3/4): 245–257.

Eltony, Mohamed Nagy (2009). "Oil dependence and economic development: the tale of Kuwait." *Geopolitics of Energy* (May).

Fiorio, Carlo and Massimo Florio (2011). "Would you say that the price you pay for electricity is fair? Consumers' satisfaction and utility reforms in the EU15." *Energy Economics*, 33(2): 178–187.

Fitzsimmons, Michael (2010). "Foreign oil dependence: the root cause of America's economic pain." *Seeking Alpha* (November 28).

Frank, Robert H. (2007). *Microeconomics and Behavior*. New York: McGraw-Hill.

Gabriel, Mark (2008). "Another inconvenient truth: the need for coal." *EnergyPulse*.

Haas, Reinhard and Hans Auer (1998). "The relevance of excess capacities for competition in European electricity markets." Vienna Technological University.

Harlinger, Hildegard (1975). "Neue modelle für die zukunft der menshheit." IFO Institut für Wirtschaftsforschung (Munich).

Harvey, Fiona (2006). "How to breathe easier." *Financial Times* (October 20).

Hilsenrath, Jon (2009). "Cap-and-trade's unlikely critics: Its creators." *Wall Street Journal* (August 13).

Huber, Peter W. (2009). "Bound to burn." *City Journal*, 19(2).

Hung, Joe (2010). "Coal: the contrarian investment." *321 Energy* (April 10).

Hutzler, Mary (2009). "The Pickens plan: is it the answer to our energy needs?" *IAEE Energy Forum* (Spring).

Johansson, Emma (2003). "The current status of the greenhouse markets." *Oxford Energy Forum* (May).

Kubursi, A.A. (1984). "Industrialisation: a Ruhr without water." In *Prospects for the World Oil Industry*, edited by Tim Niblock and Richard Lawless. London and Sydney: Croom Helm.

Lorec, Phillipe and Fabrice Noilhan (2006). "La stratégie gasiére de las Russie et L'Union Européenne." *Góéconomie* (No. 38).

Maassen, Uwe (2003). "The EU's proposal concerning emissions trading." *Oxford Energy Forum* (May).

Martin, J.-M. (1992). *Economie et Politique de L'energie*. Paris: Armand Colin.

Moller, Jorgen Orstrom (2008). "The return of Malthus: scarcity and international order." *The American Interest* (July/August).

Montgomery, David (1972). "Markets in licenses and efficient pollution control programs." *Journal of Economic Theory*, 5(3): 395–418.

Mork, Erling (2004). "NordPool: A successful power market in difficult times." *International Association for Energy Economics Newsletter*, 2nd Quarter.

Nadeau, Robert (2008). "The economist has no clothes." *Scientific American* (April).

O'Sullivan, Arthur and Steven H. Sheffrin (2008). *Economics: Principles and Tools*. New Jersey: Prentice Hall.

Raffin, Lisa (2008). "Greening commercial facilities." *EnergyPulse* (October 14).

Rhodes, Richard and Denis Beller (2000). "The need for nuclear power." *Foreign Affairs* (January–February).

Rose, Johanna (2010). "Drömmen om rentkol." *Forskning & Framsteg* (March).

———— (1998). "Nya Krafter." *Forskning & Framsteg* (September).

Söderbergh, Bengt (2010). *Production from Giant Gas Fields in Norway and Russia, and Subsequent Implications for European Energy Security.* PhD Thesis, Uppsala University.

Stern, Nicholas (2007). *Stern Review on the Economics of Climate Change.* London: H.M. Treasury.

Stipp, David (2004). "Climate collapse." *Fortune* (February 9).

Tverberg, Gail (2010). "Huge amount of oil available, but..." *OilPrice.Com* (December 26).

Victor, David and Danny Cullenward (2007). "Making carbon markets work." *Scientific American*, 297(6): 44–51.

Victor, David and Varun Rai (2009). "Dirty coal is winning." *Newsweek* (January 12).

Wallace, Charles P. (2003). "Power of the market." *Time* (March 3).

Watts, Price C. (2001). "The case against electricity deregulation." *Electricity Journal*, 14(4): 19–24.

Woo, C.K., M. King, A. Tishler, and L.C.H. Chow (2006). "Costs of electricity deregulation." *Energy: The International Journal*, 31(6–7): 747–768.

Yohe, Gary W. (1997). "First principles and the economic comparison of regulatory alternatives in global change." *OPEC Review*, 21(2): 75–83.

Chapter 3

The World of Energy 3: A Modern Survey

"War is 10 percent fighting, 10 percent waiting, and 80 percent self-improvement."

— *Chairman Mao*

This is the last survey chapter in my so-called short course on modern energy economics, although the completely non-technical Chapter 8 should be read as soon as possible after this chapter is examined (or even earlier). Here, I should note that persons who have a fluent knowledge of energy economics are as rare as those who have a similar knowledge of quantum physics. The reasons are different, however. In discussing economics, the great (Lord) John Maynard Keynes described economics as an easy subject that is difficult; however, I do not remember many of my students having any problems with micro or macro economics at the basic or advanced levels. Things are different with energy economics. The reason is that this is a subject that has not received the attention it should have received by the decision-makers in the academic world, and while a change may be taking place, it is not taking place fast enough. The continuing lack of knowledge of energy economics is a very bad thing, and should be corrected as soon as possible.

As I tell my students, they must learn the elementary aspects of the subject perfectly, which is the reason for the first three chapters of this book being constructed the way that they are. Furthermore, a key strategy for learning energy economics perfectly involves identifying the wrong kind of knowledge — or, for that matter, the right kind that is poorly presented — and rejecting it immediately. For instance, rejecting an opinion often offered on CNN that there are many answers to most questions. This rejection on my part comes from the many years that I taught financial economics. When students might someday find themselves in close contact with very large amounts of financial assets, and perhaps will have heavy responsibilities for deploying those assets, there is no room for experimentation, mistakes or the "many answers" approach touted by CNN celebrities, some of which are very likely counterproductive or worse.

1. Oil Futures; and Speculation versus Fundamentals

"It's not mostly the money, Ingrid. It's only the money."

— *Unpublished manuscript:* The Other Ingrid

In a lecture that I once gave at the Ecole Normale Superieure (Paris), the 'bottom line' in the speculator-fundamentals controversy reduced to the following: professionals mostly say fundamentals, while amateurs and conspiracy theorists mostly say speculation. The emphasis here should be on *mostly*, because amateurs sometimes get things right, while it has happened that finance market superstars are sometimes completely wrong. (I am thinking of the Long-Term Capital Management fiasco.) Of course, when dealing with this topic, certain highly educated and well-placed men and women will sometimes say other than what they believe if they feel that their salaries and perks would be jeopardized if they broadcast the correct conclusions.

More important, though, is this business of what makes a highly competent energy professional (or energy economics student). As pointed out in the preface to my previous energy economics textbook, the best approach to the study of economics and finance is to find out what you need, and if possible, learn to like a large part of it, and, in addition, do everything possible to learn that "large part" *perfectly*.

One of the things that energy professionals and committed students of energy economics absolutely need is a small amount of financial economics. There is no point in entertaining the belief that this book will supply more than a few intimations about, e.g., *futures* markets. However, even so, all readers, regardless of their background, should do everything possible to learn the materials in this section perfectly and keep them in mind, because becoming conversant in them might make it possible to avoid the enormous amount of foolishness in circulation about the impact of derivatives (e.g., futures and options) and 'hedge funds' on the price of physical oil.

We can start by modifying an elementary example that once proved useful to my first-year students of financial economics at Uppsala University. Among other things, it was suggested that they should devote some extra attention to the terminology. Having this terminology at your fingertips is crucial for impressing colleagues, friends, future employers, bloggers, and, for that matter, enemies. *It is much more important than being familiar with a few equations.*

Millicent Koslowski is an undergraduate at the University of Pittsburgh, and a financial superstar in the making. She already has an innovative way of regarding the mechanics of her career: never buy when you should sell, and never sell when you should simply go home and take a shower! Most importantly, her radar is never turned off. She knows that what it takes to become a 'rocket scientist' is more than a perfect knowledge of the fundamentals of financial markets and a sincere belief that more money is better than less money. The most important things are an iron concentration, and something they repeatedly told her brother during his basic training in the American army: *stay alert, stay alive!*

While eating breakfast one day, her father mentioned that a friend's Uncle Charlie had phoned that friend from Genoa (Italy) and told him that all the tanker crews in the Gulf were talking about going on strike. That was enough to cause Millie to immediately leave the table, and before ten minutes had passed she discovered — with the aid of her computer — that this information had not yet reached the media. In other words, assuming that Uncle Charlie was sober, if she was prepared to place what was almost a certain bet, the means for financing her graduate studies at Harvard's School of Business had probably just made an entrance into her young life. She picked up the phone and called a mentor and former teacher, Condi Montana, who is a commodities broker. (Note: "*almost a certain bet*". Formally speaking, since this wager was not 100 percent certain, it would make her a speculator, but that is strictly an academic position.)

She informed Condi that the time had arrived to buy some futures contracts for crude oil. This is sometimes called 'going *long* in *paper* barrels', as compared to the '*wet* (i.e., physical) barrels' aboard the oil tankers that might soon be lying idle (physical oil is formally referred to as the *underlying*). "How many?" inquired Condi, and so Millie informed her of the developing situation in the Gulf as explained by somebody's Uncle Charlie, and told her to use her judgment.

Ms. Montana immediately replied that she was going to buy 100,000 barrels for Millie, which meant 100 contracts, because *all* oil futures contracts are for 1,000 barrels; and since a *maturity* had to be specified, she was going to choose 30 days: after 30 days, if these (long) contracts had not been *offset* with a sell contract (i.e., a *short*), and thus her *position closed*, Millicent would be the proud owner of 100,000 barrels of physical oil in the middle of Texas or somewhere inconvenient, and would also have the pleasure of paying for them. (*This is an important observation, and it should be noted and remembered.*) In entering into this arrangement (i.e., *opening a position*), Millicent has the right to call herself a speculator if she wants, *but she is NOT a speculator in physical oil*. In any event, speculation is not what Millicent aspires to, nor does she want her name associated with that line of work, even if she happens to be doing it. Readers should be clear on this point, because unfortunately a lot of people are not. Moreover, strictly speaking, she may not be a speculator in paper oil! If the information that is in her possession is the real deal, there is no speculation in her purchase of that 'paper oil' (i.e., futures contracts). It is a sure thing.

Millie glanced at the latest edition of the *Wall Street Chronicle* and noticed that the price of oil futures (with a maturity of 30 days) was $70/b, which meant that Ms. Montana would be ordering about 7 million dollars worth of those assets for her friend (and client). That number — 7 million — had a nice ring to Millie, as it would to anyone living in one of the less distinguished residential districts of Pittsburgh. The procedure usually is that she would have been asked by Condi for a (security) deposit — from 5 to 10 percent of the value of the transaction, which is called *margin* — but today Condi Montana does not bother. She is quite aware that Millie does not possess that kind of money, but more importantly, she is too busy buying some barrels of paper oil for herself and her employers — a few million (or probably many more), to be exact. In addition, given the value of this information to Condi's employers, they would have no problem in lending Millie this margin if it were absolutely necessary.

Margin enters the picture in a perfectly logical way. Millie is able to buy these contracts because someone else is willing to sell, and in a highly *liquid* market, that "someone" will almost always be available. Moreover, the exchange where the transaction is culminated has a Clearing House that functions as the *counterparty* in Millie's transaction, which means that she has no dealings at all with the person on the other side of the contract (the seller in this example), just as he or she has no dealings with Millie. In order to ensure that there is no non-performance risk for buyer or seller or the Clearing House, margin is required by the latter (via the broker) to eliminate the risk that a buyer or seller will be unable to fulfill the terms of the contract in the event of an adverse price movement. (For instance, in this example, if the price of oil were to fall instead of rise, then Millicent might — *might* — owe the Clearing House money.)

Millicent also has every intention of reversing her initial position long before the expiry (or maturity) date of the contract (= 30 days in this example), and she should have no trouble doing this because of the adequate liquidity of the oil futures market — at one time referred to as "the best game in town". In fact, with luck, she might be able to close her position in a few hours, but if for some reason she does not reverse her position by the end of trading that day, then that night her contract will be *marked-to-the-market* by the Clearing House. This involves the daily revaluation of contracts by the Clearing House, so that if the price moves in favor of the buyer or seller, he or she is credited with the amount (and can obtain it immediately if so desired), whereas if the price moves against the buyer or seller, then he or she might — *might* — have to provide more margin, which *must* be provided on request. ("Might", because it may happen that there is sufficient money in a transactor's account at the transactor's broker to cover the loss on the transactor's position.)

Now, before continuing, go through the above paragraph again, and keep doing this until you know it by heart. It might be extremely important for your future.

Continuing, Condi Montana leans back in her chair and sips her coffee. Today is a big day in her life too, because she is aware that even a rumor of this strike could send the oil price upward and perhaps off the Richter Scale, and knowing that, she also understands that, like Millie, she has been presented with the opportunity of a lifetime: the chance to change her name from Condi to Condo, in recognition of the kind of property she intends to purchase in quiet places like New Zealand or the south of Argentina — localities that would be without interest to the new crews of hijacked planes, as her boss, Mr. Gekko, likes to put it.

Moreover, it hardly would have made any difference if Millie and Condi and some others had bought 10 million paper barrels. Even in a highly liquid market like the oil futures market, the price of futures (i.e., paper) contracts might increase somewhat in order for a much larger order to be absorbed, but this does not mean that a drastic interpretation of this phenomenon would have been made by brokers and traders and analysts in the major financial institutions: large buy or sell orders are often placed because of rumors, 'hunches', gossip or hangovers. The important thing is realizing that where the oil price is concerned, purchasing 5 or 10 million more barrels of paper oil would hardly have the same impact on the oil price as a decline, or perhaps suspected decline, in the availability of, e.g., 2 million 'wet' barrels (i.e., physical oil) that might be out of the picture every day for an undefined — perhaps short — period. That kind of event could unleash a buying wave that would cause all oil prices to rise.

We had an example of this recently. Because of the situation in Libya, global oil production might have fallen by approximately 1.7 percent, but the oil price escalated by 17 percent almost immediately. Now do you understand the value placed on physical oil?

Something that needs to be emphasized at this point is that Millie is buying *futures*, not *forward*, contracts. True, a futures contract can be regarded as a forward because the delivery of the physical commodity (e.g., oil) is usually specified on the contract, but as compared to a forward contract, *delivery does not have to take place*, and indeed scarcely takes place. Instead, in a highly *liquid* futures market, if the opening transaction was buying (or going *long*), then a contract can be *offset* (or reversed) by just picking up the telephone and selling (or going *short*). As an example, readers can think of the share/stock market, where, e.g., a position can be opened in the stock of the firm Easy Oil by calling your broker and buying, and an hour later your position in Easy Oil can be closed by calling your broker and selling the same amount.

The previous paragraph is something else that you need to read and reread until you understand it absolutely perfectly!

Transactions in the futures markets for oil and oil products are easy to carry out because futures contracts are standard contracts, for a specific amount of a commodity, and should delivery take place because the holder of a long contract keeps the contract until the *maturity* (or *expiry*) date, which is 30 days in this example, then delivery takes place to only a few specific locations. (In the U.S., the stipulated delivery locations are (or were) New York Harbor or West Texas.) However, it has become increasingly popular to settle contracts that are *open* at the expiry date of the contract with money, instead of taking (or making) delivery. This is called *cash settlement*, and is a simple matter as long as a price is available at the maturity date which both longs and shorts can use to determine which of the two is the receiver of the cash (the winner), and who has to pay (the loser). Customarily, that price is announced by the Exchange.

That afternoon, when Millie returned from university, she switched on the television and heard that tanker crews in the Gulf were indeed going on strike. Already the *spot* price (i.e., the price for immediate delivery) of physical oil on both the New York Mercantile Exchange (NYMEX) and (Brent oil) on the International Petroleum Exchange in London had jumped up several dollars. It was being quoted in both places as $72.5/b, but spokespersons for the oil companies were claiming that everything humanly possible was being done to reach an agreement with the tanker crews. This was not easy because floating objects had been observed in Gulf waters, and when representatives of these crews questioned their employers and asked if minesweepers were available, they were told by one of those gentlemen that "every ship is capable of serving as a minesweeper...once." That person was only joking, but this was not the kind of humor that the employees appreciated.

Millie called Condi again. The price of physical oil has gone up, she told her, and she had once read a brilliant energy economics textbook by a great teacher who claimed that when the price of physical oil rises, it was very likely that the price of oil on futures contracts — i.e., paper oil — would also increase (although this might not be true for contracts with very long maturities because of an absence of liquidity in those markets).

Yes, Condi Montana told her. I know the book, and as usual that gentleman was correct. The price of paper barrels just increased to $73/b, but there is an ugly rumor going around that the strike may be settled very soon, perhaps in a few hours. In addition, although this is not always the case, the fact that the futures price is below the price of physical oil is not a good sign at the present time.

"Sell my contracts now," Millie said. "Dump them all." Everything considered it had been a sweet ride, but it was time to take the money and run! Although the ride had ended in a mere 6 hours, Millie increased her 'wealth' (before taxes) by about \$300,000 [= 100,000(73 − 70)]. Normally there would also be a broker's fee, but not on this beautiful day, because Millie's information had made Condi rich, and the owners of Condi's firm even richer. Something that should be noted here is that the price of futures and physicals do not have to be equal when a transaction is initiated, and the same is true when a position is closed, assuming that it is *not* closed on the expiry (i.e., maturity) date of the futures. (I provide in my previous textbooks, particularly Banks (2001), an explanation of the last statement.)

One more question needs to be asked here: why did Condi choose a contract maturity of 30 days? One reason could be that the strike mentioned by Uncle Charlie may not take place, but even so the tanker crews obviously did not like sailing around in waters in which there were mines, and they wanted more money for taking this risk. Accordingly, this issue was not off the table, which could be reflected in an upward spike of the oil price at any time. Assuming that the oil market demonstrated more 'bull' (or rising) than 'bear' (or declining) tendencies at that time, then holding a contract with a 30-day maturity seemed to the very experienced Condi to be a good choice.

Now let us turn this delightful exercise around. An acquaintance calls Condi Montana from Rome and tells her that a large oil field has just been discovered in Western Egypt next to existing oil fields in Libya. In other words, in order to get the oil to market, hardly any new pipelines will have to be constructed. Supply up, price down, as Condi (and Millie) learned in Economics 101. Condi immediately went short (sold) a very large number of contracts for herself, expecting to close her position by the end of the day with a buy that was much lower. She then called Millicent Koslowski and gave her the news. She suggested that Millie should also make a substantial investment — for instance, sell 100,000 barrels, or 100 contracts, with maturities of, e.g., 30 days.

Readers can complete this example after reading the next paragraph. Assume that the exercise had a happy ending (i.e., the price falls), and explain in detail why Millie and Condi made money!

However . . . where is the oil? How can you sell something that you do not have, which is what my students asked me the first time that I lectured on futures markets? The answer is that if you open a futures position by selling oil (i.e., going short), you can close it at any time before the expiry date by buying the same amount (i.e., going long). In the situation being described, the 'ownership' or location of physical oil is irrelevant. The important thing is a 'liquid' market. If, however, the contract is kept open until the closing of the exchange on the expiry date, then conventionally 100,000 barrels would have to be purchased from some source and delivered to a designated delivery point.

Of course, it might be that the contract could be *cash settled*, in which case, if Millie held contracts until the expiry date, she would have to either pay something or receive something. The way this might go is as follows. If for some odd reason Millie's long contracts in the first example above are not offset by closing time on the last trading day for these contracts, her position is declared closed by the exchange, and the price for her contracts is recorded as the closing price for long contracts with a 30-day maturity that are eligible for trading as near as possible to closing time. In order to find out whether Millie pays or receives money, we need another price, and that (*reference*) price is provided by the futures exchange or an adjunct of the exchange; for example, the exchange's Clearing House, which handles its accounting and acts as a kind of *middleman* — i.e., a buyer to sellers, and a seller to buyers.

That price might be the spot price of physical oil if there is a spot market for oil, or it might be a price put together by the exchange on the basis of activities in the futures market earlier that day, or something of this sort. The key thing with the reference price, though, is that persons using the exchange who do not want to deal in physical oil — or, for that matter, do not want to think about buying or selling physical oil — are not put in a position where they must concern themselves with physical oil if for some reason they did not close their contracts before the expiration date on the contract. Furthermore, and logically, the reference price chosen should ensure that persons who would have made money had they offset their contracts around the expiry date, do not lose money if they do not offset their contracts, but take advantage of the cash settlement 'option'. The details of constructing the reference price are not considered in this elementary presentation.

That brings us to *hedging* — i.e., a sort of insurance against unpleasant (or undesirable) price movements up or down. And observe: we are not talking about hedge *funds*, because in reality these are *speculative* funds, and it has happened that there were a few spectacular failures. A few years ago, with the economies of the U.S. and Europe in deep trouble, perhaps one in six or seven hedge funds failed or lost a lot of money. Of course, some of these may already be back in business under another firm name and in some other part of the country. After all, it did not take the directors of Long-Term Capital Management many months after their fund 'tanked' before they were merrily practicing their trade in new offices.

In any event, suppose that two weeks ago you concluded that the oil price was going to rise to $80/b or higher, and so you went down to the local 7–11 and bought 2,000 barrels of *physical* oil for $70/b, which was the price for oil at that time. But now you are not so sure anymore that the oil price is going to rise, and so you find yourself with an overwhelming urge to hedge your investment. How would you initiate this particular risk management exercise?

One way is to call Condi Montana and tell her that you want to sell (i.e., go *short*) two futures contracts (= 2,000 barrels). Then, if the price falls, what you lose on the physical oil that you are holding will be approximately gained on a futures transaction, because later you can close your futures position with a 'buy' at a lower price than the price at which you went short. Moreover, if it appears that certain so-called experts were right and the bottom is going to fall out of the oil market, the market is usually liquid enough that you can sell the small amount of physical oil that you are storing in your home, while perhaps keeping your futures contracts, which — since you opened your position by going short — become more valuable with every decrease in their price. Or, if indications are that things are going to go the other way, *reverse* (i.e., *offset*) your short futures position by going long two contracts, and retain your physical oil. Maybe the price will break the $100/b barrier, and in the ensuing buying spree, you can make some serious money. If you cannot make up your mind, you can call Millie, who will soon be on her way to a corner office on Wall Street.

At this point, readers should take a deep breath and make sure that they understand the previous paragraph.

Before turning to the delicate topic of speculation (gambling) versus fundamentals (supply and demand) for the oil market, it can be mentioned that Condi Montana works as a broker (instead of a trader) because she wants to avoid some of the stress that is endemic in high-pressure financial institutions. But earlier she did *proprietary trading* in an investment bank, which means that she traded for them, and in return received a large salary and — if things went well — a nice bonus. And things usually did go well, because, as Gordon Gekko explained to Bud Fox in the film *Wall Street*, the key thing in his business is information, and the trading departments in the major investment banks have access to a very large amount of information. At the same time, they employed people who knew how to interpret and use this information.

Something else: in the investment bank where Ms. Montana worked, at no time did anyone call themselves speculators. Her firm makes it clear to its employees that it does not hire *speculators*, nor want them on the premises, although speculating is what these employees mostly do or advise their clients to do. The title on their business cards is "trader", and when asked, they sometimes tell people that they do proprietary trading. In appearance they resembled the ladies and gentlemen in front of the screens in the film *Wall Street*, and they have comparable backgrounds. Condi Montana, when she was working for an investment bank, would be a good example. The traders in her department at the bank really and truly understood the dynamics of oil markets, because traders who don't are encouraged by their superiors — at all investment banks and other financial institutions — to transfer their activities to bar stools at the nearest pub, and let others who know what they are doing take their place in front of a computer in one of the firm's trading rooms.

Exactly what the persons on CNN and Bloomberg know about the oil market at the present time is uncertain, but Condi and her colleagues understand that we have come to a point in history where demand is likely to outrun supply (unless shale oil does what some people think that it will), and OPEC has its act together. That makes it very plain what actions they should always be prepared to take. In other words, if Mr. Bill insists that the oil price is going to go down, and he wants to put his money where his mouth is, then Condi should talk to him the same way she would talk to her rich uncle — if she had a rich uncle.

In my course on oil and gas economics at the Asian Institute of Technology (Bangkok) several years ago, two oil price escalations were discussed at great length: one circa 1979–1980, when the Shah said goodbye to Iran; while the next was in 1990–1991, during the run-up to the First Gulf War. Readers should learn something about escalations of this nature, and what they mean for the global macroeconomy. Several years ago attention was concentrated on the movement of the oil price to $147/b. The finale of that breathtaking escalation began in 2008, and somewhat earlier a journalist in *Le Monde*, Jean-Michel Bezat (2008), had some very disturbing information to present to his readers about the intentions of King Abdullah of Saudi Arabia where the supply of oil was concerned. Among other things, His Majesty said that when there were new oil discoveries, they should be left in the ground for the children (*les enfants*) of the kingdom. That was not news to many of us, because 35 years earlier, another king of Saudi Arabia had said almost the same thing.

Let us look again at several of these upward oil price movements, namely *spikes*. The first mentioned above (1980) raised the price well above 100 percent before it swung back, and if you examine a plot of oil prices during that period, you would use the expression "spike" to describe what was taking place: the oil price jumped up and fell back in a fairly short time. The second drama, at the beginning of the First Gulf War, was very definitely a spike of about 100 percent. The question asked at that time was: why did things move so fast? At no time did it appear to professionals like Condi — before or after the results were in and analyzed — that these spikes could be turned into sustainable price rises, even if she and her colleagues at every investment bank on Wall Street bought futures contracts and contracts for physical oil day and night. *The objective market conditions were not right!*

The lesson here was the same one that Condi received in graduate school, and later on when she began to earn her living in the financial district: except for very special occasions, it makes sense to believe that the market is *bigger* and smarter than any individual or comparatively small group of individuals. Those spikes were caused by special events, and the players who understood the mechanics of those events and got in early made a bundle if they did not try to prolong their windfalls, while many of those who came in late took a fall. They took a fall because, as Condi and her colleagues understood, there was a great deal of easily obtainable oil in the crust of the earth, and in addition oil producers possessed considerable spare capacity (at least 5 or 6 million barrels per day). It has also been claimed — though not by reliable sources — that there was considerable 'cheating' by OPEC producers who ignored OPEC quotas, although I regard this as wishful thinking. In any event, the price had to fall.

The oil price rise that began in 2003–2004, and accelerated in 2008, was not a spike. It was a slow but sustained increase in the price of an invaluable commodity. Condi understood that very early. It was a sustained price rise in which the latter phase continued for several months, and she made a lot of money for her firm — and herself — because she understood that demand was outrunning supply. Specifically, OPEC restrained supply, and the faster the price increased, the more reluctant some OPEC countries became to increase their supply of oil. *You would have behaved the same way if you had been in their place!* Condi was paid by her firm to trade — or if you want to call it speculate, please do so — but the reason she made so much money was because she knew more about oil market fundamentals than traders/speculators whose grasp of fundamentals was inadequate.

In the economic circumstances prevailing in 2008, speculators, anti-speculators, traders, neighborhood betting syndicates, moonwalkers, day-trippers or anybody with an urge to make some quick cash went long in oil (by which I mean that they purchased oil futures, which play a (perhaps psychological) part in pricing physical oil). This was the period when the billionaire investor T. Boone Pickens predicted that oil was on its way to $200/b, and he might have been correct if the macroeconomic and financial market meltdowns had not commenced. It should also have been clear at a fairly early stage that this (very) high oil price could lead to some very bad macroeconomic news.

Note the statement above about speculators and others going long. Condi Montana analyzed the situation correctly, and she and others like her in the trading rooms went long. The question then becomes: who went short? One answer is traders and speculators and others who initially did not believe that the oil price was going to rise, or who listened to or paid the wrong people for investment advice, or who were willing to bet that the moment of truth would not arrive as fast as it did. Just as important, there were others who, for some reason, believed that if the market started to move in an unexpected way, they would be warned by their firm's analysts, or somebody's analysts, and could get out the emergency exit before suffering a great deal of damage. Many traders went short because they had been badly burned earlier when they went long, and later, some of these short speculators might have closed their positions and went net long. In that case, who goes short for them? Maybe some of Millicent's classmates, along with some people at various investment banks, who had been watching developments in the oil market and impetuously concluded that the sweet ride was over and they should close long positions.

Suppose there is nobody to go short — *everybody* wants to go long. Then the price races up until a big sub-set of speculators and others change their minds, explaining to themselves and everybody around them that all good things eventually come to an end. Perhaps the oil price reaches $147/b, at which height it slows down, and a large group of traders and others decide that the good times are going to stop rolling and it is time to 'jump ship'. Why would they decide this? Maybe because they heard that the macroeconomy was on its way into the 'tank', which meant that it was very likely that the bottom was going to fall out of the market for physical oil, and therefore oil futures.

Important in this oil escalation drama was the visit of President Bush to Saudi Arabia in May 2008, when he requested some assistance from King Abdullah with the oil price, preferably in the form of more output and also investment in new capacity. The well-known outcome of that episode was that his host graciously thanked him for his concern, and wished him a safe flight back to Washington. Please note that the President went to Saudi Arabia to talk to the King of that country, and not to Wall Street to plead his case to 'masters of the universe'. One reason he did that was because his Secretary of the Treasury, Henry Paulson — a Wall Street veteran — informed him that it was fundamentals in the form of insufficient oil that was the villain in this price escalation. Mr. Paulson had earlier been the CEO of perhaps the most successful investment bank in the world, Goldman Sachs, and he almost certainly had talked to the oil experts at that establishment, several of whom probably had more inside (and outside) information about oil markets than anyone else in the world.

In 2009, while almost every trader in Europe was asleep, and traders in Asia were brushing the sleep out of their eyes on the way to the shower, an English trader gave a demonstration of how the oil price can 'momentarily' be influenced by nonsense. That gentleman went long 16 million barrels of paper oil, or 32 times the normal amount transacted during that particular period. The price of North Sea oil reacted by climbing about $2/b. Exactly what Mr. Trader expected to achieve is difficult to say, but his colleagues were not in the mood to follow his lead, because at that point in time, they knew that relative to demand, there was plenty of oil available. Yes, there was a spike — or a 'mini-spike' — and some buying by a few other traders, and so perhaps a few traders made a little money, but it did not take long for reality to prevail. As a result, a certain gentleman who had decided to do some (unauthorized) trading in the middle of the night reportedly cost his firm 10 million dollars before his positions were 'unwound'.

That brings us to the bottom line in the fundamentals versus speculation dispute: *logically, if X depends on Y, and Y depends on Z, then X also depends on Z.* To the extent that the oil price is influenced by speculators and traders, and speculators and traders obtain their clues from fundamentals (e.g., demand, reserves, OPEC policies), then the weight that speculators and traders exert on the oil price is mostly due to fundamentals. Mostly? Well, that is a figure of speech, since there are undoubtedly a few speculators and traders who play 'hunches' instead of fundamentals, but they tend to have a minimum of 'juice' in this very complicated market, even if occasionally a few of them are lucky.

According to some information once published by Murray Duffin, an important contributor to the site *EnergyPulse*, it is possible that there was no shortage of oil during the second quarter of 2008, and speculators — or *noise speculators* as they would be called in academia — accounted for the rise in the oil price above $110/b in that same year. Maybe they did, but that figure sounds wrong to me, although it does not really make any difference. The key thing is that without demand outrunning supply, and OPEC playing the game the way that smart people are supposed to play it, the sustained oil price escalation during 2008 that accounted for enormous revenues for some OPEC countries could not have taken place, and the rise in price during 2008 would have shown up on the charts as a spike. I can add that according to the United States Energy Information Agency, OPEC raked in more than 900 billion dollars in 2008. *This could never have happened if they had not restricted the supply of physical oil!*

In his testimony before a congressional committee, a gentleman by the name of Michael Masters took what was virtually a sacred oath that speculators were responsible for the big oil price upswing in 2008. He undoubtedly acquired some sympathizers, though his logic was very different from that in the above discussion, where the point is that the price rise originated with fundamentals in which demand threatened to outrun

(OPEC-constrained) supply, and to which speculators reacted. *There is no need to claim that speculators are always losers in the long run, and for the most part are impotent in the short run, but successful speculators specialize in understanding and reacting to the fundamentals.* Mr. Masters also seemed to be in favor of not interfering with the market except to ban or restrain speculation, which would be something like banning the musical background in the dance scenes in a Fred Astaire or Gene Kelly film, or perhaps limiting it to the contribution of a tambourine and kettle drum.

OPEC weighed in on discussions about the oil price rise by claiming that speculators were running wild, although to my way of thinking, OPEC's (sophisticated) supply policy is perhaps the cornerstone of today's oil market fundamentals. Even if Mr. Masters and his distinguished interrogators did not understand this rather unique situation, intelligent traders and the analysts and managers with whom they work did.

It also needs to be stressed that the decision makers at OPEC know — as I also know — that without speculation, the liquidity in futures markets could plummet, which in turn could — could, not would — lead to a situation of the kind shown in a French TV film last year, in which mules (or horses) were seen pulling luxury automobiles through the streets of Paris. I am not suggesting that OPEC movers-and-shakers are thrilled at this prospect, but if the oil price were to exceed $300/b — i.e., the price that oil reached in that French film — those gentlemen would probably find the Paris street scenes even more charming than they did in previous visits to that magnificent city, since they would be viewing those street scenes from their suites in the most luxurious hotels.

And finally, it should be noted that a partial or full reduction in speculation would have drastic (negative) consequences for the hedging of (oil) price risk — a hedging by consumers and producers of physical oil. If Mr. Masters had been aware of this, and what it would mean for future investment and production in the oil sector, he might have tried to be less dramatic in his condemnation of speculation.

At the beginning of this section, readers were informed about the importance of terminology if it is their intention to successfully deal with real people in the real world. Some readers might therefore notice how the above discussions did not include any of the complicated mathematics or abstract terminology of financial economics. For instance, there was nothing about the *efficient market hypothesis*, which among other things has to do with how prices from the futures markets should provide some indication of future market performance. Put another way, under ideal conditions, these prices should represent future supply and demand fundamentals.

Under ideal conditions! To me that sounds like a classroom proposition, but what it means is the presence of large numbers of rational buyers and sellers — i.e., buyers and sellers who utilize the available information to the best of their ability. That might be asking for a great deal, because shortly after taking office, the present governor of the U.S. central bank (i.e., the Federal Reserve System) said that the futures market indicated an adequate supply of oil over the next two or three years. In other words, the two- and three-year futures prices were not much larger than the price at the time he made that statement.

Here we have a problem, because there is not enough liquidity in contracts of that maturity (i.e., 2 or 3 years) to exploit the supposed wisdom of the market. This point is not adequately understood, despite its importance. What might be called optimal market wisdom is available only with the interaction of large numbers of market actors, with this interaction providing a highly liquid market, and preferably where these actors are mostly rational and not 'noise traders', and trading is in contracts with comparatively short maturities! Given that background, a futures price can then be thought of as a sort of 'average' expected price, but better than any alternative. Moreover, as they sometimes say on Wall Street, "the market knows more than any one player", although in the example above it did not know more than Millicent and Condi. Of course, in that example, these two 'stars' had access to crucial private information.

Mathematically, we sometimes see the expression $E(P_{t+n}|P_t) = F_{t+n,t}$. This means that the expected value at time t of the price of the physical good at time $t + n$, is equal to the futures price of the physical good at time t, given a price at time t of P_t. $F_{t+n,t}$ is the price of a futures contract (opened at t) for the good that has a maturity of $t + n$.

Unfortunately, it can be shown that theoretically this expression only applies in the case of risk neutrality. A risk-neutral individual is one who is only interested in the expected value of the outcome. It probably does not make any difference to me if I flip an unbiased coin for one dollar, and the outcome is zero or two dollars, since the mathematical expectation is

$1.0 [= (0.5 × 0) + (0.5 × 2)] given a large number of 'flips'. A similar calculation applies if I flip a coin and the stake is $50,000. Then we have $E =$ $(0.5 \times 100,000) + (0.5 \times 0) = \$50,000$. However, I am not risk-neutral when facing a possible loss of $50,000 on a single flip. Here I need an advantage — for instance, a 'doctored' coin that gives me a 90 percent chance of winning instead of 50 percent, where 90 percent is a rough measure of my *risk averseness*. (Even better is access to a phone with an Uncle Charlie at the other end of the line.)

Moving back to the now discredited expression above, if we assume that in addition to pure gambling, speculators also perform an insurance function for hedgers (who would likely be buyers or producers of physical goods), it seems right and proper that these speculators should be given some reward for providing this service. John Maynard Keynes (Lord Keynes) seemed to take a special interest in *short hedgers*: i.e., hedgers who are afraid of the price falling, and so they sell futures contracts. The speculators who are providing them with insurance would therefore have to go long, and so speculators as a group *might* have to be compensated for providing this service.

It is appropriate at this point to remember that the (unadorned) term *backwardation* applies to the case where futures prices for contracts maturing in the distant future are below prices of near-term futures contracts, and of course the present spot price. This is also sometimes called an *inverted price structure*. The normal price structure — which is called *contango* — then displays a pattern of increasing futures prices, with all of these above the present market price. It happens that with oil, backwardation has often signified that users of a commodity (e.g., oil) think that it will be scarce in the future, and they rush to buy more, thus driving up the present price. The situation for a day or so at the beginning of the First Gulf War, when the American army opened its offensive, provides a useful example of this.

When I was writing this chapter there was considerable talk about contango, and very little about backwardation. The reason was that if the futures price is considerably above the present price, it might be possible to do some very profitable speculation — the kind of speculation in which very little risk is involved. Unfortunately this involves storing oil, and knowledgeable people who devoted a great deal of time to thinking about this subject said that storage space for oil was at a premium virtually everywhere. In fact, vessels were being used for storage rather than transportation. It has been suggested though that the large oil companies are not too worried about this: they are as skilled at the storage game as they are at removing oil from the ground. They are definitely skilled at a lot of things, considering that the enormous profits of these companies at the present time must make their directors and shareholders very happy.

By this I mean they are not making the mistake of doing what consumers of oil want, which is to put more effort into finding and — especially — producing oil.

2. The Controversy about Global Warming

"The mind that has feasted on the luxurious wonders of fiction has no taste for the insipidness of truth."

— Samuel Johnson

The people who want to take action against anthropogenic (i.e., man-made) global (or climate) warming (AGW) appear to have both quantity and quality on their side, by whom I am talking about high-level scientists and not merely concerned dilettantes. By the latter I mean journalists of the kind who deal with this topic in Swedish newspapers, regardless of what they believe or do not believe about 'warming'. This is why some of us decided to accept — or at least not to oppose — arguments in favor of anthropogenic global warming. The AGW guru, Al Gore, may have certain faults, but his way of dismissing persons working the other side of the street can be very attractive. Besides, it might be difficult to feel comfortable around persons who think that gentlemen like the anti-AGW author Paul Johnson, or the Danish gadfly Björn Lomborg, are valuable contributors to a discussion of this sort, with its complex scientific implications.

If the world were as rational as portrayed in most conventional economics textbooks, this section would be quite unnecessary. But as George Monbiot once informed his readers: "The dismissal of climate change by journalistic nincompoops is a danger to us all." I think that we can remove "journalistic" from that sentence because I doubt whether, at the present time, the ladies and gentlemen of the press are much different than most of us where this topic is concerned. They too have become more sophisticated in that they are no longer willing to believe that 'scientific truths' retailed by self-appointed gurus and mentors are worthy of their attention. It might also be useful to note that while the word "nincompoops", or its equivalent, is not unknown in many academic conversations, another description might be more suitable for many of the persons who find it imperative to repudiate global warming: *well-meaning observers with an annoying tendency to accord pseudo-scientists respect that they do not deserve.*

Under no circumstances do I regard my understanding of this topic as comprehensive or special, but I feel that one item deserves to be repeated to acquaintances and students until it becomes as ingrained as the General Orders that infantry recruits were once compelled to learn in the United States Army: *There are still a few deluded scribblers in circulation who want us to believe that the overwhelming majority of scholars who say that climate warming is the real thing and must be addressed correspondingly are anti-American and/or anti free-market loony-tunes, while the miniscule number of academic or intellectual stars who insist that the talk about global warming is hysterical nonsense should be honored as paragons of scientific virtue!* I often receive opinions and information from very smart people that I have no problem believing and respecting, who assure me that global warming is nonsense, but intuitively I find it impossible at the present time to appreciate their sagacity.

As an example of the latter, I mentioned the superstar journalist Paul Johnson, whose intellectual firepower and sustained success puts him streets ahead of the know-nothings identified as 'AGW doubters' by Mr. Monbiot. It is occasionally very easy to enjoy some of the brilliant things that Mr. Johnson has written, and strangely enough this also applied to an article of his in the British periodical *The Spectator*, in which he tells us to "pay no attention to scientific pontiffs" (in the matter of global warming) — unless, I suspect, they are *ersatz* scientific pontiffs. What I particularly liked about that unexpected advice was that it furnished a modicum of proof that Johnson's high intelligence and access to the corridors and restaurants of power did not make him a wiser human being than those of us who, for one reason or another, have come to roost much lower on the social scale.

To make a long story short, Johnson regards many recognized scientific pontiffs as snotty neurotics who, because of their shortcomings in dress and/or manners, have no right to interfere in matters outside their narrow specialities. His principal negative role models seemed to be the late Oxford University scientists Henry Tizard and Lord Cherwell, both of whom were scientific advisors to the UK prime minister Winston Churchill during World War II, but who, when summarily banished to academia after the war, morphed into bad-tempered misfits.

Tizard is a man whose life and longings are a complete mystery to me, but I know — which Johnson apparently does not — that Cherwell risked his life during the First World War to show that a spinning aircraft could be pulled out of a dive, and he was also a key player in the design of the UK air defence in the crucial years before the Second World War. (I will not bother to go into here what could have happened if that air defence had failed.) Johnson's idea of a real scientist — or "boffin", to use his language — is Björn Lomborg of Copenhagen Consensus fame, who is a total non-participant in the genuine scientific literature on any level, and whose recent appointments in the great world of Danish higher education suggest to me the kind of gratuitous welfare handouts that have characterized Swedish higher education from time to time.

As for 'his' Copenhagen Consensus, this is — or perhaps was — a conclave of well-placed academics who were brought to wonderful Copenhagen on several occasions to discuss topics about which they knew little or nothing, and given their backgrounds and specialities, cared less. The only consensus that could be associated with the participants in this half-baked song-and-dance was that travel and lodging at the expense of Danish taxpayers is even more gratifying than drinking beer in Copenhagen's Tivoli on a summer evening.

Among other things, Johnson said that the United States has done more research on "so-called" climate warming than the rest of the world combined (which is almost certainly true), and this was why — he claimed — President Bush refused to comply with the Kyoto Protocol. Ostensibly, that very expensive research failed to establish a definite link between climate warming and man-made emissions.

Perhaps this described the situation when Johnson's precious composition went to the printer, but it definitely was not the case sometime later, when President George W. Bush said that "Science has deepened our understanding of climate change and opened new possibilities for confronting it."

It has also opened new "possibilities" for understanding certain related prospects that, according to Sir David King, the UK government's chief scientific advisor, might eventually have the same ruinous impact on life and property as a succession of large-scale terrorist attacks. By that he was undoubtedly alluding to physical security and the overall economic outlook. This does not mean that the Chief Executive became a partisan of the Kyoto 'talkathon', or that he accepted the scam known as 'emissions trading', but for one reason or another he decided that he had enough on his plate without challenging the opinions of the majority of qualified climate scientists who reject scepticism in this matter.

One final observation needs to be made here. George Monbiot labels the climate warming sceptics as "tools of the fossil fuel lobby". I am not sure that he is correct with this designation, because according to the profit reports for the large oil and gas firms over the last few years, their executives do not need the assistance of a "lobby" to go to sleep at night with thousand-watt smiles on their faces. On this point it is interesting and unfortunately discouraging to note how climate warming sceptics have a tendency to flaunt various strange beliefs, one of which inevitably focuses on what they think is the plenitude of energy resources. Mr. Lomborg, for example, once declared that we do not need to start worrying about an oil shortage in the present century. For more on these and similar matters, this might be the right time to turn to Chapter 8 in this book.

After an earlier scrutiny, I considered removing this section from the present chapter, although given the position recently taken on this topic by people like Bill Gates — the richest or next richest person in the world — it is clear that a decision of that nature would have been premature. By that I am not talking about research grants or plane tickets, but the gradual acceptance by many important politicians in the United States, both Democrats and Republicans, that global warming might deserve serious consideration. In addition, I think that mainstream economic theory would have no problem proving that the very well-off (as a class) would be more discomfited by the melting of glaciers at Swiss ski resorts and the flooding of waterfront real estate in Carmel (California) than the poor, even if many footloose plutocrats were still able to afford apartments in, e.g., Dubai that were once on the block for 5 million dollars or more, or, at the other end of the scale, cosy hideaways on the great South Side of Chicago or in Soweto. Wehrmacht

Sergeant Christian Diestl, in Irwin Shaw's brilliant war novel *The Young Lions*, spoke of the U.S. as "untouched and untouchable", but as things now stand, some extremely choice properties in North America would be in the danger zone in the event of a severe climate meltdown.

In considering themes for consideration in this chapter, it would not be easy to overlook or reject *The Stern Review on the Economics of Climate Change*. Several years ago Lord (Nicholas) Stern (of Brentford) appeared in Stockholm in order to clarify for a large audience the conclusions reached in his widely advertised analysis of potential climate change calamities (see Stern, 2007). Some of us were not invited to attend his presentation, because it was probably believed by the organizers that had we been present, there might have been what is sometimes called an 'incident'. This was definitely not certain, because persons like myself would have been willing to exercise a maximum of restraint in order to avoid embarrassing Lord Stern and his hosts with the kind of information that they definitely did not want to hear, which is that many leading economic theorists regard his work on this topic as scientifically meaningless.

Meaningless, and in the light of the finished product known as the 'Stern Review', pedagogically superfluous. To my way of thinking, that document — or at least the very small portion that I forced myself to scrutinize — has no relevance whatsoever for conscientious economics teachers, and particularly for countless economics students who require our guidance where their reading materials are concerned, but who unfortunately may have been informed by advisors and supervisors that the Stern Review is state-of-the-art knowledge. There is also this matter of innocent bystanders who pay taxes in order for — among other things — systematic and professional attempts to be made to ascertain the extent and mechanics of global warming, and if necessary, to suggest or devise efficient programs for reducing its possible dangers. The analytical fragments of the Stern Review are for the most part typical of the articles, notes and comments that once filled a periodical called *The Review of Economic Studies*, which played an important part in the education of graduate students in mathematical economics at the University of Stockholm and the Stockholm School of Economics.

After scrutinizing a few pages of the Stern Review, the first thing that came to my mind was that if it is true that we are now living in the most dishonest period in modern times, as a Canadian billionaire has claimed, then the so-called research done by Stern and his team might turn out to be a prime contender for one of the most outrageous attempts at deception of the present century. In wisdom, though perhaps not verbosity, it belongs in the same category as a poor man's imitation of *Mein Kampf*, and this is perhaps the reason why Professor Richard Tol said that if it were presented to him as a Master's thesis, he would grade it "F" (for failure).

But please take note that I am not criticizing nor denigrating climate scientists or anybody who takes the position that global warming is a clear and present or nearly-present danger, because this may well be true. For instance, I am completely uninterested in a recent poll conducted by the Pew Research Center in the United States, in which climate warming turned out to be in 20th place among 20 anxieties that persons who responded to the poll were asked to rank.

Instead, I am merely commenting on a pseudo-scientific piece of research whose ennoblement can be attributed to careerists in various university faculties, as well as environmental bureaucracies where the principal concern of many employees is exposure or prestige or money or some of the beautiful things that money can buy. Here I am not indirectly referring to large amounts of subsidized travel and accommodation, but career rewards that can be accessed by skilful lecturers and conference attendees as the result of superficially comprehending a fashionable topic, and turning it into something that will hold the attention of what was once called 'the broad masses', and today might be identified as the TV audience.

There are no analytical surprises in the Stern Review, at least for me. Given the limitations of economic theory, there is only one sensible approach by economists to the project, which is to find a model that can be adjusted and/or refined and/or extended in such a way as to allow Professor Stern and his collaborators to carry out rambling discussions of the economics (or pseudo-economics) of climate change over a stretch of 700+ pages. The model that was chosen by Professor Stern, and which might have been chosen by me had I been in his position, or chosen by my graduate students, or for that matter by party animals posturing themselves and courting attention at the bar of an Uppsala University student club, could only be the one published by Frank Ramsey (1928) in his famous article "A mathematical theory of saving".

What is the logic behind this choice? First of all, the Ramsey model is complicated but not too complicated. Even if students are put off by its mathematics, the basic intent of such a model should be recognizable to a majority of advanced economics undergraduates pursuing their education in institutions where environmental studies have almost the same status as theology. It might also be worth mentioning that the Cambridge 'don' Ramsey was a certified genius, particularly in philosophy, where he was often compared with Ludwig Wittgenstein.

Equally as important, the Ramsey model can be summarily enriched in case someone visiting a lecture on global warming decides to play ego games with the lecturer, in the course of which they proclaim the inappropriateness of the basic Ramsey construction. If the basic model is polished up, a skilful lecturer might be able to successfully convince an audience that it is capable of providing some genuine insight into whether future benefits from 'present' efforts to prevent climate change outweigh the costs of this action. The model — though not the real world — reduces this puzzle to determining the discount rate that should be used to evaluate future benefits.

A simple example might feature, as a benefit, being able to continue enjoying the marvellous beach life on the west coast of Sweden in future summers, while the estimated (or presumed or postulated) present cost of achieving this goal might involve altering Sweden's energy architecture in such a way as to include more wind turbines and natural gas, while promoting a nuclear retreat. While some readers may be uncertain, I think I know how this absurd strategy would play out, because the cost of electricity in Denmark — which is generally considered the 'promised land' of wind-based energy — is often the highest in Europe, while the cost of electricity in nuclear-intensive Sweden was once lower than that of any other country except Norway, and the output of carbon dioxide (CO_2) is very low in Sweden.

For readers who want to learn more about a trade-off of this nature, I can refer them to an article in the IAEE Energy Forum by Mary Hutzler (2009), in which she examines a proposal by T. Boone Pickens for the massive employment of wind and natural gas in the United States, where it would be used to generate electricity (wind) and provide an alternative motor fuel (natural gas). I have tried to convince large numbers of persons of the possible futility of this approach; however, their fear of a situation in which the United States must depend on Middle Eastern countries for increasing amounts of expensive energy has tended to make them over-sensitive. Energy security is much more important than most of those persons believe, but this requires careful analysis and/or planning, and not just the eager adoption of fashionable predilections.

As it happens, Mr. Pickens eventually announced that he was abandoning his plan to construct a 'wind corridor' from the Rio Grande to the Canadian border. I was informed of his reasons for doing so, but almost immediately forgot them, because to my way of thinking, nobody with any important work to do could possibly be interested in the espousal or abandonment of that unworkable escapade. Its likely purpose was to capture some of the money that will almost certainly be in circulation due to an increased interest in 'climate change' by decision makers in the U.S. and elsewhere.

Another project that Mr. Pickens expressed considerable enthusiasm for several years ago involved an expansion in the production and use of domestic (U.S.) coal. This option was not especially popular among environmentalists and many others, but Mr. Pickens also supported the underground sequestering of carbon dioxide emissions from this coal, or if possible, routing them into the depths of the ocean. Until recently coal was a comparatively inexpensive resource, which is why so much of it was used, but I have seen estimates in which sequestration could raise its cost by up to 50 percent. Even so, a small amount of coal gasification and/or liquefying deserves consideration by decision makers, because in ideal conditions, like other alternatives, it may have a place in the 'optimal' energy program to which political or corporate decision makers should always be prepared to devote at least some of their valuable time and enthusiasm.

There is also the scheme whereby generators of electricity and manufacturers of various sizes can buy 'emissions permits' from, e.g., other firms that possess the knowledge (i.e., technological competence) as well as the financial resources that will allow them to economically reduce their production of CO_2. This is apparently a big thing in certain circles, and seems to have won the approval of some members of the U.S. Congress, but the opinion here is that if it were not for the almost complete unfamiliarity with mainstream economic theory that marked the relish for cap-and-trade programs, it would qualify as a deception. For instance, one of its most enthusiastic promoters — the former CEO of British Petroleum, Lord Browne of

Madingly — has changed his tune, and called emissions trading inefficient. The same is true of some economic theorists who devised and/or initially supported this scheme.

In any event, and this is important, instead of manufacturing physical capital that could be used to produce items such as *present* necessities and luxuries, including investment goods (i.e., machines and structures), some capital would (directly or indirectly) be produced (or obtained) or eliminated for the purpose of suppressing or preventing CO_2 emissions. In other words, the present consumption of necessities and luxuries would be reduced, and in theory this 'disutility' would be balanced or more than balanced by the increase in amenities such as cleaner water and air that would appear sooner or later — which might be true in a textbook world.

Ramsey-type equations are designed to provide a 'scientific' approach to comparing a future utility with a present disutility — e.g., enjoying the beaches on the west coast of Sweden in the future in return for a present expense that might be reckoned in money, or leisure, or comfort, or (as is so often the case here in Sweden) just picking up your favorite newspaper and on the first page being treated to an unscientific disquisition about climate change by a clueless politician or civil servant. Of course, neither those equations nor any derived since Adam and Eve can carry out the mission to which they have been assigned by Lord Stern, but that is irrelevant. When many persons see or hear about the mathematics used by Frank Ramsey and Lord Stern, they draw the conclusion that it is on the same level as that employed by Albert Einstein or Werner Heisenberg. My opinion here is that assumptions of this nature are precisely what is wrong with academic economics at the present time.

In a comment made on one of my favorite sites — EnergyPulse (www.energypulse.com) — I mentioned an American academic from an upmarket university who was wandering through Europe like the Ghost of Halloween Past, preaching his version of environmental redemption to naïve segments of the concerned academic masses. As to be expected, his belief in Lord Stern's work is almost total, and the warmth with which he has been received is largely based on the belief that anthropogenic global warming (AGW) is the real deal — which it may be — and that it can be brought under control by remedies suggested directly or *en passant* in the Stern treatise, or derived by other economists specializing in climate studies.

As to be expected, some troubled citizens are almost totally enraptured when confronted with the opinions expressed by Lord Stern, as well as those mouthed by lesser 'lights'. Take, for example, Joan Ruddock, who was Parliamentary Under-Secretary of State for Energy and Climate Change in the UK. She dismissed the criticisms of Stern by academic stars like Professors Tol, Partha Dasgupta and Martin Weitzman because — according to her — these economists suffer from "a fundamental misunderstanding of the role of formal, highly aggregated economic modelling in evaluating a policy issue."

I am afraid that I must reject your contention that they do not understand, Ms. Ruddock: those scholars definitely get the message! You and I lack the required insight! You because your education in economic theory is inadequate, and me because the only interest I have in what you call "aggregated economic modelling" is the slender amount necessary for me to write this book or a few articles for the 'web'. A special problem here is that by aggregated economic modelling, she might mean 'aggregate econometric modelling'. I taught econometrics in Sweden and Australia and, while working for the United Nations in Switzerland, found it useful for conveying certain ideas — as long as it was kept on an elementary level. On an advanced level it is little better than witchcraft.

I would like to end this section by saying that the Stern Review contains a number of the features that characterized important books like *The General Theory of Employment, Interest and Money* by John Maynard Keynes, or *The Theory of Games and Economic Behavior* by John von Neumann and Oskar Morgenstern. These are famous books that almost every advanced student knew something about when I was in graduate school, although very few had actually read them, particularly the very complicated latter work.

There was a time when I knew a great deal about these two books, and may even have told some inquiring student or colleague that I have read them from cover to cover, but if I made this assertion, it is for the most part a departure from the truth. From one source or another I learned a great deal about the contents of Lord Keynes' book, but the only person I have met who convinced me that he had read most of the other book was R.G.D. Allen, professor of mathematical economics at the London School of Economics. His book *Mathematical Economics* is presently almost unknown in the great world of academic economics, and probably elsewhere, but as far as I am concerned, it is still, in many respects, the best mathematical economics textbook. Macroeconomics and game theory play an interesting part in his book, but I doubt whether there would be a place for Lord Stern's work if that book were being written or revised today.

3. Energy Units, Terminology, and Some Mathematics

"Time's glory is to calm contending kings, to unmask falsehood and bring truth to light."

— *William Shakespeare*

This is a long and very important section, and one reason for its importance is its absence in many other books on energy economics. It might also seem logical for some of the present section to be placed at the beginning of the first chapter, but as I have found out the hard way, that is often the wrong place for numbers and symbols if the intention is to have the rest of the chapter or book read. As mentioned earlier, serious readers will make every possible effort to absorb most of the materials in this chapter. If desired, though, some of the math later in this section can be overlooked, but only *some* of it; the rest should definitely not be bypassed. For those of you who might find yourselves in a position to present your ideas on energy to decision makers at various levels, much of this section is essential, because there could be a *professional* — i.e., social and economic — penalty for not having this material at your fingertips.

Here I can say that regardless of how this section appears at first glance, anyone who has made any calculations having to do with the changing of currencies in their local FOREX or airport, or for that matter on quiet street corners in Gifu (Japan) or Schwabisch Gmund (Germany), will have no difficulty following the discussion.

I want to start with some observations that should be of use the next time you read this chapter, because I presume that persons who desire a thorough acquaintance with energy economics will read this chapter several times, as it contains information that 'serious readers' cannot ignore. I regard this chapter as analogous to the troop-leading procedures handed out to students at the Infantry Leadership School at Fort Ord (California), and covering about half the size of a postcard with small print. I still have my copy of that document. For *this* document, you can remember that the *energy* used *daily* in 2009 was about 333 billion kilowatt-hours ($= 333$ Gkwh, where G signifies 'giga' or billion). An interesting exercise here would be determining how this output of energy is divided between the developed and the developing world, because increasingly one sees predictions that by the end of the present decade, the global population could be close to a billion persons larger than at present, and this will mean a much larger demand for energy resources! Taking a slightly longer view, by 2050 the global population could be 9 billion persons, or even more. But that is the future, and now we turn to current business.

1. Calculating the real price of oil. The *nominal* (or *current* or *money*) price of a good — e.g., oil — is its market price on a given day. The *real* price, however, takes into consideration the rate of inflation from some previous date. With this topic we generally limit attention to the real price of the good for a single country, although we could put together some kind of approximation for a group of countries, or even the entire world, and also introduce exchange rates.

In order to calculate the often cited real price of oil (P_r) at time t, we use the relationship $P_r = [\mathrm{CPI}_b/\mathrm{CPI}_t] \cdot P_t$, where the calculation is made with respect to the consumer price index (CPI) for a given country at a time b, which is usually defined as the base year (b). This relationship can be made clearer by writing it as follows: Real price in year t, or $P_r = [\mathrm{CPI}(\text{base year})/\mathrm{CPI}(\text{year } t)] \times P_t$, where P_t is the nominal price at time t. Sometimes the base year is in the early 1970s, before OPEC burst onto the scene with a tripling of the oil price, but I have not found it particularly useful to discuss the real price of oil using 1970 as the base year in my lectures, since it often leads to misunderstandings. As far as I am concerned, the most useful period for studying the oil price are the 10 years between 1999 and 2009, during which time OPEC put its long-run strategy into action.

Taking numbers at their face value, with no attempt at interpretation, it is easy to come up with the conclusion that the maximum real price (P_r) for the U.S. over the period 1973–2005 was in 1981. Perhaps this is true, since the Iranian Revolution took place around that time and the oil price surged, but individuals who feel that their standard of living is threatened by rising oil (or energy) prices after 2004–2005 might feel that another base year is preferable — one that results in a greatly increased real price for the years 2006–2008.

It can also be pointed out for calculus enthusiasts that a total differentiation of the above gives $g_r = g_t - g_{ct}$, where g signifies a growth rate for P_r, P_t and CPI_t, since the growth rate of the base year CPI (i.e., g_{cz}) is of course zero. It often seems that $g_{ct} > g_r$, and so you might be told that g_r (the real price of oil) is actually descending at the present time. This may or may not be true, but the nominal (i.e., *current*) West Texas Intermediate (WTI) oil price just now is $103/b, and the 'Brent' price (which seems to be at least as important as the WTI price these days) is close to $110/b. Nominal or real, these are high prices, given the lamentable economic situation for a number of countries. When students or conference participants ask me how to deal with a 'two-price' situation (WTI and Brent), I suggest employing some sort of average.

2. The reserve-production ratio. As will be clarified in the chapter on oil, geology and economics suggest a limit to the amount (q) that we should remove from an oil (or gas) deposit in a given period. This is a very important observation! With R reserves, write $\Delta R/R \leq \mu^*$, where μ^* is a number that geologists or petroleum engineers tell us is appropriate, and in this context is the maximum rate at which, on average, an oil deposit should be exhausted during, e.g., a year. However, for the present analysis, $\Delta R = q$, i.e., present output, and so we immediately obtain $q/R \leq \mu^*$. In the chapter on oil, careful attention will be paid to the inversion of this relationship, or $R/q \geq 1/\mu^* = \Theta^*$, because until recently the general discussion of oil shortages usually involved the reserve-production (R/q) ratio. In my earlier work I often emphasized that — *ceteris paribus* — a sharp decline in the R/q ratio is not a good thing; however, when the R/q ratio seems stuck at a certain value, while the oil price is escalating, the conclusion should be drawn that something else is much more important than the R/q ratio. For me, that *something else* is OPEC.

Incidentally, we often have to entertain strange tales about the R/q ratio, in the context of which the expression "dynamic" is used to claim that reserves are adequate; however, if we start at the present date and ask about the development of this ratio, we might write $Re^{\alpha t}/qe^{\beta t}$, and since it is very clear that β is larger than α, the dynamics do not seem to work in favor of the oil optimists. Readers should also remember that the key thing where this

172   *Energy and Economic Theory*

topic is concerned is the date or approach to the date at which the global q peaks, because for various reasons (e.g., psychological), when that happens, the lifestyles of a good many conscientious citizens might be regarded as being in significant danger. (q might also mean 'liquids' as well as crude oil!) Things like investments in structures and machines, as well as financial investments, are often considered in the light of the future availability of energy, and the announcement of an unambiguous peaking of the global supply of oil on primetime TV would not be encouraging for most categories of, e.g., investors, entrepreneurs, managers, etc. (And please remember, the talk now is increasingly about 'liquids', and not just crude oil.)

3. Units and equivalencies. Table 1 hardly requires any explanation, but it should be looked at carefully anyway because it is a part of the terminology that you want to have at your disposal. Fortunately, these designations should be easy to memorize.

Next it should be noted that 1 t is the designation for one metric ton, or 1 tonne, which equals 2,205 pounds (lbs). We also have a short ton, which in most countries is simply called a ton. One short ton (ton) = 2,000 lbs, and so 1 t = 1.103 tons. Finally, there is a long ton, which is 2,240 lbs. As most readers know, 1 mile = 1,609 meters = 5,280 feet, and so 1 meter is approximately 3.28 feet = 39.37 inches, giving us 1 inch = 2.54 centimeters. Furthermore, 1 kilogram is approximately 2.2 pounds (lbs), and if we have to convert Centigrade to Fahrenheit we use the formula {Fahrenheit = (9/5)Celsius + 32}, remembering that Celsius now seems to be used instead of Centigrade.

When working with energy we are often explicitly or implicitly interested in heat, which is usually measured in British thermal units (Btu) and

Table 1.   Commonly Used Prefixes

Prefix	Symbol	Power	Meaning	Example
kilo	K (or k)	10^3	thousand	kW (kilowatt)
mega	M (or m)	10^6	million	MW (megawatt)
giga	G (or B)	10^9	billion	GJ (gigawatt)
tera	T	10^{12}	trillion	TJ (terawatt)
peta	P	10^{15}	thousand-trillion	PJ (petawatt)
exa	E	10^{18}	million-trillion	EJ (exawatt)

Table 2.   Conversions: Joules–Kwh–Btu

	Joules (J)	Kilowatt-hours (Kwh)	Btu
1 joule	1	0.278×10^{-6}	0.948×10^{-3}
1 Kwh	3.6×10^{6}	1	3.412×10^{3}
1 Btu	1.055×10^{3}	0.293×10^{-3}	1

joules. 1 Btu is the amount of heat needed to increase the temperature of 1 pound (lb) of water by 1 degree ($= 1°$) Fahrenheit. (After reading the materials below, you should convert pounds and degrees Fahrenheit into kilograms and degrees Celsius.) One metric ton (1 t) of bituminous coal has an energy content (on average) of 27,700,000 Btu, and the reader can convert this to joules using Table 2. 1 t of crude oil has an average energy content of 42,514,000 Btu, and 1,000 cubic feet ($= 1$ kcf $= 1$ kft^3) of natural gas has an average energy content slightly in excess of 1,000,000 Btu. (The exact figure is given below, but 1,000,000 is a useful approximation for calculations.)

What we can do with this information is to write equivalence relationships between different energy sources. You will be given an introduction to this below, and you should pay close attention to these very simple calculations. Note in particular that a watt is one joule per second, which is why it is possible to say that power — which is measured in kilowatts — is the *rate* at which energy is expended, or the rate at which energy can be expended.

It was often said that Btu(s) would be completely replaced by joules (J), but this is not likely. It would not seem natural at the present time to quote natural gas prices in dollars per joule instead of dollars/Btu, although joules are a member of the international system of units (SI units). The kilowatt-hour (see Table 2) and megawatt-hour — which are units of energy (and not capacity) — are very popular and well-known units, and readers should get used to working with them. A megawatt-hour is a million watt-hours, and a gigawatt-hour is a billion watt-hours.

It is worth remembering that 1 million tonnes of oil-equivalent ($= 1$ mtoe) can be converted to Btu, should this unit be relevant. A handy transformation is 1 barrel of oil ($= 1$ b) $= 5,800,000$ Btu. We also have 7.33 barrels (on average) enclosing 1 t of oil. The most popular unit for measuring the consumption and production of oil is barrels per day ($= $ b/d).

For example, the output of the OPEC countries the last time I checked was about 28 mb/d, and exports perhaps 25 mb/d, and this can be converted into another popular unit — millions of tonnes per year (mt/y) — by multiplying by 50. Thus, 25 mb/d = 1,250 mt/y. Where this 50 is concerned, we get it from a simple dimensional analysis: 1 (barrel/day) × 365 (days/year) × (1/7.33) [tonnes/barrel] = 50 (tonnes/year), and thus 1 mb/d = 50 mt/y. Here it can be mentioned that if you are really and sincerely interested in impressing friends and neighbors and employers and students, it is a good idea to memorize some of the above numbers.

Power is defined as the rate of doing work. The best-known unit for measuring power is the watt (W or w), which is equal to one joule (J) per second. These units will be examined more closely in the chapter on nuclear energy, but just now the conversions in Table 3 seem useful.

Finally, in summary form, some useful equivalencies are:

- 1 barrel of crude oil = 42 U.S. gallons, and weighs 0.136 Metric tons (t).
- 1,000 cubic feet (= 1,000 ft^3) of natural gas = 28.3 cubic meters, or 1 m^3 = 35.33 ft^3.
- 1 kilowatt-hour (kWh) of electricity = 3,412 Btu = 860 kilocalories (kcals).
- 1 tonne (1 t) of bituminous coal = 27,700,000 (= 27.7 × 10^6) Btu (on average).
- 1,000 ft^3 of natural gas = 1.035 × 10^6 Btu = 2.61 × 10^8 calories.
- 1 tonne (1 t) of hard coal = 4.9 barrels of crude oil (on average).
- 1,000 cubic feet of natural gas = 0.178 barrels of crude oil (on average).
- 1,000 kWh of electricity = 0.588 barrels of oil (on average).

A thermodynamic staple is probably in order at this point. To begin, it needs to be recognized that heat is a form of energy, and one of the most

Table 3.   Conversions: Watts–hp–Btu/hr

	Watts (W)	Horsepower (hp)	Btu/Hour
1 watt	1	1.342 × 10^{-3}	3.41
1 hp	0.746 × 10^3	1	2.54 × 10^3
1 Btu/hr	0.293	0.393 × 10^{-3}	1

important scientific discoveries had to do with the equivalence of heat and work (where work is roughly anything requiring physical effort). Once we accept this, the logical next step is the law of the conservation of energy, which is also called the First Law of Thermodynamics.

This states that the total energy in the universe is constant and that, when energy is transmitted from one body to another or converted from one form to another, the total quantity of energy (and its mass equivalent) in the system after the conversion will be the same as before. This is the thing you need to remember and recite to friends and neighbors in a superior tone of voice if you get the opportunity. The best thing about all this is that standard units for power, energy and work have fixed mathematical relations to one another, as you may realize if you think about the discussion just above.

A problem, though, is that heat differs from other forms of energy. The (chemical) energy in, e.g., fossil fuels can be converted to heat with little or no loss, and from heat into work — via, e.g., a moving part in a turbine. But this last conversion will usually involve a considerable loss, which takes the form of a descent in temperature of the heat involved to that of the surroundings. Once the heat obtained from the fuel has descended to the 'ambient' level, it is no longer capable of doing work.

It is at this point that we encounter one of the most mysterious expressions in thermodynamics, which is *entropy*. Heat is sometimes spoken of as low-grade energy, because it is not convertible into any other form, and it becomes of less use as its temperature falls. In fact, when it reaches the ambient level, it is of no further use. However, escaping this degradation of energy might be possible through gaining access — in an economical way — to, e.g., solar energy, which is endless. Notice the term "economical"; this is often overlooked by solar enthusiasts.

4. An example. Now for an example I first employed when teaching energy economics at the University of Stockholm, which uses some of the above equivalencies. Suppose we have two light bulbs. One of these produces a great deal of illumination and has a power rating of 500 watts, while the other is considerably weaker and has a rating of only 50 watts.

If the heat energy in coal is totally and perfectly transformed into electrical energy (i.e., with 100 percent efficiency), then 3,412 Btu are required to generate a kilowatt-hour (kWh) of electrical energy (where the kWh is the unit in which electrical energy, as distinguished from *power*, is measured).

The power rating of the bulbs, 500 and 50 watts respectively, informs us of the rate at which the energy potential of the coal is consumed, and so if the 27,700,000 Btu in a tonne of coal is transformed into electricity in a *perfect* system, then it could provide exactly $27,700,000/3,412 = 8,118$ kilowatt-hours of electrical energy. In other words, in a perfect system, the stronger of the two bulbs — which consumes power at the *rate* of 500 watts ($= 0.5$ kW) — could function for $8,118/0.5 = 16,236$ hours. The other bulb requires $8,118/0.05 = 162,360$ hours to consume a tonne of coal.

In reality, the efficiency with which fossil fuel can be converted to electrical energy is well under 100 percent. An efficiency of about 32 percent seems typical for much of the industrial world, and so, on average, it would require $10,662 (= 3,412/0.32)$ Btu to obtain a kWh of electrical energy. The 10,662 Btu/kWh is called a *heat rate*. This can be put another way: 1 kWh(e) $= 3.12$ kWh (fossil fuel). The UN and OECD also calculate, e.g., 1 kWh(e) $= 2.6$ kWh (oil). It was stated above that a tonne of coal has an (average) energy content of 27,700,000 Btu. Similarly, a barrel (b) of oil has an average energy content of 5,800,000 Btu. Averages of this sort permit us to discuss things like coal and gas in terms of an oil-equivalent.

The talk above has mostly been about oil, but as noted previously, the expression "oil" will increasingly be replaced by "liquids": oil, biofuels, natural gas liquids, and perhaps gas and coal to liquids (GTL and CTL), etc. Furthermore, although it is not completely certain, the global output of crude oil may have peaked, or is in the process of peaking. A question that readers might find useful to examine is how easy will it be for items like biofuels to replace conventional crude oil.

Readers can now turn to Figure 1, noting that system output was changed from 'million barrels of oil-equivalent per day' to 'terawatt-hours per year'. In this figure we have an output of 11 Mboe/d of electrical energy. Note that 1 tonne of coal is equivalent to $27,700,000/5,800,000 = 4.77$ barrels of oil-equivalent. To get the heating value we write $11 \times 10^6 \times 5.8 \times 10^6 = 63.8 \times 10^{12}$ Btu. This can easily be converted to kilowatt-hours $[= (63.8 \times 10^{12})/3,412 = 1.869 \times 10^{10}$ kWh/d]. Notice too that these are 'per day' values. Because these figures tend to be quite large, another unit — the Terawatt-hour (TWh) — is often more useful, where 1 TWh $= 1$ trillion watt-hours. Accordingly, the above figure becomes 18.69 TWh/d, or

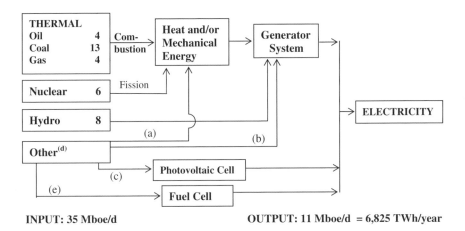

INPUT: 35 Mboe/d   OUTPUT: 11 Mboe/d = 6,825 TWh/year

Figure 1. A Comprehensive Electrical System: Inputs and Output

Notes: (a) Solar-Thermal: Biomass, Geothermal, Solar; (b) Wind, Tidal, Wave; (c) Solar; (d) Negligible; (e) Hydrogen-Oxygen.

Source: Shell Briefing Service, 1986.

6,825 TWh/y as shown in Figure 1. By the same token it would have been easy to use megawatt-hours (MWh) or gigawatt-hours (GWh) if these fit better into the discussion.

5. As simple as the above is, some readers may feel uneasy. While electrical power is defined as a 'rate', it is not always explicitly associated with a time dimension. For instance, the 'rating' of a power station is likely to be in megawatts. However, in the example with the bulbs, we saw that a large bulb 'milked' the given unit of coal of its energy potential more rapidly than a small bulb, which trenchantly suggests that the dimension for power is energy per unit of time. As it happens, a watt is one joule per second (which is easily recognized as a rate) or 3,600 joules per hour; and since 1,055 joules is 1 Btu, 1 watt is 3.412 Btu/hour (which is more easily recognized as a rate by those of us accustomed to working with the Btu). Observe that 1 kW = 1,000 J/second, where J signifies joules.

This example can be extended by treating power as analogous to velocity, and energy as analogous to distance. The total distance travelled by a vehicle is velocity (in miles per hour or kilometers per hour) multiplied by time (hours), just as the total energy converted is power (J/time or kilowatts) multiplied by time. The power rating of the vehicle determines (*ceteris paribus*) its speed, while the power rating of the bulb determines its illumination. In both cases, the larger the power rating, the more work that can be done, where the units in which work is measured can be converted into, e.g., kilowatts. Put more succinctly, the major uses of energy are for the production of work or heat: the flow of energy is called work when it exerts a force, and heat when this is not the case. Work and heat are alternative modes for the flow of energy. Something we are aware of is that the more work the bulb or car can do, the higher its acquisition cost, and often the higher its operating cost.

6. Now for an extension of the simple mathematics in the first chapter of this book. It will be an elaboration of something for which I have very little respect, the Hotelling model, but it will be a good introduction to the analysis employed in the subsequent chapters on oil and natural gas. To begin with, we want to maximize the following expression:

$$V = \sum_{t=0}^{n} (p_t q_t - c(q_t) q_t)(1 + r)^{-t} = \text{Max}. \tag{1}$$

Readers not used to summation signs should write this out for $t (= n) = 3$. The same should be done for the equation below, and then explain your work to someone.

Take note that $c(q_t)$ — i.e., average cost — is not a constant, and in addition, usually increases as q increases. Now, add a constraint to account for the using up of reserves S over the n periods for which planning takes place. Replacing V with V^*, we get:

$$V^* = \sum_{t=0}^{n} (p_t q_t - c(q_t) q_t)(1 + r)^{-t} + \tau \left(S - \sum_{t=0}^{n} q_t \right). \tag{2}$$

Write total cost for a period i as $C(q_i)$, and then differentiate partially with respect to q_t for $t = 0, 1, \ldots, n$. We now get for a typical period i: $[p_i - C_i'(q_i)](1 + r)^{-i} = \tau$. In this expression, note that $C_i'(q_i)$ is the marginal cost when output is q_i. Now we can write:

$$[p_0 - C_0'(q_0)] = [p_1 - C_1'(q_1)](1 + r)^{-1}$$
$$= \cdots = [p_n - C_n'(q_n)](1 + r)^{-n} = \tau. \tag{3}$$

This result has a Hotelling-like connotation, but, fortunately, a more universal significance. Intertemporal profit maximization involves equalizing present values by 'manipulating' planned outputs — "planned" because *future* prices and costs are estimated. The steps leading up to this result are valuable, and all students who can handle the differential calculus are expected to learn them perfectly. Something else worth noting is that the average cost, $c(q_t)$, increases as q_t (output) increases, and if we had something else then the so-called *second order conditions* might not be satisfied.

I want to finish this section with another mathematical demonstration. Specifically, I want to utilize an observation by Robert Feldman, chief economist at Morgan Stanley. He states that time lags between supply and demand that characterized the famous hog cycles of the 1930s are now at work in "energy" (by which he meant, of course, oil). These time lags can create imbalances that lead to very large price swings. These lags can be detected in Equations (4) and (5) just below, although what is not shown in these expressions are the root causes of fluctuations. These are speculative tides of bullishness and bearishness, which have their origin in actual and/or expected price trends, the outcome of OPEC meetings, and possibly the forecasts of high-status organizations such as the International Energy Agency (IEA) and the Energy Information Agency of the United States Department of Energy.

We can now derive a simple equation which might have some pedagogical value; however, it is not an essential part of this exposition. Notice the term "pedagogical". What it specifically means is that every student of elementary economics should learn how to solve simple difference equations. First we have a (flow) demand curve for oil, $h_t = a_0 + a P_t$, and a (flow) supply curve, $s_t = b_0 + b P_{t-1}$, with $a < 0$ and $b > 0$. Note that the supply curve is lagged, while the demand curve is called h instead of the usual d, and I will add trend terms to these relationships to get $h_t = a_0 + a P_t + \alpha_t$ and $s_t = b_0 + b P_{t-1} + \beta_t$. Introducing the equilibrium condition $s_t = h_t$, we immediately obtain the following first order difference equation:

$$P_t = \frac{b}{a} P_{t-1} + \frac{\beta_t - \alpha_t}{a}. \qquad (4)$$

(For α_t and β_t, we might have $\alpha_t = \lambda_x + \alpha_1 t$ and $\beta_t = \lambda_y + \beta_1 t$.) In any event, if we begin with a price of P_0, then the solution to the above equation is obviously:

$$P_t = \left[P_0 - \left(\frac{\beta_0 - \alpha_0}{a - b} \right) \right] \left(\frac{b}{a} \right)^t + \left(\frac{\beta_t - \alpha_t}{a - b} \right). \tag{5}$$

The expression to the left of the plus sign in Equation (5) represents a cobweb model, while the expression to the right is a trend term for the price. On the basis of earlier statements, at the present time β_t is less than α_t, and so P_t is trending upward. Basically, in Equation (5), there is a trend term around which there are oscillations in both price *and* inventories. In addition, if supply lags demand, and there is an upward spike in demand, inventories will decrease even if the lag is very short. For an optimal presentation, a flow model of the type above is inadequate. The trend terms in the equations suggest that we are dealing with the long-term, while even the discussions on CNN and Bloomberg suggest the relevance of inventories for short-term developments. Thus, we need another model — one in which inventories are explicit. That means a stock-flow construction of the kind shown and discussed later in this book.

4. Loose Ends: Mr. Malthus, Options, and Refining and 'Crack Spreads'

On October 22, 2009, Professor Riccardo Basosi of Siena University (Italy) opened a workshop in which energy was discussed on a considerably more abstract level by him and Franco Ruzzenenti than in the present book. The discussion by Professor Basosi began with thermodynamics and ended with 'complexity', where the latter topic involves, among other things, the rather delicate subject of sustainability. My book, *Scarcity, Energy and Economic Progress* (1977), alludes constantly to these issues.

With sustainability, we find ourselves considering long-run situations in which technology may be unable to work the miracles that we have always been assured it can and will work in the short or maybe the medium run. Accordingly, if we cannot have those miracles when we want or need them, then we might have to start thinking about compensations or trade-offs, or, in plain language, just adjusting to the kind of situations that we thought or hoped we would never encounter. This is a topic that is not particularly easy to contemplate and discuss except on the most advanced intellectual and/or scientific levels, because one of the most powerful driving forces of modern societies is the belief that tomorrow will be better than today, and under no circumstances could our actual or expected prosperity be interfered with by *technological limits* of one sort or another.

The research that Professor Basosi and his colleagues have commenced is going to be an integral part of the energy economics of tomorrow, and in the very long-run perhaps the most important part, because what might be termed the established philosophy of (economic) growth is going to be called into question. Readers who desire a largely non-technical insight into these matters can turn to the last chapters in the book by Earl Cook (1976); although less comprehensive, the final chapters of Donald E. Carr (1976) also deserve examination. The interesting thing here is how much more important these comparatively unknown books are today as compared to when they were published, because in, e.g., 1976 the mention of resource scarcities — or a limit to growth — was considered too unorthodox for polite academic consideration.

Something else that was and still is considered out of place in serious forums is the work of (Reverend) Thomas Malthus, who several hundred years ago insisted that runaway population growth could result in very bad economic and political scenarios that neither technology nor markets could ameliorate. If you add things like a shortage of space and various social tensions in major population centers, declining educational standards in many countries, pandemics, and possible climate woes that may or may not be due to human activities, then it is clear that the kind of economics presented in this book will require extensive augmenting. As Jorgen Orstrom Moller (2008) suggested, "Mr. Malthus is a man that we would not like to meet in person. He is an unwanted guest, one well worth working to avoid." In the very long run, however, an encounter with that gentleman might be inevitable. By mid-century the global population will probably be around 9 billion persons, and while this might be acceptable, mid-century is not the end of the world. In the decade after that, another billion persons may be added.

We now move to a somewhat more pleasant (though not more important) subject. We know about futures contracts due to the work and worries of Millicent Koslowski and Condi Montana as described in the first section of this chapter, and we proceed by moving ten years into the future and scrutinizing another situation. But first, a reminder about futures. Unless it is offset, a futures contract provides the buyer with the obligation to take delivery of a specific amount of a commodity such as crude oil, and provides the seller with the obligation to deliver that specified amount at a future date and at the price of the contract on that date (i.e., the expiry date of the contract). As we know, however, after a position has been opened it can either be kept open until the expiry (or maturity) date, when delivery or cash settlement takes place, or be closed by a reverse transaction — which is almost always the case. As you may remember, this is what Millicent did in the earlier discussion of futures: she *opened* her position by going long (buying) futures contracts with a 30-day maturity in the morning, and *closed* it that afternoon a much happier person as a result of closing her position by a short (futures) sale at a higher price than when she opened.

Options provide an extra dimension, because options can be exercised either before or at the expiry date, with delivery coming into the picture, or, surprisingly, can be abandoned to expire worthless at the expiry date, without the transactors having to be concerned with making or taking delivery. (Options can also be sold.) Now we begin a new discussion by looking at an unusual situation that involves finger food.

To be exact, wedding bells will soon be ringing. The socially prominent Evelina Grundy of Aberdeen (Washington State) and Algiers (Louisiana) is about to tie the knot with a former Gunnery Sergeant in the U.S. Marines, Hoover Lee Clarke, who is presently a popular Roper County (Alabama) sharecropper. There are, however, a few problems.

The main problem is money for the wedding and wedding reception, since in the Grundy family the custom is for the bride and groom to share expenses when it comes to elaborate nuptials. As bad luck would have it, things have not gone well for Hoover. The man who pulled his plow departed to take a job at Hamburger Heaven, and while the Grundy family fortune was founded on timely purchases of houses and lots during the Civil War, later generations of Grundy men dissipated it by untimely purchases and sales of the same items. The bottom line is that these two lovebirds must throw themselves on the tender mercies of Hoover's oldest son, Hector Lee Clarke. Once a teacher of nuclear physics, Hector is now a quantitative analyst (or 'quant' or 'rocket scientist' as they are sometimes called) at Canino & Soprano, a well-known Wall Street (New York) investment bank.

Hector accepts the assignment with great reluctance, but after a few days of thinking about it, he and his wife, Lucy Sweet Clarke, become enthusiastic. Hector's father and his other male relatives have always considered him a blight on the family name, since most of them were moonshiners, pool sharks, 'do-ah' choristers, stand-over men, and hoboes. Now he and Lucy have a chance to show those 'good old boys' what the good life in a duplex apartment on Park Avenue is all about, and in addition, demonstrate to them what it means to be a successful 'quant'.

The wedding will be on a Friday, and the reception is in the form of an open house that lasts from Friday evening until midnight on Sunday. Unfortunately there might be hundreds of guests, most of them invited by Evelina's seven brothers. Food is going to be a problem, but not as much of a problem as paying for it. The recent oil price escalation came so unexpectedly that it confused the judgment of Hector and several of his colleagues, and they bought the wrong assets. As a result, those gentlemen were informed by Mr. Soprano that if they forgot about their bonuses and annual increase in salaries, then they might escape the notoriety associated with 'security' escorting them to the front door, and dumping them and their personal effects into the gutter. Hector's memory, which he always boasted of as being perfect, suddenly became extremely bad, although he did remember to come to work earlier every day and stay later.

The cuisine at the wedding reception will consist of finger food, ordered from the best caterers — huge amounts of it — and it will be washed down with French wine. Evelina and Lucy Sweet Clarke decided on finger food because when Hoover Lee is in his cups, he eats everything with his fingers except his soup: his soup he drinks directly from the bowl, and so there will not be any soup at the reception. When planning for the party starts on January 30, finger food is selling for $42/kilogram, but neither of the women have any idea as to how much it will cost on April 1, when the reception begins. If only Hoover Lee had married a Grimaldi or someone affiliated with the Court of St. James, Lucy finds herself thinking, but as Hector's brother Harry was fond of asking, "Why be in a war without learning something from it?" The war that Lucy and Hector had been a part of since they got married was fought in trading rooms and offices, and corridors and restaurants of financial power in places like New York, London, Frankfurt and Tokyo. It was a war to which only people who knew how to make large amounts of money were invited. Recognizing that she needs some advice in this case, Lucy picks up the phone and calls a friend, Ingrid Ewing, who is an investment banker. "I'm going to make this thing work," she said aloud, and make it work without adding to the professional burdens putting lines in the face of her husband.

On the basis of what Ingrid tells her, Lucy purchases some options contracts which give her the right to buy 1,000 kilograms of finger food by April 1 — or earlier if she desires — for $50/kilogram. These are named *call options*, which give the buyer the right to buy the underlying entity (as compared to *put options*, which give the buyer the right to sell the underlying entity). The $50/kilogram is called the *striking* (or *strike*) or *exercise* price. When the initial transaction takes place, Lucy must pay a *premium* to the seller of the option, who is also called the *writer*. (The premium is the price of the option.) For the purpose of this exercise, let us take the premium as $5/kilogram, with the *expiry* (or *maturity*) *date* of the option contract April 1. This premium money (= $5,000 for 1,000 kilograms of finger food) is transferred immediately to the option writer via Ingrid Ewing, who occasionally does some consulting for a brokerage, and also has important contacts at a local options exchange.

Notice the position of the writer of the option. On or before April 1, he must be prepared to deliver 1,000 kilograms of finger food to Lucy for $50/kilogram. (*Before* if the option is exercised!) If the market price at which he purchases finger food — *in the event Lucy exercises her contract* — is $75/kilogram, or for that matter $750/kilogram, he must still obtain these victuals. Thus, theoretically, his losses are unlimited. But on the other hand, if the market price is less than $50/kilogram when Lucy must have her finger food for the reception (e.g., $40/kilogram), then instead of exercising the option contract, Lucy tears it up and buys her finger food in the open market (for $40/kilogram). As for the option writer in this situation, he breathes a sigh of relief: he is $5,000 to the good (since he keeps the premium money), and his concern with the possible future behavior of Lucy is over.

Lucy is even happier. She was prepared to pay $55/k for her finger food (= exercise price + premium), but now only has to pay $45 (= $5 premium + $40 market price) because the price of finger food on the open market luckily turned out to be $40/kilogram. She ends up paying $45,000 for 1,000 kilograms of finger food, and as for the option contract, that becomes confetti at the marvellous reception she gives, because she did not exercise that particular document.

Yes, Ingrid Ewing does not make many mistakes, Lucy thought (as the music began to play at the reception for Hoover Lee and Evelina's wedding). When this reception business first began, Lucy had plucked up her courage one morning at breakfast and asked Hector for some advice as he was stuffing another piece of toast into his mouth. He immediately suggested hedging the finger food price with a forward contract or a futures contract, but when Ingrid Ewing was informed of that solution she almost laughed aloud. There was no futures market for finger food, and since there were so many parties and receptions taking place around April 1, a finger food supplier might have asked Lucy for $70/kilogram for a forward agreement (i.e., a promise to deliver 1,000 kilograms to her apartment on April 1). But, argued Ingrid, since the finger food supplier in this case kept talking about a likely market price for finger food of $50–$60 per kilogram, it would be best for Lucy to pay a relatively small premium ($5/kilogram) in order to keep her options open. As things fortunately turned out, keeping her options open meant that Lucy only paid $45/kilogram.

Of course, if Hector becomes peevish because of his precious advice being rejected, and insists on Lucy using a futures market, Ingrid suggests that Lucy might find it useful to immediately go long in futures for a delicacy called Long Island Trout, which is consumed in large amounts (together with finger food), around April 1. The money that Lucy should be able to make on that investment could then be used to buy finger food at the prevailing market price. However, as Ingrid points out, the *basis risk* on this transaction might be considerable. By basis risk Ingrid meant that hedging one commodity with another does not always work. Before there were futures contracts for crude oil, it often happened that crude was hedged using futures for heating oil. Sometimes this worked, but sometimes it didn't, and because sometimes it didn't, Lucy and Evelina chose the options route, and stopped communicating with Hector on this matter.

That leads immediately to another question: in the discussion at the beginning of this chapter, why didn't Condi Montana suggest that Millicent Koslowski speculate on the price of oil using options instead of futures? The answer is that given the particular situation — i.e., information from somebody's Uncle Charlie — it was less complicated to use futures. And even if there had been a futures market for finger food, it might have made more sense to deal in options. People who occupy corner offices on Wall Street, like Ingrid Ewing and now Millicent Koslowski, know about these things, and are paid well to know them.

A few other comments about options are in order. Like futures, options are usually listed and traded (bought and sold) on an exchange by members of that exchange, but they are also sold 'over the counter' (OTC) by financial institutions, in which case they are not traded. The price of the option (i.e., the premium) usually depends on three things: the market price of the underlying entity, the amount of time that the option has to run before expiry, and the volatility of the market price of the underlying entity. This can be a somewhat complicated matter, and readers might find it useful to examine my finance book (2001) for a more thorough view of the topic.

Now for a few words about *crack spreads*, which is an expression that we see quite often, and which deserves much more attention than it has received, especially by those of us who teach energy economics. We can begin by saying something about the refining of oil. In my previous energy economics textbooks I avoided this subject, but it is extremely important. For instance, when the oil price escalated in 2008, some oil producers in OPEC claimed that it was because of the lack of refining capacity.

An oil refinery is an installation for turning crude oil into a *slate* of various products. By convention these products are divided into three cuts or fractions: gas and gasoline (light products); middle distillates of various types; and fuel oil and residual cuts (i.e., heavy products). Usually the most valuable oil products are the *white* products: domestic gases, aviation fuels, motor fuels, and some feedstocks for the petrochemical industry. (For example, naphtha, which is extracted from both the light and middle ranges of distillate cuts.) The other middle distillates are kerosene, paraffin and light gas oil, while the rest of the refinery output consists of heavy lubrication oils and residues.

This system functions as follows. Crude oil is pumped into a tall distillation tower that is pressurized and is hotter at the bottom than at the top. The various oil products have different boiling points, with those that are the lightest having the lowest. When the crude enters the tower, the heaviest part remains in liquid form and falls to the bottom. The rest is vaporized, but the various constituent products return to liquid form as they reach the lower temperatures higher up the column. As a result, they can be piped away. Given the dominant preference for light (i.e., white) products, the most valuable input is light oil, and here Libyan oil is the most preferable.

Libya has the largest oil reserves in Africa, but its production tends to be low, and the sad fact of the matter is that a great deal of expensive upgrading has been necessary in order for existing refineries to handle heavier grades of crude. As this takes place, smaller refineries are at an increased disadvantage relative to the refineries operated by the oil majors, who can finance this upgrading with profits from the sales of crude.

The *crack spread* has to do with the profit margin associated with refining crude oil, and this expression generally considers movements in the price of an oil product such as motor fuel relative to the cost of crude oil. To paraphrase some remarks in an excellent introduction to this topic by Richard Bloch (2011), this oil is refined to make it less crude. As alluded to above, the firms doing this are 'independents', or more successful large integrated companies such as Exxon and Shell. For the independents, bad news is often the outcome, because, as noted, one of the key factors in obtaining a profit is making expensive investments that provide the technology required to deal with unexpected changes in the pattern of demand, as well as changes in the quality (e.g., weight) of the crude input.

As you found out in Economics 101, refineries are like other businesses in that they are first and foremost concerned with selling the products mentioned above, as well as others, at a price that exceeds the cost of the crude. (Of course, there are also the costs of the production factors (e.g., labor and capital) required to transform the crude into oil products, and as a result, the expression *refinery margin* seems more appropriate to me.) It is easy to deduce that when the oil price increases, it costs more to drive your car, and in Sweden it appears that the price of motor fuel rises the same day that the price of crude increases. The oil price increase due to the troubles in Libya raised motor fuel prices by well over 10 percent in both the U.S. and Western Europe.

The crack spread is often described as the difference between the price received for an oil product and the cost of 'cracking' the raw materials into oil products. The way this issue is often approached is by examining something called the 3-2-1 spread, where three barrels of crude oil are transformed into two barrels of gasoline (petrol) and one barrel of heating oil, taking into consideration the cost of the crude and the prices received for gasoline and heating oil.

Suppose though that the intention is to get a different set of oil products, for example, gasoline and jet fuel. Assuming a comprehensive knowledge of the relevant technology, determining the cost is not a difficult calculation; but when amateurs on this topic like myself think about refining, I do not attempt to go beyond the 3-2-1 crack spread. I suspect that the 3-2-1 spread provides a kind of opportunity cost, by which I mean that if, e.g., you are in possession of a barrel of oil and want to turn it into oil products, the 3-2-1 spread will provide an *alternative* to doing what is required to obtain various other spreads (or output mixes) whose profit margins might be uncertain. (To take an example from financial economics, if you are in possession of $1,000 and are considering an investment in some financial asset, the purchase of a government bond provides an opportunity cost, or a guarantee of a certain return (or *yield*), because there is a risk associated with a private bond or a share.) This does not mean, however, that resorting to the 3-2-1 spread will always guarantee positive profits. Among other things, the expansion of the refinery sector in the Middle East will provide plenty of headaches for refineries in other parts of the world.

5. Final Statements

> *"Things must change if they are to remain the same."*
>
> — *Italian saying*
>
> *"Nothing is ever settled until it's settled right."*
>
> — *Rudyard Kipling*

It is to some extent sad that energy economics has not received the warm welcome in the halls of academia that it deserves, but that may be quite in order. A few decades ago, anyone who studied this subject, in any university in the world, was forced to deal with trivia a large part of the time. When I began teaching this subject, the focus in virtually every university was on the Harold Hotelling model of exhaustible resources, which meant that a minute or two after taking the final examination for the course, students were as blissfully ignorant of the subtleties of energy economics as they were a minute or two before the start of the first lecture. The problem was that they were never given the special rhythm, the special swing, needed to master this subject.

An attempt was made in Chapters 1–3, and will be made in the remainder of this book, to examine some elementary need-to-know topics in energy economics. Needless to say, it cannot be guaranteed that these topics will be recognized as the most important by all readers. There might also be some dissatisfaction with the brief treatment — or absence — of some themes. For instance, one of the things not dwelled on in this book is the macroeconomic details of energy price shocks. A considerable amount of statistical (i.e., econometric) work has apparently been done on this topic over the years, but the belief here is that it may not be worth a great deal of attention. Econometric equations cannot substitute for straightforward answers in, for example, plain English.

It is also not certain that researchers have arrived at the right conclusions about the very special nature and outcome of energy (e.g., oil) price shocks, although some of the conclusions about the effects — or likely effects — on the economy have a modicum of sense about them. I am thinking in particular of an article by Lutz Kilian (2008), in which he claims that the "surge" in crude oil prices since 2002 can be explained by large increases in the demand for crude oil. This is only a partially correct observation, because a large portion of that surge can be attributed to a restraint in investment and production by OPEC countries that was not witnessed earlier, and may well dominate the oil scene in the future. The correct expression here is *resource nationalism*, and since that terminology has started to make the rounds, it is only appropriate that readers of this book incorporate it into their vocabulary. They might also be able to incorporate it into their econometric equations with the help of some dummy variables, but that seems to me to be complicating something that in reality is best served by a thorough non-mathematical analysis, based on the history of some key oil producers and/or exporters.

Perhaps a more unambiguous wake-up notice is useful at this point. The Middle Eastern OPEC countries, perhaps together with Russia, have the discipline, knowledge and intention to defend an oil price close to and eventually above the present price, whatever that price happens to be, and *regardless of anything they may say or do in the short run*. The enormous export incomes they are still realizing suits them fine, just as it would suit you and me fine if we had access to a slice of it. The OPEC countries definitely prefer this to the dreary alternative, which would have involved going to the large importers with their hats in their hands and trying to explain to them that the price of a barrel of oil should exceed the price of a barrel of Coca-Cola.

Wouldn't you feel the same way if you were in their place? Wouldn't you continue doing the *simple things* that are necessary to obtain high prices, regardless of what speculators actually or *ostensibly* do or do not do? In case you have a problem with the expression "simple things", it means the same kinds of things that were done by wholesalers (i.e., generators) in California and/or states close to California when electricity was 'deregulated': although in theory deregulation was supposed to increase the availability of electricity, it happened instead that unforeseen 'glitches' caused it to decrease. (Many of these glitches apparently had their origins in strategic behavior by wholesalers that later came to be described as *gaming the system*.) Similarly, the once promised increase in output by Saudia Arabia of almost 3 mb/d of oil by the end of 2009 was what Jean-Paul Sartre would have called "a fire without a tomorrow" — since those of us who had studied this matter understood that this increase in output, if it came about, would not be sustainable.

The thing to be appreciated here is that an increase in output was never going to come about. A 'sustained' upward movement in the oil price began in 2003–2004, and an upward acceleration in the oil price began early in 2008, and between January and March of that year OPEC may have reduced output by more than a few barrels. Global oil demand tended to be higher than global oil supply, but instead of closing the gap, OPEC scrutinized the continued growth in demand, particularly in Asia, and decided that it was in their own best interests not to supply increased amounts of oil. Instead, they dreamed up — or supported — a fairy tale about the activities of 'speculators'.

What about new oil discoveries? Several of these have been announced recently in the media, by which I mean one in India and one in deep water somewhere south of Houston (Texas). According to my primary school mathematics, the India 'strike' will not make much of an impact on that country's goal of energy independence — a goal without the slightest chance of being realized. As for the strike in the American portion of the Gulf of Mexico, it will be badly needed by the time the first barrel of oil from it is sold, by which I mean that large-scale production from that source — if it really happens — is many years in the future, and in any case will not be capable of substantially altering the global oil supply.

Oil and gas play a very large part in this book, as well as in my previous textbooks (e.g., Banks, 2007). For this reason, it might be a good idea to refer to an article in the *Financial Times* by Carola Hoyos entitled "Mideast oil to play bigger role in global growth" (January 22, 2006). Several of the points touched upon by Ms. Hoyos were considered by the present author 15 or 20 years earlier, and here it is meant the possible shifting of the center of gravity of world oil refining to the Middle East. I also circulated at great length my opinions about the role that petrochemicals would play in the Middle East, and to a lesser extent, so did Professor Morris Adelman of MIT.

This comes under the heading of what I call in my lectures "OPEC's Strategy". Moreover, that strategy is very simple, and reduces to the following: OPEC intends to export (and perhaps produce) as little oil as possible. It ends with: OPEC intends to export (and perhaps produce) as little oil as possible, regardless of what they say or do! I occasionally repeat this mantra a number of times during a single lecture, and I intend to continue this practice. I sincerely hope that other teachers of energy economics do the same thing, but I am not an optimist.

The logic in play here is an extension of the work of three brilliant economists: Professor Gunnar Myrdal, who was one of my teachers at the University of Stockholm; Professor Hollis Chenery, who organized a small conference I attended in Paris in the 1980s, and whose book (together with Paul Clark) I used when I taught at a UN Institute in Dakar (Senegal); and, of course, the superb article by Professor A.A. Kubursi (1984) entitled "Industrialisation in the Arab states of the Gulf: A Ruhr without water".

A short comment about those gentlemen is highly relevant. Gunnar Myrdal, Nobel laureate in Economics, and known and famous throughout the world, was unbeatable in any seminar room or conference, or for that matter at the Nobel Banquet in the Stockholm City Hall in 1974. His belief was that the study of development economics should be based on a study of the economics and sociology of successful economies. The two countries he recommended for study were the United States and Sweden.

Hollis Chenery, a professor at Harvard, went almost unnoticed by the Nobel Academy, which was another example of its characteristic misunderstandings or short-sightedness, although he was a leading mathematical economist in the study of economic development (and on the 'applied' level, could be ranked above the first winners of the Nobel Prize in Economics, Ragnar Frisch and Jan Tinbergen). More remarkable, however, is the neglect by economists of Professor Kubursi (of McMaster University in Canada), although the big mistake here might be the failure of Professor Kubursi to adequately market his theories of economic development to the 'academy'. The only place that I have seen or heard his name called is in my books, articles and lectures, although it is not impossible that he discussed his work in OPEC councils, because somebody (or persons) in those councils listened to or read what he had to say, and we see the results today. By that I mean high oil prices, and an expansion of refineries and petrochemicals in some OPEC countries.

The explicit observation by Professor Kubursi — and implicit in the work of Gunnar Myrdal and Hollis Chenery — was that instead of exporting oil in its crude form, if the development process is taken to an optimal conclusion, that oil should be used in OPEC-owned refineries, and a large amount of the refinery output should be used in the production of petrochemicals. Robert J. Beck, director of the Oil and Gas Journal Research Center, who spent much of the 1970s in Iran, said that the Shah of Iran stated on several occasions that oil should be saved to produce petrochemicals. (He also was making arrangements to introduce nuclear energy to Iran.) More importantly, simple mathematics leads us to conclude that investing in facilities to produce refinery products and petrochemicals in, e.g., the Middle East without having enough crude to utilize these facilities for *X* years is a serious mistake! (The book edited by Professor Jean-Marie Chevalier (2009), and used at his institute at the University of Paris (Dauphine), deserves mention here.) Something else that is meticulously overlooked is the value of those facilities for training members of the workforce of a developing country.

As I indicated earlier in this book, Professor Morris Adelman of MIT (who definitely is *not* a friend of OPEC) and his colleague, Martin Zimmerman, make clear the enviable position of certain OPEC firms, particularly those who obtain their refining (and therefore petrochemical) feedstocks at a fraction of world prices (Adelman and Zimmerman, 1974).

Among other things, it is going to mean acceptable profits for a very long time for these firms, as well as impressive national incomes for the countries in which these firms are located. Moreover, even if desired outcomes are not immediately achieved, major oil-producing countries can look forward to the high 'returns' that result from transforming inexpensive refinery products into high-priced petrochemicals. This was pointed out by the Nobel Prize (in chemistry) winner Sir Harry Kroto many years ago with regard to agricultural chemicals (i.e., fertilizers).

It cannot be overemphasized that since energy costs are the key burden for chemical industries, the combination of inexpensive energy and state-of-the-art technology will ensure that the center of gravity of the global petrochemical industry will move toward the 'least-cost' Middle East. According to the time-honored theory of *comparative advantage*, that is where it belongs if the countries of that region can provide sufficient output (which is not certain).

"Center of gravity", though, does not mean complete domination. At any time the global petrochemical industry is a mixture of small and large, low and not-so-low cost, new and old, etc., and so *theoretically* the OPEC output would be adjusted so that product prices would be high enough to keep some of the less favorably endowed plants in operation in order to supply a certain portion of the total demand. Even so, many firms in this line of work have become unpleasantly aware of the realities brought about by the cheaper methods of production at the disposal of countries that no longer want to be a hostage to unfavorable oil or gas prices. Exactly how traditional firms will react to this challenge is uncertain, particularly in the short run, although in the long run many of them have no choice but to cut-and-run, to use one of President George W. Bush's favorite expressions.

After reading my above thoughts on this subject, and once again reviewing the paper by Professor Kubursi, I would like to say that while you may not find the observations directly above attractive, it is impossible to deny the bottom line: OPEC strategy is going to turn on producing and exporting as little crude oil as possible, and their business and political acquaintances throughout the industrial world will simply have to get used to and adjust to this arrangement.

Some further references to refining should perhaps be noted, since deficient refining capacity (and therefore deficient capacity for producing motor fuel) is often associated with high motor fuel prices. It is easy to get the impression that there is not enough refining capacity to refine (at low cost) the output of crude oil, and so there would be no problem for motorists if only more refineries were constructed. Unfortunately, however, it is not as uncomplicated to build and operate a refinery as it is to build and operate a fast food outlet. Refining is one of the riskiest of all industrial pastimes. Refining and red ink have a way of going together, and with only a relatively small number of exceptions, the winners in this business tend to be the large integrated oil companies who have *upstream* profits (from crude production) that enable them to carry substantial *downstream* (e.g., refining) losses should they occur. To repeat, the center of gravity of refining and petrochemicals logically belongs in the Middle East, and it is very possible that this is where it eventually is going to be found.

The bad news for refiners usually begins with large shifts in demand. Refineries produce kerosene and fuel oil that give light and heat; gasoline and diesel fuel which are inputs for transportation; lubricating oils; lighter products that are building blocks for the petrochemical industry; and asphalt. Refineries typically are configured to produce a certain 'cut' of these outputs, and it can happen too often that suddenly demand for that 'basket' declines while demand for another increases. Refineries that want to stay in business then have no choice but to make costly investments (i.e., *upgrade*) in order to accommodate the new demand. It can also happen that demand falls for the product(s) to which refineries are most intensively committed, or the cost of inputs — particularly light or heavy crude oil — unexpectedly increases, or they are not sufficiently alert to compete with other establishments in a game where mistakes or misjudgments are exceptionally costly. Even firms with good management that have the financial resources to make large investments can miss out on their timing.

To make matters more complicated, there are large expenses in the offing that have to do with environmental issues. It was once claimed that as many as 20 of the approximately 124 refineries in the U.S. that produce gasoline and diesel may elect to close their doors rather than do the expensive upgrading required to meet the more demanding environmental laws that could be scheduled to take effect in the near future. This is not a welcome development in a country where it has been many years since the last refinery was constructed.

It is always possible to say that certain people who should have received Nobel Prizes were deprived of them for one reason or another, but where quantitative development economics is concerned, I have no doubt at all that the late Hollis Chenery was the champion. Unfortunately, his use of linear programming and input-output analysis did not go over too well with the rank-and-file engaged in teaching and studying development economics; but for those of us who taught from his articles and the book that he co-wrote with Paul Clark (1962), it was clear that the time would come when the big oil producers of the Middle East would not be eager to ship their oil in unprocessed form. That time has now arrived.

It has been noted by several observers that the unique feature of recent oil market developments is the near-term capacity constraints existing in some parts of the petroleum industry — e.g., refining in the U.S. — and the gradual decrease in excess sustainable crude oil production capacity in virtually every major oil-producing country except, perhaps, Saudi Arabia. ("Perhaps", because in both Saudi Arabia and Iraq, phenomena like water flooding have reportedly led to an increase in the natural decline of deposits.) The complacency displayed toward this ominous situation is nothing less than remarkable, although behind the scenes I am sure that the heads of any number of central banks are informing their principals that they should do everything possible to prevent the kind of 'anomalous' events (such as serious political flare-ups in or near an oil-producing country) that would remove a few million more barrels of oil (per day) from the market, because that is all it would take to send the oil price to a level where ugly macroeconomic and/or political consequences could follow.

In addition, at the present time, the oil price is much more volatile than it was in the recent past, largely because the low investment of the past few years has kept storage and transport facilities from expanding as fast as output, and as a result, has increased the possibility that bottlenecks could appear that lead to wild price swings.

Items like oil and gas have been stored and transported for many decades, but even so, things have occasionally happened that lead some of us to question the wisdom of markets, although this does *not* mean that we are prepared to propose an alternative, or, for that matter, to listen to proposals on this subject. For example, the logic behind storing gas is simple. Demand is not constant over a year, and usually rises during the winter. Having gas available to meet this demand can mean substantial profits for the owner of a storage facility. Storage is not costless, though, and given the amounts of gas usually involved, there is ample opportunity to make catastrophic errors of judgment.

This brings us to the curious developments in the U.S. natural gas industry. What has happened of late is that energy companies have filled salt caverns, acquifers, depleted oil wells and other receptacles with gas. Preliminary figures indicated that in 2009, the amount of stored gas reached $3,800\,\text{Gft}^3$, which was the highest ever. As has already been mentioned, and will be emphasized later in this book, the ratio of inventories to consumption is usually a key variable for determining price. Now, with overflowing inventories but production remaining steady, natural gas prices might fall below present low prices, which is very bad news indeed for gas producers.

A real dilemma then appears as the result of continuing to produce natural gas in the present amounts. This strange behavior cannot be taken up in detail in this book, but in light of the statement above about the "wisdom of markets", Richard Bellman's 'Principle of Optimality' comes to mind. This states that "An *optimal* policy has the property that, whatever the initial state and initial decision are, the remaining decisions must constitute an optimal policy with regard to the state resulting from the first decision." What is taking place now may have very little to do with an optimal policy. Instead, while the continued high rate of flow of U.S. natural gas is good for consumers (in the short run), it is bad for producers — except possibly those who hedged their output with long-term contracts and derivatives. However, it might also turn out to be bad for consumers because investment in new gas sources might decrease, and present output could be involuntarily 'shut in'. In addition, and more ominous, large producers might form a producers' organization — a 'GAS-PEC'. This option has been discussed for years by the large gas producers, including Russia.

One of the most influential physicists of the 20th century, Niels Bohr, once said that "true expertise comes only after making all possible mistakes," but even so, it might be wise for most of us to appreciate that it would not be a good thing if we become indifferent to the mistakes that are possible concerning the availability of oil and natural gas, because the correcting of mistakes is something that cannot take place overnight. A perfect example here is the plenitude of mistakes leading up to the macroeconomic meltdown that has plagued a large part of the world over the last few years.

The next topics that I discuss are also sources of controversy, by which I of course mean renewables, alternatives and nuclear energy. According to *BusinessWeek* (March 6, 2006), President George W. Bush could be described as a poster boy for renewables, specifically plant-based (or grain-based) ethanol, wind power and photovoltaics. I hesitate to label this an accurate description of the former president, but in any case it is impossible not to conclude that there is still considerable work to be done before an optimal 'refinement' and deployment of these and other items can be realized. At the same time I am certain that eventually we are going to have the energy assets that we need and deserve, because we absolutely must have them. Hopefully, they will appear earlier rather than later.

According to several commentators on the important sites *EnergyPulse* and *Seeking Alpha*, biofuels have been 'oversold', and in another statement dealing with this topic, Jim Beyer argues that "ethanol makes sense as a fuel additive, but the economics are much less favorable as a significant fuel component itself." I prefer to put it this way: in the long run, biofuels, wind power, solar, oil sand and tar sand oil, and other unconventional energy sources are not only promising, but probably indispensable. The quandary is that there are some gross misunderstandings associated with their promotion and use that could lead to considerable misinterpretation by the general public and their elected representatives. For example, a gentleman writing in a recent issue of the (UK) *Financial Times* assured the readers of that influential publication that we are approaching a period when renewables will be the flavor of the century, while "unsequestered" emissions from fossil fuels, along with nuclear energy, will be a lost cause. I strongly suspect that only money — and/or the promise of a great deal of it — could explain that quaint and completely useless conviction.

There are many claims that the real advantage from ethanol will appear when the underlying fuel is cellulose-based (e.g., switchgrass and wood) instead of grain- or plant-based (e.g., corn), but generally it is believed that it will take several more years to obtain the right degree of proof. In reality, proof is no longer necessary, but a great deal of complicated research needs to be done before starting to build the refineries that will produce enough biofuels — ethanol (and perhaps also methanol) and others — that will be capable of dealing with a decreased supply of oil from the main oil producers, should that occur.

A key problem here is obtaining an optimal portfolio of renewables and alternatives, by which it is meant an amount of these items supported by science and technology, and not fantasy and wishful thinking. If we take ethanol, for example, there is no question that some of it must be produced. Exactly who or what will determine the exact amount is a mystery just now, although I suspect that leaving the decision *entirely* to the market may not be wise. Instead, I prefer to leave it up to the readers of this book, who in turn can take their findings and/or opinions to their employers or their political representatives, because there is too much questionable advice in circulation at the present time.

At the same time it should be recognized that in many aspects of energy economics, *social profit* is as important as *private profit*, and so measures like production tax credits for biomass and similar initiatives should be resorted to without hesitation in the battle to reduce an excessive dependency on fossil fuels. It might also be useful if governments consider functioning as buyers for some of the new products, as the U.S. government did with microchips: they provided a market that encouraged producers to move rapidly up the learning curve in order to lower costs (and raise profits) by exploiting economies of scale and doing additional research.

Wind has been mentioned on several occasions in this book, and it is rewarding to note that in the U.S., for example, its use increased by 35% in 2005, which amounted to 2,500 megawatts of additional power. The U.S. is a very large country and its wind resources are considerable, but having noticed the deceleration in the growth of wind-based power in Northern Europe — especially Scandinavia and Northern Germany — it should be made clear that wind power is not the nostrum that some observers believe it to be. There are also a number of wind installations in North America with very low capacity factors.

Unfortunately is a word that I use a great deal in this book, and probably even more in my lectures, but where the present topic is concerned there are many things that are ignored that deserve very close inspection. Denmark is often regarded as the promised land of wind energy, since it generates (on average) 26 percent of Danish electricity. The question that has never been asked, however, is why it does not generate twice that much or more? There are several answers here. One is that if windmills were located in certain locales in Denmark, they would not turn very much. The capacity factor in those districts would be 9 or 10 percent instead of perhaps 25 percent. Danish voters would not be very satisfied with a situation of this nature for very sound economic reasons having to do with the price of electricity. Moreover, although not widely known, coal plays a major role in the supply of electricity in Denmark.

Furthermore, without the ability to utilize Swedish hydro- and nuclear-based power, it is doubtful whether more than a few percent of the electricity in Denmark would be generated by wind. Swedish hydro and nuclear power functions as a 'back-up' or 'stand-by' for the Danish electricity system. The most interesting thing about this situation for me, as a resident of Sweden, is that *ceteris paribus* my electric bill has increased as a result of Danish (and German) demand, whereas the electric bills of the Danes have decreased as a result of the relatively inexpensive electricity generated in Sweden being available to keep the lights burning in 'Wonderful Copenhagen'.

Under no circumstances, however, should the development and eventual large-scale deployment of a full arsenal of renewables and alternatives be obstructed. There is no place in the world where governments and voters are friendlier to renewable energy than in Scandinavia, and in addition, technological know-how in these countries is world-class, but even so there is a visible stagnation in the rate at which renewable energy is entering the energy mainstream. For instance, the 1,600 MW of new generating capacity that will be installed in Finland could possibly have been provided by wind and other renewables or alternatives (instead of nuclear power), but hardly at a price that enlightened Finnish voters would have been glad to pay, given the cost advantages of the right amount (= the optimal amount) of nuclear power in the long run.

As noted in an editorial in *Science* (July 30, 1999), "Affordable energy is the lifeblood of modern society. Without it, the network of transportation, agriculture, healthcare, manufacturing, and commerce deemed essential by many of the world's inhabitants, would not be possible."

Not mentioned in this résumé is entertainment and various other forms of relaxation. When these extremely important activities are ushered onto the scene, then it becomes quite clear that many of our fellow citizens are not particularly enthusiastic about adjusting their behavior in such a way as to promote low-energy lifestyles, nor adopting lifestyles that would facilitate materially reducing discharges of CO_2 or other greenhouse emissions into the atmosphere. Changing this situation in the short run is unlikely, although the technological means are now becoming available that could eventually bring this about in the medium to long run. Paying for this technology might be something that will not come about without extensive political conflict.

The governments of many countries have confronted the very (politically) sensitive issue of private transportation by engaging in half-hearted attempts to convince motorists that it is in the best interests of themselves and their descendants to make sacrifices that may be necessary in order to enable the provisions of the (very overrated) Kyoto Protocol to be honored, but inevitably their entreaties are tuned out as quickly as possible. After all, it would be difficult — to say the least — for a number of politicians to portray themselves as icons of self-denial. This has certainly become true in Europe, and in particular applies to those politicians dreaming night and day of highly paid non-jobs in Brussels.

According to Michael Farrell, director of the program for global environmental studies at the Oak Ridge National Laboratory (Tennessee, USA), the (average) estimated increase in the global temperature will be 2.5 degrees (Fahrenheit) for the present century, while even if the provisions of the Kyoto Protocol are fully carried out, the increase will be only slightly less. In addition, Sidney Borowitz — a New York University physicist — calculated several years ago that the atmospheric concentration of CO_2 was 358 ppmv (volume parts per million), and increasing at a rate of 1.5 ppmv. He considered this to be without precedent over the past 160,000 years. If these assessments are reasonable, or nearly reasonable for that matter, then it might be a good thing if many of the loving references to the Kyoto Protocol are toned down as much as possible, and a new program for efficiently reducing climate warming is set into motion by some influential and charismatic person.

"Influential and charismatic" is a description that fits the multi-billionaire Bill Gates. He has taken an interest in global warming, and one of his suggestions for doing something about it turns on what he calls "energy miracles", which include miracles in the production of electricity from new nuclear reactor designs. The truth is, though, that it is not technological miracles that are of primary importance, but political and psychological miracles, because present nuclear designs — as well as those that are certain to appear in a decade or two — will be quite sufficient. What is not sufficient is the willingness of governments in some regions to find a place for this nuclear equipment in the energy programs that occasionally blossom in their rhetoric.

The thing to always keep in mind when dealing with the bad news that could come about because of prolonged energy shortages, is that we are not talking about occasional blackouts, or the possibility of irksome increases in motor fuel prices because a pipeline somewhere got in the way of some rockets, but economic disasters that in earthquake terminology belong at or above the top of the Richter scale. The recent President Bush was often awarded the bad-guy role in the environmental drama, but it might be a good idea to remember that although his father once had a similar attitude toward ozone depletion and acid rain, when the very conservative UK Prime Minister Margaret Thatcher insisted that something be done to diminish this hazard, he felt compelled to go along.

Moreover, in a speech to the Royal Society, Baroness Thatcher said that "We may have unwittingly begun a massive experiment with the system of the planet itself." She was talking about global warming, but what she did not say was that the experiment might be impossible to stop or slow down. When I wrote my coal book I could not picture the price of coal going above $55 per metric ton. Now it is $130/t because the demand is accelerating. In particular, China and India are major buyers of thermal coal. (Major exporters in order of magnitude are Indonesia, Australia, Russia, Colombia, South Africa, and finally, the U.S., but U.S. exports are increasing.)

Perhaps a healthier activity than looking at or thinking about the above information is merely to understand that many countries absolutely must have a new energy system, and if bringing it about involves a correct or incorrect belief about global warming, then smile and let it happen. Another thing worth considering is that really bad climate news may be on the way, regardless of how much carbon is bought, sold or piped into the depths of the earth or the oceans. In a rational world, our political masters would devote at least part of their working hours to considering this possibility. According to David Stipp (2004), some of this may be happening, and what it at least partially involves is dealing with the very large population movements that may take place in the worst of circumstances.

There is so much uncertainty in dealing with the topics in this chapter, that it is difficult to avoid showing an excessive amount of intolerance for the low degree of honesty that we often find on the part of many people dealing with, e.g., electric deregulation. This is sometimes regarded as 'attitude' by academic deregulation insiders; however, as U.S. Congressman Peter DeFazio remarked at the beginning of the electric deregulation escapades on the U.S. West Coast, "Why do we need to go through such a radical, risk-taking experiment?" As you might remember, the answer Congressman DeFazio gave is that there are people who are going to make "millions or billions" from this malicious practice.

Some of Congressman DeFazio's potential millionaires and billionaires still believe that there is a place for them on the deregulation gravy train, while many of those already on board want to upgrade their tickets. I have no problem with this. My problem is with the so-called energy experts in California and elsewhere who failed to discern that when regulated utilities are replaced by unregulated oligopolies, the exploitation of market power by these oligopolies is exactly what their textbooks told them would happen.

Now that politicians have been mentioned, it is a very small step toward the subject of energy independence, which is something that a year or two ago was constantly referred to in the media. With all due respect to politicians and their advisors and supporters, I have a very strong feeling that the mere idea of energy independence is to a certain extent illogical, because, as generally outlined, there is an inadequate reference to cost. (Professor Eric Smith of Tulane University calls energy independence an oxymoron.) During the Second World War many countries were forced into an independence mode, particularly where raw materials were concerned, and for the most part they found the defects of this departure so great that they abandoned it as soon as possible after the war ended. Sweden is a good example here, because unlike the U.S., it was necessary for Sweden to replace almost all oil-based motor fuel during WWII. The main replacement was 'wood gas' (or gengas), which was used in both civilian and military vehicles, and only a small amount of *gengas* production still takes place today. As for its general desirability, I have never heard anyone claim that a large-scale resort to this expedient at the present time would be capable of outweighing its disadvantages, although contemporary technology may now be capable of upsetting that judgment.

This might also be the place to mention once again that Len Gould once pointed out in the important forum EnergyPulse (www.energypulse.net) that voters in oil-importing countries might prefer war to foregoing the large quantities of motor fuel that they believe they deserve. I have no reason to doubt this, because if voters in the U.S. (and elsewhere) were really interested in buying the concept of energy independence, and thought it 'doable', their purchases of energy-intensive goods and services would be of a very different magnitude and configuration from those we noticed before the onset of the recent macroeconomic and financial market unpleasantness. Unless I am badly mistaken, the cost of energy independence in the short to medium run — assuming that it was made known — would be unacceptable to a majority of American voters, although in the long run a 'transformation' of one sort or another should be tolerable. "Necessary" is probably the correct term, though it is too early to take a position on this complex topic.

The mention of voters preferring war to abstention from a fraction of their customary motoring immediately brings up the question of how community leaders receive their instructions from the electorate.

As Kenneth Arrow (1951) suggests in his famous book *Social Choice and Individual Values*, in a capitalist democracy there are ideally two avenues by which social choices are made: voting (which means voting for certain programs advanced by candidates running for public office) and the market mechanism. The problem of course is that voters/consumers as well as politicians may lack the knowledge to appraise and rank choices that are critical for satisfying both their present and their likely future energy preferences. Without this knowledge, the kind of rationality that is indispensable in certain situations might be impossible, because as Kurt Rothschild (1946) noted, "Unless economic units act in conformity with some rational pattern, no general theory about what would follow from certain premises would be possible."

Professor Arrow evidently found it difficult to accept this proposition; however, I can understand his reluctance, because his book was written during those wonderful post-war days in which there was growing prosperity and overwhelming optimism in the U.S., and when it was less important than at the present time whether voters and politicians really understood what they were doing or what was going on around them. There was plenty of oil, plenty of energy, plenty of faith in the future, and it was unnecessary for politicians to make preposterous statements about things like the unsuitability of nuclear technology.

As intimated above, if there were a better understanding of economic theory, energy independence might be an unwelcome suggestion. This is because assuming a burden of this nature would probably require behavioral patterns that are unacceptable in times of peace, at least in the short run. But even so, given the possible shortages of energy that could take place in the not-too-distant future if the supply and demand trends of the past few years persist, a comprehensive energy policy will almost certainly have to be crafted by many governments, even if total implementation can and should be delayed. Although not broaching the question of energy independence, Lisa Raffin (2008) seems to believe that the way ahead in matters of this nature is the one suggested by Sheila Sheridan of the U.S. Green Building Council: "Doing a little at a time, and eventually making a difference."

I do not see how this philosophy can be rejected, if only because it concurs with my approach to appraising the various difficulties that are certain to be encountered if there is an ambitious attempt in the U.S. or elsewhere to move toward a full-fledged state of energy independence, although it might be found optimal to stop short of that particular goal. What is needed here is a 'Cold War' rather than a 'Hot War' approach, more or less in tune with advice put into circulation by Chairman Mao (of the People's Republic of China): "...10 percent fighting, 10 percent waiting, and 80 percent self-improvement." It might also be wise to stay on the alert for lies, nonsense, provocation and various forms of trickery. For instance, some delegates at the supposedly important Cancun Climate Meeting (in 2010) apparently signed a petition to ban dihydrous monoxide because, ostensibly, it was a key ingredient in climate change, and could be fatal if inhaled. As it happens, however, dihydrous monoxide is the chemical term for water.

In the scientific community, systematic thinking on the energy future (and not necessarily on energy independence) seems to focus on diversity. (Note the word "seems", because a large part of the general public is not particularly interested in the opinions of what they call nerds, geeks and wonks.) In some situations, diversity has a great deal to offer, if only because of the annoying technical or political shortcomings that characterize the entire roster of energy sources, including some that in the long run will be found to be extremely valuable. To be explicit, an energy structure that features an optimal nuclear base supporting a large and heterogeneous structure of the *right* (i.e., optimal) *renewables* has a great deal to recommend it. (And note, optimal does not mean huge!) Hopefully, in the long run this is precisely what we are going to get, although as mentioned earlier in this book, and occasionally in my other energy economics books, one of the great challenges in energy economics at the present time, and perhaps the greatest challenge, is to determine the exact or near-exact optimal assortment of renewables and alternatives. This is something that will require a great deal of creative thinking, and readers of this book have an important part to play in this process.

6. Appendix: An Easy Derivation of the Annuity Formula

Originally my intention was to present a fairly long book that was essentially free of mathematics. This was because, although a teacher of mathematical economics, I no longer find it useful to dress up important topics with superfluous mathematics in order to make them attractive to the editors of 'learned' journals — which are journals where, according to a certain economist, even important articles are read by an average of only 12 persons (excluding students, who of course constitute a captive audience).

One of the first equations in this book featured the annuity formula, following which there was an example of its use in a two-period situation. What I want to do now is to generalize this two-period example to T periods. I begin by noting that two equivalent arrangements for paying a debt of PV ($=$ present value) entered into at the beginning of the first period is to pay $PV(1+r)^T$ at the end of T periods, or to pay via annuities A at the end of each period, beginning with the *end* of the first period and ending at the end of the last period! Thus we get:

$$PV(1+r)^T = A + A(1+r) + A(1+r)^2 + \cdots + A(1+r)^{T-1}.$$

Multiplying both sides of this expression by $(1+r)$, we obtain:

$$(1+r)[PV(1+r)^T] = A(1+r) + \cdots + A(1+r)^T.$$

Subtracting the second of the above expressions from the first yields:

$$[(1+r)^T]PV[1-(1+r)] = A - A(1+r)^T.$$

From this we get Equation (2) in Chapter 1, which was:

$$A = \left[\frac{r(1+r)^T}{(1+r)^T - 1} \right] PV. \tag{A1}$$

Please observe that in this exercise, PV is used instead of P_0, which might have been designated as the price of an asset. As an exercise here, you can take PV as the cost of a private jet or a condo in Monaco, and calculate the annual payments — or for that matter the monthly payments. (Do you have any suggestions for calculating the monthly payments if this arrangement is suggested by someone important?) As noted earlier, this expression can also be derived using some elementary calculus, beginning with a fundamental (neo-classical) economic concept: the capital cost of an investment is the uniform return per period that an asset must earn, in order to achieve a net present value of zero. In other words, the asset price is the present value of future net yields (i.e., revenues minus costs). A is also sometimes referred to as the *levelized cost*.

Notation in this derivation is changed somewhat in order to correspond to standard usage. Taking I as the asset price (i.e., the investment cost), P as the capital cost per period, and r as the market discount rate, we can write for T periods:

$$I = \int_0^T Pe^{-rt}dt = \frac{P}{r}\left(1 - \frac{1}{e^{rT}}\right). \tag{A2}$$

It takes very little manipulation to obtain $P = re^{rT}I/(e^{rT} - 1)$. Remembering that we can approximate e^{rT} by $(1+r)^T$ for small values of r, we get Equation (A1), though with a different notation. The discount rate here is the market interest rate, because in the neo-classical world, there is no risk/uncertainty on the part of lenders and borrowers, which means that the risk-free interest rate is always appropriate. This is not the kind of recommendation that needs to be taken seriously outside a seminar room, and instead of worrying about it, readers should concentrate on distinguishing between the investment cost and the capital cost, because if you examine certain academic contributions, you might note that the two are often confused.

The exercise above will be given in another chapter. The derivation of A1 (though not A2) is one of the things that I kindly ask my students to learn perfectly, and if they find that impossible, then I often find it impossible to give them a passing grade.

Key Concepts and Issues

backwardation	long and short
Clearing House	Millicent and Condi and futures contracts
contango	NYMEX and OPEC
Copenhagen Consensus	option value
ethanol	paper oil
gas storage	price of oil (e.g., the spot price)
gigawatts and megawatts	real price/nominal price
hedging	refining and petrochemicals
Hotelling's model	upstream and downstream

Questions for Discussion

1. Go immediately to Chapter 8 and read it carefully. Do you understand most of it?
2. Describe in detail how Millicent made a lot of money.
3. What is 'going long', margin, 'going short', maturity, liquidity, closing a position?
4. How does a futures market function?
5. Should President Bush have gone to Wall Street instead of Saudi Arabia?
6. Speculation or fundamentals?
7. How do you feel about the global warming controversy?
8. What is the Stern Review? Who is Joan Ruddock and what does she say about Stern Review critics?
9. Do you pay for electricity in kilowatts or kilowatt-hours? Answer by presenting a simple example!
10. Why would I say 1 kWh(e) instead of 1 kWh?
11. 1,000 barrels of oil is the energy equivalent of how much gas?
12. The late Mr. Malthus is not a very popular man to many economists. How do you feel about him?
13. Suppose that you were a large gas producer. How would you feel about joining a GAS-PEC? Suppose you were a large oil producer, but not a member of OPEC. If asked to join, what would you say? Why?

14. If you hear that the reserve-production ratio of oil in the world was 100 (= 100 years), would you feel that there would be a shortage of oil in the near future? Suppose it was 40. What is your opinion now in thinking about the future price?

15. From where does OPEC get its impressive market power?

16. The famous John D. Rockefeller was not particularly interested in upstream oil operations. Downstream was his 'thing'. Can you comment on why?

17. Argue for energy independence! Argue against energy independence!

18. Argue in favor of going to war for oil! Argue against going to war for oil! In the very popular television series "The West Wing", a candidate said that he would never go to war for oil. How does that sound to you?

19. 'Red ink' (or bad news) seems to go together with the refining of oil. What seems to be the problem here?

20. "Dirty coal is winning", two observers said. Discuss briefly!

21. A BTU is the amount of heat necessary to increase the temperature of one pound of water by one degree Fahrenheit. What numbers do we have for converting these units into one kilogram and one degree Celsius, respectively?

22. My belief is that OPEC is now in a position to determine the global price of oil. What is your belief where this subject is concerned?

Bibliography

Adelman, Morris A. and Martin B. Zimmerman (1974). "Prices and profits in petro-chemicals: an appraisal of investment by less developed countries." *Journal of Industrial Economics*, 22(4): 245–254.

Allen, R.G.D. (1960). *Mathematical Economics*. London: MacMillan & Co.

Arrow, Kenneth (1951). *Social Choice and Individual Values*. New York: John Wiley.

Badal, Lionel (2009). "What the IEA doesn't want you to know about peak oil." *Seeking Alpha* (September 6).

Baltscheffsky, S. (1997). "Världen samlas för att kyla klotet." *Svenska-Dagbladet*.

Banks, Ferdinand E. (2008). "The sum of many hopes and fears about the energy resources of the Middle East." Lecture given at the Ecole Normale Superieure (Paris), May 20, 2008.

_____ (2007). *The Political Economy of World Energy: An Introductory Textbook.* Singapore and New York: World Scientific.

_____ (2004). "A faith-based approach to global warming." *Energy and Environment*, 15(5): 837–852.

_____ (2001). *Global Finance and Financial Markets*. London, New York and Singapore: World Scientific.

_____ (2000). "The Kyoto negotiations on climate change: an economic perspective." *Energy Sources*, 22(6): 481–496.

_____ (1987). *The Political Economy of Natural Gas*. London and Sydney: Croom Helm.

_____ (1980). *The Political Economy of Oil*. Lexington and Toronto: Lexington Books.

_____ (1977). *Scarcity, Energy, and Economic Progress*. Lexington and Toronto: Lexington Books.

Basosi, Riccardo (2009). "Energy growth, efficiency and complexity." Conference Paper; Center for the Study of Complex Systems, University of Siena (Italy).

Bauquis, Pierre René (2003). "Reappraisal of energy supply-demand in 2030 shows big role for fossil fuels, nuclear but not for non-nuclear renewables." *Oil and Gas Journal* (February 17).

Bell, Ruth Greenspan (2006). "The Kyoto Placebo." *Issues in Science and Technology*, 22(2).

Bezat, Jean-Michel (2008). "Petrole: le pouvoir a changé de camp." *Le Monde* (May 20).

Bloch, Richard (2011). "The crack spread: theoretical oil margins near 2-year high." *Seeking Alpha* (March 25).

Boiteux, Marcel (2007). "Les Ambiguite's de la Concurrence" (Lecture).

Carr, Donald E. (1976). *Energy and the Earth Machine*. London: Abacus.

Casazza, Jack A. (2001). "Pick your poison." *Public Utilities Fortnightly* (March 1).

Chenery, Hollis and Paul Clarke (1962). *Inter-industry Economics*. New York: Wiley.

Chevalier, Jean-Marie (2009). *The New Energy Crisis: Climate, Economics and Geopolitics*. London: Palgrave MacMillan (with CGEMP, Paris-Dauphine).

Cohen, David (2009). "Mr Market gets it wrong again." *321 Energy* (May 29).

Constanty, H. (1995). "Nucleaire: le grand trouble." *L'Expansion* (pp. 68–73).

Cook, Earl (1976). *Man, Energy, Society.* San Francisco: W.H. Freeman and Company.

Cooke, Ronald R. (2009). "The clean energy act is not going anywhere." *321 Energy* (July).

Costello, Kenneth (2003). "The shocking truth about restructuring of the U.S." *The Electricity Journal*, 16(5): 11–19.

Eltony, Mohamed Nagy (2009). "Oil dependence and economic development: the tale of Kuwait." *Geopolitics of Energy* (May).

Frank, Robert H. (2007). *Microeconomics and Behavior.* New York: McGraw-Hill.

Gabriel, Mark (2008). "Another inconvenient truth: the need for coal." *EnergyPulse.*

Goodstein, David (2004). *Out of Gas: The End of the Age of Oil.* New York and London: Norton.

Hamilton, Carl B. (2009). "Sahlin slarvar med sanningen om Barsebäck." *Svenska Dagbladet* (January 22).

Harlinger, Hildegard (1975). "Neue modelle für die zukunft der menshheit." IFO Institut für Wirtschaftsforschung (Munich).

Henderson, James M. and Richard Quandt (1995). *Microeconomic Theory: A Mathematical Approach.* New York: McGraw Hill.

Holmes, Bob and Nicola Jones (2003). "Brace yourself for the end of cheap oil." *New Scientist* (August 2).

Huber, Peter W. (2009). "Bound to burn." *City Journal*, 19(2).

Hutzler, Mary (2009). "The Pickens plan: is it the answer to our energy needs?" *IAEE Energy Forum* (Spring).

Kilian, Lutz (2008). "The economic effects of energy price shocks." *Journal of Economic Literature*, 46(4): 871–909.

Kubursi, A.A. (1984). "Industrialisation in the Arab states of the Gulf: A Ruhr without water." In *Prospects for the World Oil Industry*, edited by Tim Niblock and Richard Lawless. London and Sydney: Croom Helm.

Lorec, Phillipe and Fabrice Noilhan (2006). "La stratégie gasiére de las Russie et L'Union Européenne." *Góéconomie* (No. 38).

Macdonald, Gregor (2009a). "Oil production: Brazil making the wisest choice of all." *Seeking Alpha* (August 18).

_____ (2009b). "Should Mexico stop exporting oil?" *Seeking Alpha* (June 24).

Martin, J.-M. (1992). *Economie et Politique de L'energie*. Paris: Armand Colin.

Moller, Jorgen Orstrom (2008). "The return of Malthus: scarcity and international order." *The American Interest* (July/August).

Morse, Edward L. (2005). "Oil prices and new resource nationalism." *Geopolitics of Energy* (April).

Nadeau, Robert (2008). "The economist has no clothes." *Scientific American* (April).

O'Sullivan, Arthur and Steven H. Sheffrin (2008). *Economics: Principles and Tools*. New Jersey: Prentice Hall.

Raffin, Lisa (2008). "Greening commercial facilities." *EnergyPulse* (October 14).

Ramsey, Frank (1928). "A mathematical theory of saving." *Economic Journal*, 38(152): 543–559.

Rhodes, Richard and Denis Beller (2000). "The need for nuclear power." *Foreign Affairs* (January–February).

Roques, Fabien, William J. Nuttall, David Newbery, Richard de Neufville, and Stephen Connors (2006). "Nuclear power: a hedge against uncertain gas and carbon prices." *The Energy Journal*, 27(4): 1–24.

Rose, Johanna (1998). "Nya Krafter." *Forskning & Framsteg* (September).

Rothschild, Kurt (1946). "The meaning of rationality." *Review of Economic Studies*, 14(1): 50–52.

Söderbergh, Bengt (2010). *Production from Giant Gas Fields in Norway and Russia, and Subsequent Implications for European Energy Security*. PhD Thesis, Uppsala University.

Stern, Nicholas (2007). *Stern Review on the Economics of Climate Change*. London: H.M. Treasury.

Stipp, David (2004). "Climate collapse." *Fortune* (February 9).

Tverberg, Gail (2010). "Huge amount of oil available, but..." *OilPrice.Com* (December 26).

Victor, David and Varun Rai (2009). "Dirty coal is winning." *Newsweek* (January 12).

Wallace, Charles P. (2003). "Power of the market." *Time* (March 3).

Watts, Price C. (2001). "The case against electricity deregulation." *Electricity Journal*, 14(4): 19–24.

Webb, Michael G. and Martin Ricketts (1980). *The Economics of Energy*. London: The MacMillan Press.

Woo, C.K., M. King, A. Tishler, and L.C.H. Chow (2006). "Costs of electricity deregulation." *Energy: The International Journal*, 31(6–7): 747–768.

Yohe, Gary W. (1997). "First principles and the economic comparison of regulatory alternatives in global change." *OPEC Review*, 21(2): 75–83.

Chapter 4

An Introduction to Oil Economics

On January 2, 2008, at approximately the same time that an offbeat article by Christopher Helman entitled "Really, really cheap oil" (*Forbes*, October 2, 2006) was belatedly brought to my attention, the price of oil on the New York Mercantile Exchange (NYMEX) touched $100 per barrel (= $100/b) for the first time in modern history. That price continued to increase until by late summer of that year it was slightly over $147/b, and in the wake of the severe recession that touched most of the world (and which was partially due to the record oil price), it declined to below $35/b. The oil producers organization (OPEC) found this unsatisfactory however, and they took steps to immediately redress the situation in their favor. Their efforts would have to be judged impressive, because due to the troubles in Libya, when I began writing this chapter, the price of (WTI) oil was $113/b, while the price of Brent oil was $122/b. There are simple but not widely circulated explanations of price movements in the oil market, and while several have already been presented, several more will be presented later. The thing to be emphasized though is that a statement was made in 2008: a fundamentally different oil era has arrived in which OPEC has *real* as opposed to *theoretical* power, and it is in the interest of oil consumers to give it the closest attention. (Note: WTI = West Texas Intermediate.)

A great deal of Mr. Helman's composition provided false impressions. "New" oil may be "coming from everywhere" he claimed, but under the most favorable of conditions, it is unlikely that it will be able to change supply and demand realities in world oil markets after the departure of the present macroeconomic doldrums, and the same might be true of "new" technology. His work contained an ill-advised attempt to discount the growing energy demand of China, although he was correct in noting that it might make sense (for price reasons) for Saudi Arabia to reduce its production in order to make room for non-OPEC output. The potential of the Caspian and perhaps other 'new' producing regions were in all likelihood overestimated when demand is taken into consideration, and there was a drastic mistake in the conjectured spare capacity — or perhaps I should say *effective* spare capacity — of the global oil production system: it was well under the 5 mb/d that one of Helman's gurus believed prevailed at the time Helman's article was published. Financial speculation (in the oil futures market) was basically misinterpreted and/or theoretically incorrect. These and similar issues were briefly examined in previous chapters of this book, where non-technical language was used to outline the departure from mainstream economic logic that still characterizes much of the commentary on petroleum markets.

It is sometimes believed that the principal explanatory factor for the very large oil price rise in 2007–2008 was the increase in the *growth rate* of Chinese (and Indian) demand, which could only be supplied by greater imports. For instance, China now has at least 100 million vehicles, and there are estimates that by the year 2020, this figure may be very much larger. However, even if there is only a modest increase, the difference will be more than covered by the increased sales of cars, vans and trucks in India. Readers should also be aware of the growing oil consumption in leading exporters. CNN recently called attention to a growing demand in the Middle East, where the largest fraction of OPEC oil is produced, and in addition where all governments evidently agree that an excessive reliance on the export of crude oil is both economically and politically imprudent. "Resource nationalism" is an expression that has been introduced to describe the likely behavior of many OPEC countries, and perhaps also Russia.

Underlying the present contribution is the belief that the supply of (conventional *and* unconventional) oil will be unable to expand at a fast enough rate to put a sustained downward pressure on the price of oil, and eventually there could be a traumatic piece of bad news in the form of an unambiguous peaking of the global oil supply instead of the 'flattening' of the conventional oil supply that seems to be taking place now. Of course, an unhealthy freight of ill-tidings could be featured in association with an ugly oil price scene well before global production declines. I was lecturing in Paris in 2008 just prior to the oil price escalation, and there was no talk of peaking in

French publications. Instead, there were dramatic references to the demand for oil outrunning the supply.

And speaking of dramatic references, if you run into trouble with the math below, take a break by going to Chapter 8, which contains no math, but the arguments you need to captivate your audience.

1. 1931–1974: Examining a Fundamental Relationship for the Pricing of an Exhaustible Natural Resource (i.e., Oil)

My lectures on oil often begin by citing an important article on resource economics by Professor Robert Solow (1974) — a Nobel Laureate in economics and a brilliant teacher — which refers at great length to a mathematical relationship that is associated with Professor Harold Hotelling (1931). I studied and taught this relationship for many years; however, as a result of later developments in the oil market, I now contend that it is (and was) pedagogically misleading, and consequently I do not encourage my students to discuss or even mention it in classroom situations.

Instead, I usually begin my work by formulating an orthodox profit-maximizing relationship of the kind featured in undergraduate economic theory, and which has already been introduced in the first chapter of this book. To be explicit, we can write:

$$V = \sum_{t=1}^{N}(p_t q_t - c_t q_t)(1 + r)^{-t} + \tau \left[R^* - \sum_{t=1}^{N} q_t \right]. \qquad (1)$$

In the first parenthesis we have profits in period t, which are defined as usual as revenue (price times quantity, or $p_t q_t$) minus 'cost' (*average* cost times quantity, or $c_t q_t$). In the third parenthesis we have the given amount of the resource (e.g., oil), R^*, at the beginning of the period designated $t = 1$, distributed over N periods ($q_1 + q_2 + \cdots + q_N \leq R^*$). The second parenthesis, $(1 + r)^{-t}$, merely discounts the profit in period t: profits in later periods have less value than those of the same amount in earlier periods. In conventional presentations N is taken as given, and in the Hotelling article (and many others) c_t is regarded as a constant (e.g., $c_t = c$) that is equal to both average and marginal cost for the N periods. In any case the intention is to maximize V, which is discounted profit, and usually requires calculus. I see no point in this complication now.

Furthermore, p_t is the *expected* price for all periods except $t = 1$, for which period it is the *actual* price, and the implicit assumption is that these prices as well as the initial amount of the resource (R^*) are correctly forecast at the beginning of the current period. τ (*tau*) is a Lagrangian multiplier, and denotes the scarcity value of the resource: e.g., τ is zero if R^* exceeds the amount of the resource extracted during the N periods (because then the resource is not scarce). Put another way, $\tau = \partial V / \partial R^*$.

What I want from my students is that they can write this expression down and have a vague idea of what it signifies. The details can be delayed until they have studied some elementary calculus, because another approach will also be used below.

We obtain the Hotelling result from Equation (1) by differentiating V partially with respect to the values of q_t, and setting these values equal to zero ($\partial V / \partial q_1 = 0$, $\partial V / \partial q_2 = 0, \ldots, \partial V / \partial q_N = 0$), keeping the cost constant. Then, moving from *infinitesimal* to finite changes, we get $\Delta p / p = r$ for successive periods, with both sides of this expression given in 'percent'. Please note however that by taking this approach, we explicitly replace p in the previous expression with the 'net' price, or price minus the marginal cost ($p_t - c$). We learn that this net price increases at the rate r.

Whether the directors of BP and Exxon would be impressed by all this is uncertain, although I suspect that if politely asked, they would inform you that c_t is NOT constant and equal to the marginal cost for *real-world* oil deposits. Instead, deposit pressure decreases — and costs tend to increase — as cumulative output increases (and thus reserves decrease). By costs I am specifically referring to average and total costs, and, where relevant, marginal costs.

Accordingly, in these circumstances, we would NOT obtain $\Delta p / p = r$ from a properly formulated real-world profit-maximizing exercise. Put another way, we do not have the same fundamental laws operating in the real world as we do in class and seminar rooms, where teachers are sometimes inclined to present showy bunkum to their students. For example, when discussing real-world oil deposits, $\Delta p / p = r$ is useless. I can also mention that in the 1980s some OPEC directors openly cursed the Hotelling relationship for its inaccuracy. They had actually believed that in this best of all possible worlds, the oil price was going to increase annually by whatever the value of *r* happened to be. Something that needs to be appreciated here is that deposit pressure also falls due to 'natural depletion' (as well as physical production), which in mainstream capital theory would be called 'depreciation by evaporation', and is mentioned later.

Two extremely important events took place during the period now being discussed. The most important of all was the formation of the strong oligopoly OPEC (or the Organization of the Petroleum Exporting Countries). During my many jogging tours past the University of Chicago, it was impossible for me to avoid thinking of some of the goofy talk originating at that institution about rational thinking in economics. I discussed this in many of my lectures and publications, and the conclusion I came to — particularly in my book on oil — was that if rational thinking prevailed, OPEC would eventually come to the conclusion that their optimal strategy is to produce as little oil as possible. Perhaps a number of economics professors in or near the Chicago area at that time came to another conclusion, and the best known of them — the late Milton Friedman — predicted in an American news magazine that OPEC would collapse and the oil price would fall to $5 a barrel or less.

A similar belief existed on the part of Professor Michael Beenstock, now a Professor of Economics at the Hebrew University of Jerusalem, but in 1973 an important expert on oil issues for the UK Treasury Department. He was best known at that time for insisting that the price of oil at the turn of the century would decline to its 1972 level, which in real terms was almost the case. What he did not know, however, was that that preposterous price concentrated the minds of OPEC executives, and a few years later a 'slow motion' oil price escalation began that ended with a price of $147/b in 2008, which happened to be ruinous for the global macroeconomy.

As I constantly argue, it was not that price that demonstrated OPEC's power, but the fact that when the oil price descended to $32/b as the demand for oil fell, OPEC had one of their famous emergency meetings and, unlike the situation on previous occasions, was able to immediately take steps that resulted in the price increasing by $40 a barrel. *This needs to be appreciated!* Similarly, in 2011, with bad economic news in almost all of the major oil-importing countries, OPEC was able to export enough oil (and perhaps some oil products) to obtain a total revenue of about a trillion dollars.

The second event that can be cited here involved a quadrupling of the price of oil during the period 1973–1974. Needless to say, this brought about a panic in certain official circles, which resulted in talk about sending marines and paratroopers to the Middle East to seize and operate a large part of the oil-producing assets in that part of the world. At one time I was convinced that there was a good chance of this taking place, but later, after an intensive study of game theory, I concluded that talk of this nature, at that point in time, was intended as a threat rather than a preparation for action. What it will be in the future if the oil price should go into orbit remains to be seen.

By way of repairing Equation (1), $c_t = c$ (a constant) is rejected, and I suggest that we write $c_t = c_t(q_t; R_t)$, where R_t is the amount of reserves at time t. *Ceteris paribus*, R decreases as production takes place, and both marginal and average costs increase as it becomes necessary to exploit less 'well-endowed' portions of a deposit (or well).

How can we describe the development of R? One way is to start by deriving a logistic function of the type that I have often derived for my students, but which many of them found to be too abstract. This derivation is given in Appendix 2 of this chapter. Much better is the arrangement I often used in my energy economics courses in Grenoble (France) and Stockholm (Sweden), in which we do not have to delve into the minutiae of the above Equation (1), by which I mean having to make assumptions that tell us how average and/or marginal costs change when production takes place.

As was alluded to before, a new drilling boom may have started in North America in which the object is to use advanced technology to obtain oil and gas found in large shale beds. This advanced technology turns on exploiting horizontal (instead of vertical) wells into which water or some liquid is pumped in order to *fracture* sedimentary rock. I can recall conferences in the United States and elsewhere in which horizontal drilling was advertised as an approach that would revolutionize the access to oil; however, this did not happen because the oil — unlike perhaps the gas — was not there.

Given these circumstances I think it best to wait before accepting that the basic oil and perhaps gas situation in the U.S. (or in Canada) is in the process of a *major* change, and also wait before accepting the estimates of some natural gas producers about the richness of the natural gas deposits they own. There are several ways that these deposits can make producers super rich. One is for them to extract and sell large quantities of natural gas, and another is to sell deposits that they have convinced naïve potential buyers to purchase because these deposits ostensibly contain more natural gas than can be found anywhere else on the face of the earth.

One of the key variables in the following discussion is the reserve-production ratio ($R/q = \theta$), although it will not be used — as is customary — to discuss the 'length of life' of an oil or gas deposit, or field, or for that matter a region or even the entire world. Something readers should appreciate here is that R is a 'stock' (like an inventory) whose dimensions are 'units', while q is a 'flow' whose dimensions are units per time period. This distinction is important, and readers should immediately provide examples!

If the value of θ for a deposit — meaning a field or even perhaps individual 'wells' — falls below e.g., θ^*, which can be called the *critical* reserve-production ratio, then it is likely that the deposit is being worked too hard and it is losing a great deal of its productive power. In other words, the cost of additional production might increase by a large amount. As will be indicated in the following simple numerical example, when R/q declines to θ^*, then optimally θ^* will be allowed to determine production.

Suppose that we begin with a deposit that has 225 units of oil (or natural gas), and we desire to lift 15 units per year. If $\theta^* = 10$, the initial R/q ratio is 15, so there is no problem. (Note also that R/q is equal to $R/\Delta R$, since $q = \Delta R$, and so $1/\theta^* = \Delta R/R$, or the rate at which reserves are decreasing. Saying that θ^* should not be lower than 10 is the same as saying that the rate at which reserves are removed should not exceed a tenth of the available amount of reserves in a given period.)

Continuing, it is clear that we can remove 15 units every year for 5 years (measured at the end of the year) without violating the above criteria. During this period, the R/q ratio ($= \theta$) falls from 14 (at the end of the first year) to 10 at the end of the fifth year. (This can be denoted by the sequence 210/15, 195/15, ..., 150/15.) But if we continued to remove 15 units per year, then we would be removing more than one-tenth of the deposit per year, and presumably the long-run productivity of the deposit would be damaged (as would be made clear by the rising cost of extraction).

After the fifth year in this example, θ^* is allowed to determine production. In order for θ not to fall below 10, production in the sixth year cannot be larger than 13.64. This may not be completely obvious, so consider the situation during the sixth period (i.e., year). The R/q ratio is constrained by a critical ratio of 10, and so $(R_5 - q_6)/q_6 = 10$. With R_5 (reserves at the end of period five) equal to 150, we get $q_6 = 13.64$. This discussion can be easily generalized as:

$$\theta^* \leq \frac{R_t}{q_t} = \frac{R_{t-1} - q_t}{q_t}. \tag{2}$$

And this can be written as:

$$q_t \leq \frac{R_{t-1}}{(\theta^* + 1)}. \tag{3}$$

In the example given here we can calculate the maximum production for the 'next' year, or q_7. R_6 is $150 - 13.64 = 136.36$, and so with $\theta^* = 10$ we easily obtain $q_7 = 136.36/11 = 12.4$.

In this important subject of energy economics, it pays to deal in genuine economic science instead of the kind of nonsense that I presented the first time I taught the subject, and that apparently is still taught in many universities. As a result, a simple relationship such as Equation (3), and the steps leading up to it, which are a formalization of a brilliant paper by Andrew Flower (1978), are well worth knowing. Similarly, if you do not mind a couple of simple integrals, consider the following.

Once again R is reserves (a stock), while q is production (a flow). In addition, n is the rate of growth of consumption (which in the above example was zero), while g is the rate of growth of reserves (which was also zero in the above example). T and t refer to time, and θ is the reserve-production ratio R/q.

More numbers might be useful at this point. For example, suppose we begin with reserves at 150, and the intended output is 15 units per period, which gives us a reserve-production ratio of 10 to start the production cycle, but while $n = 0\%$ for the period, g is equal to 5% (or 0.05). It is easy to calculate the reserves and the reserve-production ratio at the end of the period. Reserves are $150 - 15 + (150 \times 0.05) = 142.5$. Reserves have fallen and so has the R/q ratio, which is now $142.5/15 = 9.5$. The unpleasant reality is that even though reserves were increasing by 5% a year, given the consumption ($= 15$ units), that was not enough to keep the R/q ratio from falling below θ^*.

Now let us take an example where we start with a value of $\theta = 40$, and $q = 15$. R to begin with must be equal to 600. With the same values for n and g as in the previous example, the same type of calculation as before will give us $R = 600 - 15 + (600 \times 0.05) = 615$. Accordingly, as opposed to the previous example, θ has increased from 40 to $615/15 = 41$. The conclusion in this situation is that the more reserves the better, if only because they provide options where consumption is concerned.

The next step is to generalize these results using some calculus, although first, readers should complicate the above with a few values of $n > 0$ and see how things work out. We start the next exercise by formulating an expression for R at time $t = T$, having commenced with a value of $R(0)$ at time $t = 0$. Then we have:

$$R(T) = R(0) - \int_0^T q(0)e^{nt}dt + \int_0^T \frac{dR}{dt}dt. \tag{4}$$

We should recognize here that $q(t) = q(0)e^{nt}$, and $g = 1/g(dR/dt)$. Now we write:

$$R(T) = R(0) - \int_0^T q(t)dt + \int_0^T gR(t)dt. \tag{5}$$

This can immediately be differentiated to give $dR(T)/dT = -q(T) + gR(T)$. We can continue by noting that since $\theta(T) = R(T)/q(T)$, a differentiation will give:

$$\frac{d\theta}{dT} = \frac{\theta}{R}\frac{dR}{dT} - \frac{\theta}{q}\frac{dq}{dT}. \tag{6}$$

The penultimate move in this derivation is to put dR/dT into Equation (6), and to use the quotient $\theta = R/q$ where it is appropriate. This leads to:

$$\frac{d\theta}{dT} = \frac{\theta}{R}[-q(T) + gR(T)] - \theta n = -1 + \theta(g - n). \tag{7}$$

Thus, for $d\theta/dT \geq 0$, we need $\theta(g - n) \geq 1$. For the values in the previous example, with $n = 0$ and $g = 5\%$, for θ to increase we must have $\theta > 1/0.05 = 20$. Readers should attempt to give a verbal interpretation to the above, perhaps using a few numbers.

2. 1960–1973: From OPEC's Formation to the First Oil Price Shock

This is an important interval, though often ignored by many academic energy economists. In 1960 a group of oil-exporting countries followed an example set by oil producers in Texas (U.S.), who had formed a cartel for the purpose of limiting output, and therefore keeping the price of oil from falling to a few cents a barrel. In case readers are curious, the man often called the best brain of the 20th century, John von Neumann, once said that collusion by firms is the only way that the joint profits of a cartel can be maximized (otherwise, as you may remember from your courses in game theory, producers face the problem of "conjectural reaction" — that is, having the optimal oligopolistic production arrangement upset by competition). OPEC's ambitions were also in the direction of control as well as cooperation, because the wording in the 'Letter of Intent' they published in 1967 definitely implied that when they felt the time was right, they would take complete responsibility for the oil produced within their borders.

Then, in 1962, Dr. M. King Hubbert expanded some of his previous work by insisting that oil production in the 'lower' 48 (states) of the United States would peak in 1970 or 1971. This contention was ridiculed by the same kind of people who claim today that a global peak will not take place; however, the actual peak in the U.S. came late in 1970. Output peaked at about 9.5 million barrels per day (= 9.5 mb/d), and almost immediately began to decline (instead of forming a plateau). It declined (for the entire U.S.) until the super-giant Prudhoe deposit in Alaska came on stream, at which time it went up. But output never attained its previous peak. Instead it reached 7.5 mb/d, and began to fall again. Today, output for the entire U.S. is increasing because of the exploitation of shale deposits, although many of us believe that there is a limit to this process.

Hubbert's thinking must have gone as follows. The more oil (*reserves*) in a particular plot of earth, i.e., a *deposit*, the easier they are to extract. After a while though, oil becomes more difficult to extract, which is primarily due to a decline in *deposit pressure*, which in turn might be influenced negatively by an increase in *natural depletion*. Eventually a natural production limit is approached that is set by the influence on deposit pressure of the amount of the resource (i.e., oil) remaining in the deposit. As designated earlier, the resource ceiling could be called R^*. This is the estimated quantity (in barrels) of reserves in a deposit, regionally or globally, depending upon the object of the exercise. And note: reserves are less than *oil in place* — sometimes a great deal less.

If you look at the production curves for oil — or frequency curves as they are sometimes called — and particularly those for the most productive deposits in the world, what you will see is that they increase shortly after production is initiated, reach a peak, and either decline or form a plateau. In the latter situation, they decline later. We see these arrangements with almost all the major deposits or producing areas.

Consider for example the Cerro Azul Number 4, located near Tampico, Mexico, which was the most productive oil well ever seen until then, and probably for a long period after. When it 'blew in' on February 10, 1916, a column of oil rose 600 feet in the air. Its initial production rate was reputedly 260,000 b/d, which was unbelievable. Moreover, expectations were that while its output might decrease, there would not be a collapse.

Regardless of expectations, after producing 60 million barrels, it suddenly produced nothing but salt water. The kind of price system that we talk about in microeconomic theory cannot deal with this kind of phenomenon — a situation where something completely out of the ordinary takes place. What we have in the world oil picture is a number of very large deposits producing a great deal of oil, with many of them past their peak where production is concerned and more or less on 'life support', or whose output displays a plateau. Those on a plateau will also begin to decline some day, and eventually global production will be on its downward slope.

There are some observers who claim that if we stick with conventional oil, a peak has already taken place, or at least is very near. What has happened though is that now, as compared to, e.g., a decade ago, we talk about *liquids* instead of oil, where liquids is defined as, e.g., conventional oil plus natural gas liquids. Or, more realistically, when we are talking about oil, we are actually talking about conventional oil plus liquids. Something else that deserves more attention than it has received is the peaking in 1965 of global conventional oil discovery. The important thing here is that — as pointed out by Colin Campbell — the peaking of world oil output will in some respects be an analogue of the peaking of world oil discovery, and of course unavoidable.

The eventual peaking of world oil output will likely be a very serious matter, both economically and psychologically, and not just the kind of pseudo catastrophe that best describes some early assessments of the nationalization of foreign oil-producing assets in, e.g., the Middle East. These nationalizations began in October 1973, and the reason for my assertion is that before nationalization, the foreign enterprises operating in, e.g., Saudi Arabia had planned to raise production from 10 mb/d to 20 mb/d. If production had actually been raised to 20 mb/d and maintained at that level for as long as possible, we might already have experienced a peaking of global output.

Producing 20 mb/d did not, however, coincide with the strategy of the government of Saudi Arabia, whose king clearly stated that *sustainable* Saudi production would not exceed 10 mb/d. The same promise surfaced in 2008, when the oil price soared to over $147/b, and the president of the United States flew to Saudi Arabia for the purpose of convincing the Saudi king to raise output. The answer to his request was an unequivocal but respectful "no", and here readers should carefully note that had President Bush and his skillful and intelligent advisors genuinely believed that speculators and not oil *fundamentals (i.e., supply and demand) were the cause of the oil price escalation*, he could merely have taken an interstate bus to New York, and once there, ordered Wall Street's 'masters of the universe' to cancel any socially harmful financial schemes they were practicing or contemplating.

The first oil price shock caused the oil price to increase by a factor of slightly more than five (from $2.05 to $10.35), but shortly afterward the price showed a tendency to decline. A price rise of that nature has been termed a 'spike', and there were two more pronounced spikes and a tame spike before the sustained price rise that began in 2003–2004, and accelerated during 2007–2008. The thing to appreciate here is that speculators were just as eager to become rich during those spikes as they purportedly were when the oil price went into orbit in 2008, but sadly those who stayed too long at the party took a fall, because during all the spikes in the 20th century there was plenty of high-quality crude oil in the crust of the earth, and much of it was located in the right place.

There was of course considerable price volatility on both upward and downward trends during this period, but to a considerable extent this was merely a sideshow or distraction. In a talk I gave at Uppsala University I was questioned about the dynamics of price surges, and I probably suggested that these could be approached using the same mathematics as employed when discussing non-fuel minerals or some agricultural products. But while we can easily obtain trends and assorted vibrations from some elementary difference and differential equations, this is mostly a mathematical rather than an economics exercise. It looks good on a blackboard, but does not say a great deal.

Here it might be suitable to cite a useful observation by Robert Feldman, chief economist at Morgan Stanley. He stated that time lags between supply and demand that characterized the famous hog cycles of the 1930s are now at work in "energy" (by which he probably meant oil). These time lags can create imbalances that lead to very large price swings. These lags can be detected in Equations (8) and (9) just below, although what is not shown in these expressions are the root causes of fluctuations. These are speculative tides of bullishness and bearishness, which have their origin in actual and/or expected price trends, the outcome of OPEC meetings, and possibly the forecasts of high-status organizations such as the International Energy Agency (IEA) and the Energy Information Agency of the United States Department of Energy. Of course, as Kristofer Jakobsson and his colleagues have made clear, the less said about the forecasts of the IEA the better, and I made sure that my students in Bangkok understood this.

We can now derive a simple equation which might have some pedagogical value; however, it is not an essential part of this exposition. First we have a (flow) demand curve for oil, $h_t = a_0 + aP_t$, and a (flow) supply curve, $s_t = b_0 + bP_{t-1}$, with $a < 0$ and $b > 0$. Notice that the supply curve is lagged, while the demand curve is called h instead of the usual d, and I will add trend terms to these relationships to get $h_t = a_0 + aP_t + \alpha_t$ and $s_t = b_0 + bP_{t-1} + \beta_t$. Introducing the equilibrium condition $s_t = h_t$, we immediately obtain the following difference equation:

$$P_t = \frac{b}{a}P_{t-1} + \frac{\beta_t - \alpha_t}{a}. \tag{8}$$

(For α_t and β_t, we might have $\alpha_t = \lambda_x + \alpha_1 t$ and $\beta_t = \lambda_y + \beta_1 t$.) In any event, if we begin with a price of P_0, then the solution to the above equation

Energy and Economic Theory

is obviously:

$$P_t = \left[P_0 - \left(\frac{\beta_0 - \alpha_0}{a - b} \right) \right] \left(\frac{b}{a} \right)^t + \left(\frac{\beta_t - \alpha_t}{a - b} \right). \tag{9}$$

The expression to the left of the plus sign in Equation (9) represents a cobweb model, while the expression to the right is a trend term for the price. On the basis of earlier statements, at the present time β_t is greater than α_t, and so P_t is trending downward. Basically, in Equation (9) there is a trend term around which there are oscillations in both price *and* inventories. In addition, if supply lags demand, and there is an upward spike in demand, inventories will decrease even if the lag is very short. For an optimal presentation, however, a flow model of the type above is inadequate: it requires too much interpretation. The trend terms in the equations suggest that we are dealing with the long term, while even the discussions on CNN and Bloomberg suggest the relevance of inventories for short-term developments. Thus we need another model — one in which inventories are explicit. That means a stock-flow construction of the kind that will be presented in the next section.

One thing remains in this section. This is to mention that while for several decades the key issue with OPEC was their production, their exports have now become just as important. The OPEC countries are raising the domestic 'consumption' of their oil (and gas), just as you or I would if we were in their place. In some cases they are concerned with adding value to oil with refining and petrochemical activities, rather than exporting it in the crude form. This shows that they have grasped some of the key lessons of development economics, in particular the lessons taught by people like the late Hollis Chenery and Jan Tinbergen. On this score I can mention that I recently encountered one of my former finance students, and when he informed me that he had left finance and was interested in development economics, I congratulated him and suggested that he should become pro-foundly acquainted with the work of Hollis Chenery.

3. 1974–1999: Relatively Quiet Days on the World Oil Markets

"Relatively quiet" means quiet relative to the present somewhat confused situation, where hopefully the global macroeconomy is in the early stages of a recovery from a partial macroeconomic and financial market meltdown. The rain on this parade is the movement of the oil price to over $100/b, which I happen to consider unhealthy given the unemployment situation in many countries. More alarming, if another oil price escalation begins, it might begin at about $100/b, whereas the escalation in 2007–2008 that carried the price to above $147/b apparently began earlier at about $40/b. An oil price of $147/b or above is an experience that should definitely be avoided if possible, because despite what some people might think, the global macroeconomy cannot support that price.

In 1981 the Iranian Revolution took place, which removed approximately 2 mb/d of oil from world markets, but this shortfall was quickly replaced with oil extracted from elsewhere. There was of course a short-lived and not particularly intensive oil price spike. However, in 1982–1983 another disturbing — although again largely unnoticed — event took place, in that more oil was consumed than discovered. This tendency has continued, and only incurable optimists expect it to be reversed, despite the exploitation of shale oil in the northern part of the U.S. Another price spike took place in 1991, at the time of the first Gulf War, but once again there was plenty of oil available in the world, and as peace descended on the Middle East, the price of oil moved into a prolonged decline that lasted until the final years of the century. That was when oil touched $10/b, and *The Economist* predicted a possible descent to $5/b, making the absurd claim that according to the fundamentals, this was where the oil price belonged.

The possibility that the future price of oil might remain in that range quickly concentrated the minds of the OPEC directorate, and unlike many commentators on the world oil scene, they realized that the supply of (UK and Norwegian) North Sea oil was about to peak. As a result they understood that if they played their hand carefully and correctly, the time would soon arrive when some countries in that grouping would see their most incredible dreams come true.

What we need to consider before moving to the great oil price upswing that began several years later is a refinement of the analysis that I developed many years ago for the copper and aluminium markets, when I was employed in Geneva (Switzerland) by UNCTAD. I would like nothing better than to claim full and total credit for this construction, which I can do for the diagram; however, a large part of the basic reasoning was supplied by the brilliant econometrician Professor Franklin Fisher of Massachusetts Institute of Technology, who also made one of *my* academic dreams come true when he published a note of mine that was related to this topic in the prestigious journal *Econometrica*.

To begin, it needs to be recognized that a flow model of the type above is inadequate, and also the flow models that you had so much pleasure dealing with during the elementary course in economics that you might have taken. The trend terms in the equations suggest that we are dealing with the long term, which is well and good, but the discussions on CNN and Bloomberg emphasize the relevance of inventories for short-term developments. Thus we need another model — one in which inventories are explicit. That means a stock-flow construction of the kind shown in Figure 1. All of my energy economics students, regardless of the level, must be able to deal with this model, particularly on examinations.

Derivative markets (e.g., futures) could also be brought into the analysis in a non-superficial manner, but first, readers should make some effort to absorb the following discussion, which does not require any advanced mathematics. As I tell my students, the first step in doing this is to look very carefully at Figure 1, and appreciate that it is both simple and logical! For example, it is NOT the kind of flow diagram that we see in electrical engineering textbooks, and thus Kirchoff's laws are totally inapplicable, even if we wanted to make heavy weather of the first-order servomechanism that is associated with the p-DI component of the diagram. It is in this (feedback) circuit that the instabilities mentioned earlier originate, although to adequately verify this contention some second-order or higher differential or difference equations should be formulated and solved.

Hedge funds and futures markets influence to some extent the expected price, and consequently the magnitude of desired stocks (i.e., inventories). If, for example, $DI > AI$ because it is expected that price will increase, then price will increase as an attempt is made (via an increase in flow demand) to increase stocks (i.e., AI). But at all times the key items in this short-term price formation model are stocks (i.e., inventories), which conceptually are more important than the supply (s) and demand (h) *flows*! The mechanics of this market (and also futures and options markets) are explained in detail in my previous energy economics textbooks

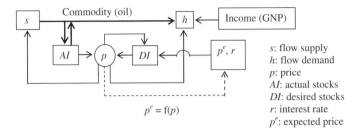

Figure 1. Stock-Flow Model of the Oil Market

(1977, 1980), where it is emphasized that equilibrium comes about when $AI = DI$, and not when $s = h$. My students must understand this perfectly, and it — and Figure 1 — must be presented and explained on every examination.

Sometimes I handle this model with a differential equation, but the only thing this does is to infuriate some readers. Instead, let us keep the analysis simple. If $AI > DI$, and thus is a non-equilibrium situation, then price falls in order to make the excessive inventories more attractive to consumers. The sequence is $AI > DI \rightarrow p\downarrow \rightarrow h > s \rightarrow AI\downarrow \rightarrow AI = DI$. What about when $DI > AI$? Then the adjustment procedure is equally simple: $DI > AI \rightarrow p\uparrow \rightarrow s > h \rightarrow AI\uparrow \rightarrow DI = AI$. Pedagogically, dealing with this issue as is done here makes more sense than writing a differential equation in which the degree of the equation is arbitrary. (Readers can now explain why we do NOT have an equilibrium if we have $s = h$, but $AI \neq DI$.)

In conjunction with Figure 1, some simple algebra might be useful, though it should not be over-emphasized. Rather than formulate and solve the differential equation that I usually use, I sometimes resort to the following simple relationships, where DI has been changed to D, AI to A, p to P, and Φ is a constant:

$$h_t = a_0 + aP_t \quad \text{flow demand}$$

$$s_t = b_0 + bP_t \quad \text{flow supply}$$

$$P_t = P_{t-1} + \Phi[D_{t-1} - A_{t-1}].$$

Two things need to be pointed out here. The first is the logic of inventory adjustment in this scheme. If, for example, we have $D > A$, then price

(P) must be raised to an extent that we have flow output greater than flow consumption ($s > h$), and thus an increase in inventories. The second is the 'triviality' of this model. A first-order difference equation will be obtained from these relationships; however, it would be a simple matter to obtain one of higher order. The important thing in this section is Figure 1, which all of my students in every university where I have taught must be able to duplicate in an examination situation, and explain employing a few symbols.

Obtaining the basic difference equation is simple. From the system and what I call the logic of inventory adjustment, we immediately obtain:

$$P_t = P_{t-1} + \Phi[D_{t-1} - A_{t-1}] = P_{t-1} + \Phi[(b_0 + bP_{t-1}) - (a_0 + aP_{t-1})].$$
(10)

Notice that there is no lag in the flow supply and demand equations. The reason is that these equations are arbitrary, and are merely intended to stress that when the issue is price formation in this kind of model, inventory behavior is crucial, and this would be true regardless of the exact configuration of the flow equations. Readers who desire something more sophisticated are referred to the first chapter of R.G.D. Allen's book, *Mathematical Economics*, which is probably one of the most important economics books ever written. A little manipulation of Equation (10) will give us the following:

$$P_t = \Phi(b_0 - a_0) + [1 + (b - a)]P_{t-1}.$$
(11)

Solving this is easy. If we have a price P_0 when the decision is made to adjust inventories, and the equilibrium price — *if there is one* — is P^*, then the solution is:

$$P_t = P^* + (P_0 - P^*)[1 - \Phi(a - b)]^t.$$
(12)

As it happens, if $\Phi < (b - a)^{-1}$, then price approaches P^* steadily. Otherwise there are damped or explosive oscillations, where in the last case there is no equilibrium.

I see no point in continuing along this line, although one thing needs to be emphasized. An instantaneous adjustment of inventories (i.e., stocks) such as we might encounter on the pages of an Economics 101 textbook, more or less eliminates the distinction between stock and flow equilibria (where, borrowing from physics, an equilibrium means a state of rest). That observation does not help us very much where discussing this topic is concerned, and so we will have to touch on speculation, which will be treated somewhat more expansively in the next section.

Average inventories of oil for the U.S., Europe and Japan from January 1991 through March 2005 came to about 775 million barrels. These were fixed inventories, and an additional 830 million barrels (called floating inventories) were in transit at sea. More commercial stocks were held in the rest of the world, but there are no figures on the exact amounts. (There might also have been a billion barrels in official inventories — e.g., the U.S. Strategic Petroleum Reserve (SPR) probably has about 800 million.) Now suppose that for one reason or another there is an increase in DI, and this is accompanied by an intention to raise AI by some fraction of one percent (1%), and in addition to do so in a short time.

In terms of Figure 1, this puts a pressure on supply (s) that it cannot easily support, given the absence of reserve production capacity in the *real* world market. (For instance, OPEC's spare capacity is almost certainly not the 5–6 mb/d estimated by the IEA and EIA, but probably 2–3 mb/d.) As a result, the price (p) will immediately increase, and perhaps by a large amount. (What we have here is the difference between short- and long-run supply elasticities.) Yes, the people in front of computers may have contributed to this situation by misjudging the developing situation in the oil market, and thus playing a small or large part in causing p^e to become something that it should not be, but the big problem was the failure of major oil producers in OPEC and elsewhere to locate sufficient new reserves — if that was possible or, to their way of thinking, desirable — and then invest in new production capacity.

By way of summation, the following should be pointed out. Three genuine price spikes have been mentioned: 1973, 1980–1981, and 1991. The most interesting spike though was a partial price spike at the end of the century, when OPEC reduced its output by 1,500 barrels per day. That showed the OPEC directorate what might be possible a few years later, if they stopped playing ego games and manipulated the oil supply the way they had intended all along when the opportunity presented itself. Moreover, I am certain that I was not the only student of the oil market to note that the

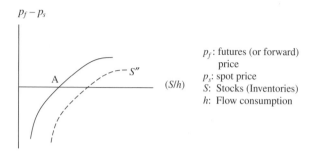

Figure 2. The Relation of Futures and Spot Prices with the Ratio of Inventories to Consumption

oil production outside OPEC was 'decelerating', and the peaking of non-OPEC (conventional) oil might take place in the not-too-distant future.

Now we can turn to Figure 2, which is a diagram that readers should look at very carefully, and to which they should give some thought. It happens to be very important for the rounding out of your energy economics education. A *normal* arrangement in this matter of holding inventories, and perhaps hedging with futures, is that holders of inventories will only increase 'inventory coverage' if the spot price declines relative to futures prices, or if they expect a very high price for the item in the future. This makes sense for several reasons, one of which is that inventory holders want to be rewarded for buying and storing a commodity, instead of, e.g., merely purchasing interest-bearing financial assets such as bank accounts or bonds.

Figure 2 is special in that to the left of point A, present prices are higher than futures (or forward) prices. This situation is called *backwardation*, and is attributed to a comparatively low inventory coverage. The usual terminology is that *there is a positive 'convenience yield' associated with a larger inventory coverage*. The opposite situation, to the right of point A, is called *contango*. As it happens, oil markets have shown a surprising tendency to be in backwardation a large part of the time, although over the past few years contango has been the rule. This terminology should be mastered!

There is no reason to complicate the above just now, although a slight rephrasing and extension of the above discussion might be useful. First, note that if producer or consumer inventories of a commodity are low, then each extra unit held in stock reduces the likelihood that, e.g., industrial operations will have to be scaled down because a crucial input is unavailable.

Remember that both producers and purchasers of industrial raw materials are bound by contractual obligations to their customers, and so inventories must be held (as a kind of insurance) as long as uncertainty exists as to whether an essential input or promised output will be physically available during the period when it is required.

Moreover, even if the expected money yield from acquiring, storing and later selling a unit of a commodity does not cover such things as its storage cost, this negative aspect might be regarded as counterbalanced by a positive *marginal convenience yield*. When the size of inventories is small relative to the amount of the commodity being used as a current input in production processes, then it could make economic sense to store more of the commodity. An effective price system should function in such a way as to ration existing stocks among perhaps many demanders of inventories, and if stocks are generally judged to be too small, we get a departure from what we usually think of as normal (i.e., contango). Instead, we could find the spot price ending up at a premium to the futures price (backwardation), and/or the price of futures contracts whose maturity dates are in the near future being higher than the price of contracts that expire later.

Perhaps the most immediate impact that variations in inventory levels have is in the *spreads* between the nearer futures contracts. Until comparatively recently, market actors apparently believed that there was a light surplus of oil in the market or coming to the market, which meant that the position designated 'A' in the diagram would be fairly far to the left. As it happened though, for various reasons this opinion changed, which is shown in Figure 2 by a shifting down of the curve. 'A' would then be located a considerable distance to the right, and the familiar backwardation that often characterizes the oil market would be intensified.

4. Natural Decline

General Douglas MacArthur once said that his favorite song was "Old soldiers never die, they just fade away". The same can be said about old oil fields. The three largest fields are still very large, but all of them could start fading away, except for the Cantarell in Mexico — it is already on the wrong side of its peak, and its output is declining at an alarming rate. These three largest oil fields in the world are Ghawar in Saudi Arabia; Cantarell in Mexico, which, though in rapid decline, may still be the second largest oil-producing field; and Burgan in Kuwait, which is second in reserves, but where claims a few years ago about the exact amount of proved reserves were aggressively questioned by outsiders, who believed that these claims were too high. Dr. Veronica Cinti (2011) notes that Saudi Arabia may have a super large offshore oil field (called Safaniya), but I think that the emphasis here should be on the word "may". Similarly, readers should make every effort to become familiar with the work of Dr. Mamdouh G. Salameh on all the intricacies of oil markets, including his most recent examination of the 'shale oil revolution' (Salameh, 2013).

A fundamental theme in this book is that oil is scarce. *It is scarce given the demand that is going to be made for oil in the not-too-distant future!* To understand the unpleasantness that oil consumers could face, it behoves readers to obtain some insight into what is known as natural decline, because there is now a constant reference to 'decline' in the press. Some mathematics must be employed in this part of the exposition, but most of it should not disturb readers who prefer a less formal type of presentation. Moreover, just below, and prior to the mathematics, are a few statements about natural decline that should be comprehensible to all readers. Of course, there is a non-technical summary of the oil market in the first chapter of this book, and some non-technical remarks in Chapter 8 that readers should examine as soon as possible. Those non-technical remarks are intended to help make you a star.

What is being aimed for in the exercise below is to say something meaningful about investment as well as production. In a more technical presentation, it would be optimal to work toward an implicit function of the type $\psi(q_1, q_2, \ldots, q_T; I_1, I_2, \ldots, I_T) = 0$, where the q's are production and the I's are investment.

Several years ago Mr. Lee Raymond — the former CEO of Exxon Mobil — gave an interview in which he emphasized the importance of the natural decline rate of oil deposits. He also expressed some disappointment that this was usually overlooked by self-appointed experts on the production of oil. Rather than turning to the technical literature, it might be best if readers consult Google, where they will find a reference to some work by the International Energy Agency (IEA) on this subject. Apparently, after a study of the 400 largest oil fields in the world, IEA concluded that (*ceteris paribus*) the average natural rate of decline is 9% a year, ranging from 4%–5% to almost 20%. What is happening is that if the inputs being used are held constant, then instead of the production of an average well remaining constant, or nearly constant, it declines by 9% on average. This is where 'natural decline' comes into the picture, and one way it can be described is in terms of the loss in capacity that would occur in a given 'asset' if no remedial or offsetting action is taken. (Personally, I think that 9% is too large.)

As for this remedial action, it can take the form of locating contiguous new reserves and drilling new wells, or perhaps taking steps to increase the output of existing wells (via, e.g., injecting water or carbon dioxide or the use of 'surfactants' to increase viscosity), and therefore turning more of the *oil-in-place* into reserves. (*Make sure to remember that globally, on average, only about 35% of the oil-in-place in deposits can be classified as reserves, and thus are immediately accessible.*) These procedures can be labelled 'investment', and they roughly have the same monetary significance as the investments required to produce, process and transport the output of an oil field.

It might be useful to add that, according to estimates that one often encounters, decline rates for Iran may be as high as 8%/y onshore and 13%/y offshore, while for Saudi Arabia the figure is ostensibly 2%–4%. (Determining the accuracy of these estimates however — particularly the one for Iran — will be left to readers.) Accordingly, if Saudi Arabia's decline rate averages 3%/y, then — via investment of one type or another — it could be true that additional oil (to the tune of several hundred thousand barrels of oil) might have to be located every year in order for a *sustained* output of 9–10 mb/d to be economically viable. A problem here is that while new deposits can conceivably be located, it is not likely that they will be as 'rich' as those now being exploited. In addition, the deposits now being exploited are old, and investments required to maintain output could become very costly because of damage sustained by fields due to (among other things) production processes which involve the extensive use of water. As a result, given the expected future demand for (and price of) oil, Saudi oil field managers may have concluded that optimal behavior on their part takes the form of minimizing the expansion of output.

At the same time I want to make it clear that the managers of Saudi oil know more about the oil in their country than many of the fly-by-night journalists who call themselves experts on this subject. Natural decline will be mentioned again below, but in order to obtain an optimal clarification, a simple model will be presented that could easily be extended to take into consideration the natural decline rate.

First of all we need to understand the insignificance of a statement such as "with all the reserves in place now, we have a 40-year supply of oil even if we do not find another drop." This statement originates with observing that the global reserve-production (R/q) ratio is 40. However, the important issue is not the R/q ratio, but when production in a field, region, or for that matter the entire oil-producing world, moves toward a situation in which it 'plateaus' or declines. As should be obvious from a consideration of the example below, oil could be present and exploited hundreds or even thousands of years in the future; however, once past the global production peak, a great deal of faith in the future of oil might be reduced.

This is not to say that the R/q ratio should be ignored, but a statement such as the above (which postulates a 40-year availability for global reserves) is scientifically meaningless. In looking at a deposit or oil field, the important thing is that if the R/q ratio falls below a certain level — probably somewhere between 10 and 15 — then, as already discussed, *if an attempt is made to continue with the existing output, the deposit could be 'damaged' in the same manner that sucking too hard on a straw will damage an ice-cream soda.*

Returning to the simple numerical exercises in Section 1, readers should confirm that there is a large amount of oil still in the ground when output decreases. Moreover, when we look at the production profiles of *actual* major oil or gas regions like the United States, what we see is that when peaking takes place (and production sooner or later begins to decline), there is still a huge amount of oil in the ground, and in addition — if economic considerations are ignored — much of this is immediately extractable. The interpretation here is as follows: *the peak is explained by economics and not geology. More is not extracted — and the peak delayed — because in the interests of profit maximization, the optimal behavior is to extract it later!* As explained in Banks (2004, 2000), and in my previous textbooks, geology essentially functions as a constraint. This is something that everyone should make every effort to understand.

But a crucial point requires more explanation, and that is the natural decline! In the main example in Section 1, sufficient investment had been made to obtain an output of 15 units/year for 5 years, but what would the situation be if there was a natural decline of 5 units a year? Then, instead of the sequence of reserves being 225, 210, 195, 180, 165, and finally 150 at the end of the fifth year, it would be 225, 205, 185, 165, and 145 at the end of the fourth year. Thus, we would arrive at what we might call the *critical reserve level* before the beginning of the fifth year. Moreover, the decline in output after that would be steeper than in the previous case.

But what would happen if we had this decline 'pattern' and the intention was to maintain an output of 15 units? This is possible, though perhaps more investment would be necessary, and in addition there would probably be a more rapid 'depreciation' of the deposit, unless associated reserves could be increased. The decline rate above is probably excessive, and was chosen to make the arithmetic simple, but what it did not show is that natural decline is often influenced by the extraction program.

In the example below only production is considered, although holding output constant as shown may influence exploitable reserves. This is not dealt with because of a lack of relevant statistics, but the IEA once claimed that globally, due to natural decline, 3.2 mb/d of new reserves must be found annually just to maintain output. This cannot be done for 'lunch money', because the management of Pemex (Mexico) says that it must spend $20 billion a year for two decades just to keep its output of 2.8 mb/d stable.

At the same time it needs to be made clear once more that when the asset under discussion is an oil deposit rather than a conventional capital good, it is still possible to think in terms of what in economic theory is called "depreciation by evaporation". What this means in the present example is that the asset is subject to a force of 'mortality' that will be taken as constant, and equal to Θ. If A is the constant annual revenue generated by the asset, or perhaps the size of an annuity derived from non-constant revenues, we can write for the value (V) of the deposit:

$$V = A \int_{t}^{t+T} e^{-(\theta+r)(\tau-t)}d\tau = \frac{A}{\theta+r}\left[1 - e^{-(\theta+r)T}\right]. \qquad (13)$$

Equation (13) might serve as a useful starting point for examining this topic if many readers were not allergic to integrals, because it indicates that the presence of natural decline (Θ) reduces the value of the deposit; but even if readers were madly in love with calculus, the presentation below is more suitable because the importance of investment is made explicit.

Suitable or not, these materials are only approximations, but as Bertrand Russell once pointed out, "all science is based on approximations." Moreover, energy economics is the kind of subject where there is not enough knowledge of the right sort in circulation. The world is filled with observers and commentators who want to believe that pessimism about the reserves of oil is socialistic nonsense, and many of these observers and commentators occupy important positions in the media.

In order to keep things simple, a constant decline rate of 20% is assumed. As in the previous example, the intention is to hold production at 15 units/year; however, the initial reserves are increased from 225 units to 'something' much larger. The reason why the size of this 'something' is not specified is that we are not concerned in this example with, e.g., peaking or the critical reserve-production ratio. The intention is to merely clarify the significance of the natural decline rate, and allude to its influence on investment. If readers want further details, they can examine my earlier textbooks, particularly Banks (2007). Now we have:

Year 1 : $15(I_1)$
Year 2 : 0.8×15 $0.2 \times 15(I_2)$
Year 3 : $0.8^2 \times 15$ $0.8 \times (0.2 \times 15)$ $0.2 \times 15(I_3)$
Year 4 : $0.8^3 \times 15$ $0.8^2 \times (0.2 \times 15)$ $0.8 \times (0.2 \times 15)$ $0.2 \times 15(I_4)$
Year 5 : $0.8^4 \times 15$ $0.8^3 \times (0.2 \times 15)$ $0.8^2 \times (0.2 \times 15)$ $0.8 \times (0.2 \times 15)$ $0.2 \times 15(I_5)$

. . . .

Year T : $0.8^{T-1} \times 15$ . . . . $0.2 \times 15(I_T)$

If we look at the above, what we see is that in Year 1, an investment of I_1 was made to obtain 15 units of output. Because of natural depletion, in Year 2 additional investment of I_2 was necessary to obtain an additional output of 0.2×15 — i.e., the decline *rate* times 15 — in order to keep the total output at 15 $[= (0.8 \times 15) + (0.2 \times 15)]$.

Mathematical induction could be useful here if the logic behind this scheme were not so simple. Let us take the decline rate as $(1 - \Theta)$, which in the example means 0.20, which in turn means that $\Theta = 0.80$. Now let us see what we have for Year 4 in symbolic terms: $15(1-\Theta)[1+\Theta+\Theta^2+\Theta^3]$. The expression in the large parenthesis can be simplified to $[(1 - \Theta^4)/(1 - \Theta)]$, and so in Year 4 we have $15(1 - \Theta^4) + 15\Theta^4 = 15$.

Nothing has been said here about the size of the I's (which represent additional investment in, e.g., wells for the purpose (in this example) of holding output at 15 units/year), but as with Pemex it could be an expensive proposition. (Something like this kind of program may be relevant for oil in Saudi Arabia, where for the time being the intention seems to be to hold output in the 9–10 mb/d range.) Note also that if we had numerical values for the value of I, we could work with an implicit expression of the type $\psi(I_1, I_2, \ldots, I_T; q_1, q_2, \ldots, q_T) = 0$. For instance, if we take $T = 5$, this expression would be $\psi(I_1, I_2, I_3, I_4, I_5; 15, 15, 15, 15, 15) = 0$. Putting together an example with explicit values of I, which also says something about the depreciation of the deposit due to additional investment, should not require a great deal of algebraic sophistication, but it would involve a degree of arbitrariness relating to the decline of the deposit that I prefer to avoid in this elementary presentation.

Moreover, it is just as well that readers attempt to comprehend this issue without being concerned with an overload of mathematics. As pleasant as it is to stand in front of a blackboard or whiteboard and write equations dealing with the present subject, most of the very important persons at international meetings like the annual Davos get-togethers for high-fliers prefer to be informed about important subjects without a resort to calculus, or for that matter secondary school algebra.

5. 2000–2020: Then, Now and Later

A considerable amount of bad economic news might be the legacy of this period. As mentioned just above, OPEC appears to have gotten its act together, because otherwise it would be impossible to have an (aggregate) oil price above $100/b, as is the case at this moment. (Aggregate means an average of WTI and Brent.) The important thing for all readers of this contribution to notice is not the sustained oil price rise from 2003–2004 to the early autumn of 2008, and in particular the acceleration of the oil price that took place in 2007–2008, but the way that OPEC managed to cut its losses in 2008–2009 by immediately reducing aggregate output after the oil price peaked and then fell to $32/b. Virtually without fanfare, output was reduced by something between 2 mb/d and 4 mb/d. That was the end of the absurd claims that OPEC was a 'paper tiger' and that the price of oil would soon collapse to a bargain basement level.

Now for speculation and speculators. The claim that I made earlier about speculation and the oil price is really all that is necessary to deal with the financial market foolishness that at one time was running wild in various forums and classrooms across the civilized world. Anyone incapable of understanding that the United States government could easily have suppressed any speculation in oil futures (or, for that matter, physical oil) that originated in the United States, could not possibly have understood the elementary technical discussion that I had planned for this part of the present contribution. Meeting the king of Saudi Arabia with his hat in his hand, and humbly asking him to increase the production of oil, was hardly the option that President Bush would have chosen if the destructive oil price escalation of 2007–2008 had its origin in speculation, and especially speculation that originated in the financial district of New York.

When this topic came up at my Paris lectures, the explanation I could have given, had I not been intent on repeating (for the third or fourth time) the materials above, is roughly as follows. *The preposterous theory that speculation rather than fundamentals (supply and demand) was behind the spectacular rise in the price of oil that began in 2003–2004 and continued to 2008, was first launched by OPEC* (just as OPEC is to thank for the present comparatively high oil price). To repeat, it was launched and put into circulation by OPEC, and they repeat it every chance they get. The speculation conjecture has as much veracity as the 'weapons of mass destruction' fantasy that President Bush used to instigate the war in Iraq. The ostensible guilt of, e.g., Wall Street and the financial districts of London and Paris was initially an OPEC fabrication, although at the present time there is a strident 'anti-financial market' backlash in one of the forums to which I contribute. I do not admonish OPEC though, because if I were in their place, I would be doing the same thing — and so would you.

The reason that OPEC could whip up this anti-capitalist hostility is because a majority of the *professionals* reputedly speculating on oil resemble the people in front of the screens in the film *Wall Street*, and they have comparable educations and prospects in a world where many employees in other sectors of the economy feel that they are under increasing pressure. Unlike their critics, most of these Wall Street winners really and truly understand the dynamics of oil markets, or financial markets, or whatever market they are involved with, and if they don't, they are encouraged by their superiors — at, e.g., investment banks and other financial institutions — to find another line of work. They also do not call themselves "speculators"! The few whom I encountered when I worked in Singapore and Sydney called themselves "traders": they trade for their firms, and in the interests of their high incomes and enviable careers, they do everything possible to avoid making mistakes.

In my course on oil and gas economics at the Asian Institute of Technology (Bangkok), I discussed several articles by a journalist in *Le Monde*, Jean-Michel Bezat, who had some very bad news to present to his readers about the intentions of King Abdullah of Saudi Arabia where the supply of oil was concerned. Once again, probably after revealing their intentions dozens of times, beginning in 1973, a Saudi king reaffirmed — though not explicitly — that his country was going to produce as little oil as possible from existing deposits, and if it happened that there were new discoveries of oil in his country, they would be left in the ground for the children (*les enfants*) of the Kingdom. I have pointed this out on many occasions, maybe even in this book, and given the opportunity I would point it out again, because it is a decisive signal where the future oil price is concerned.

6. A Summary

The materials that I put on the whiteboard during my lecture at the University of Paris (Dauphine) will have to be duplicated by my future students of energy economics if they prefer a passing to a failing grade. An item that will definitely have to be understood perfectly is Figure 1 — which was also on that whiteboard. Some of those materials are discussed at length in this chapter, and I summarize them below:

1. The papers by Harold Hotelling (1931) and Robert Solow (1974), and my discussion of their shortcomings.
2. 1960: The formation of OPEC, and the background to its forming.
3. 1962 (and 1956): The papers by M. King Hubbert stating that oil production in the lower 48 states of the U.S. would peak in 1970 or 1971.

The exact contours of this peaking would have to be given, including the circumstances under which output fell to the present level.

4. 1965: The global discovery of conventional oil peaks.
5. 1967: OPEC publishes a 'Letter of Intent'.
6. 1970: U.S. oil (50 states) peaks.
7. 1973: The first oil price shock, following nationalizations in the Middle East.
8. 1980: Iranian Revolution. 2 mb/d of oil temporarily disappear from the market.
9. 1982: More conventional oil is consumed than discovered.
10. 1991: First Gulf War.
11. 1991: Oil price falls to about $10/b, with North Sea oil (UK and Norway) about to peak.
12. 2003: Second Gulf/Iraq War. In his book *The Age of Turbulence*, the former director of the U.S. Central Bank, Alan Greenspan, claims that this war was about oil.
13. 2003–2004: The beginning of a sustained oil price rise to $147/b in 2008.
14. 2008: Oil price begins a decline that falls to about $32/b in 2009, but OPEC reduces its output by 2–4 mb/d. As a result, the oil price moves up to approximately $70/b in 2009.
15. 2010: Non-OPEC conventional oil supply (about 58% of conventional oil output) flattens.
16. 2016–2030: Somewhere in this period a definitive global peak of the conventional oil supply could occur, which would have serious consequences for the macroeconomic/financial market outcome. In a recent article in the journal *Scientific American*, the peaking date was given as 2016. This of course was an estimate, but if the global macroeconomy regains its former momentum, this might well happen. There is also the matter of what the production of conventional oil will be when the peak takes place. When I lectured in Bangkok, I discussed with my students the latest prediction of the IEA at that time, which was 121 mb/d for 2030. I took the liberty of describing that prediction as absurd. Almost immediately after, the IEA began to lower their forecasts, and an important person at the IEA sent me a mail in which he said that predictions of his organization were for consumption, and not

production. I was forced to inform him that this did not make economic sense to anyone except himself and his colleagues. I have also heard a rumor that they were prepared to publish a forecast of under 100 mb/d, which is an interesting number, because the director of the French 'major' Total has said that global production will never exceed 100 mb/d, and has offered to discuss this publicly with decision makers holding a different opinion. As to be expected, his generous invitation has been ignored. And finally, let me inform everybody that the recent war in Libya was about oil, and not about protecting civilians as the director of NATO insisted.

7. Final Statements and Conclusions

"In 20 years I predict energy wars over oil and gas resources. By the time it becomes politically profitable to react to problems in the transport energy sector, it will be too late for significant development of alternatives and too politically risky not to fight over remaining supplies."

— *Len Gould (in* EnergyPulse, *January 8, 2008)*

Someone who is alarmed by the turn of events on the energy scene is the former oil producer T. Boone Pickens, who is now a billionaire hedge fund manager, and he believes that the only way an energy disaster can be avoided in the U.S. is to fuel commercial vehicles with natural gas instead of diesel and gasoline. His position is that the supply of low-priced oil is limited, while the U.S. has or will have access to an abundance of natural gas. He apparently has accepted an estimate that the U.S. has a 100-year *domestic* supply of natural gas, and this is without accessing the supply of really exotic gas resources. Unfortunately, some of us feel that this hypothesis of Mr. Pickens has not been proven, nor should it be considered improved by decision makers.

As an aside, according to a Rand Corporation report that was prepared for the U.S. Air Force and Energy Department, liquids (oil + natural gas liquids) would have to be extremely expensive before it can be considered economical to produce coal-based transportation fuels. (Mr. Pickens has also expressed enthusiasm for a sizable CTL (coal to liquids) option.) The same Rand report concludes that coal could provide 3 mb/d of transportation fuels for 90 years. Readers should perhaps think about this estimate, because in a country where the consumption of liquids might be above 20 mb/d now, and will be considerably more when the global macroeconomy regains its traditional momentum, and a barrel of conventional gasoline requires more than a barrel of conventional oil because of refining 'mathematics', CTL may deserve to be an important part of the energy policy that the American president and his energy team might eventually propose.

In an issue of *Business Week*, there was a long article on the adventures in Russia of one of the most prestigious oil-field services firms, Schlumberger, where this multinational giant (with French roots) is frequently engaged in many countries to carry out exceptionally difficult drilling. Schlumberger has experienced considerable success, and among other things it has a reputation for leaving the politics to others, and concentrating on outperforming competitors and potential competitors.

This does not mean, however, that the presence of this and similar firms with state-of-the-art technology at their disposal will make the impossible possible, which means finding and producing oil that does not exist. Instead, the thousands of wells that reputedly will be drilled by Schlumberger might be necessary to prevent a steep decline in Russian output. This is an important observation, because Professor Gary Becker of the University of Chicago once advanced the opinion that we can find a sizable portion of our energy salvation beneath the frozen tundra of Russia. The thing to remember here is that one of the most important energy players in Russia has said that it should not be expected that Russian production will ever exceed 10 mb/d, and since domestic consumption is increasing in that country, exports will likely decrease. Shale oil and gas might, however, come to the rescue of the countries receiving Russian energy exports.

There is little or no good news in this book for those of us on the buy side of the oil market. One of the reasons is the stunning failure of key experts and players to draw the correct conclusions about the structure and mechanics of the world oil market. In addition, in considering the actions and claims of certain oil companies who are involved with activities in places like the Gulf of Mexico and the Caspian Sea, some words from the billionaire Canadian investor Stephen Jarislowsky are highly applicable:

"It's absolutely unbelievable what's going on. We're living in just about the most dishonest time in the history of man." He could have added that a large part of this dishonesty originates with so-called students of the oil markets in universities and 'think tanks', who have decided that their best career move is to take advantage of the veneration and/or pocketbooks of their devotees and clients by abandoning the restraints imposed by history and conventional logic. Instead, they turn to mythical claims that the price system and/or technology will relieve our oil anxieties.

A few provocative observations are in order at this time, beginning with one offered by Professor Michael Klare, editor of the important journal *Current History*: "If the oil from the Persian Gulf cannot be kept under U.S. control, our possibility to remain the dominating power in the world could be brought into question." A useful comment on this can be derived from the likely outcome of the present troubles in the Gulf and North Africa, which is that it may no longer be possible to ensure that the oil in those regions can be kept under U.S. control. Of course, for what it is worth — which isn't much — it may still be possible to guarantee the stipulations of the Carter Doctrine from 1980: "Every attack by a foreign power to win control over the Persian Gulf will be interpreted as an attack on the vital interest of the United States. Such attacks will be repulsed employing all necessary means, including military force."

One problem here, though, is an interpretation of the term "foreign power", which at the time that President Carter issued this warning almost certainly meant the Soviet Union. At present it could mean countries to which the oil legally belongs. There is no evidence that a definitive peaking of the global oil supply will take place in the *very* near future, but if it does, some doubt must be expressed as to whether the countries in, e.g., the Persian Gulf or North Africa would be encouraged/allowed to produce oil at the rate that they consider desirable. The energy wars to which Len Gould referred can only mean wars between some oil exporters and some oil importers, since even owners of a fleet of SUVs might hesitate to endorse nuclear war in order to enjoy the thrill of zooming down the Pacific Coast Highway at speeds close to those registered in a Monte Carlo Rally.

In another *Business Week* article (January 21, 2008), it was stated that six Gulf States control sovereign wealth fund assets of about $1.7 trillion — which puts them in the class of the largest hedge funds, or perhaps a collection of these large funds. Since the prestige and competence of a majority of hedge funds is largely an illusion, the deal-making referred to in that article belongs in a soap opera as much as it does in a serious business publication, and readers should attempt to appreciate my observation. However, something that cannot be disregarded is the ability of money generated in that part of the world to influence the price of oil by financing the diversification of Gulf States away from oil and into 'something else'. Enough has probably already been said in this book on that subject, but I would like to state again that the kind of economics studied and taught in the most *and* least prestigious faculties of economics makes it clear that the greater the pace of diversification, the less will be the effort made to produce and export (unprocessed) oil and gas. The decision makers in the oil- and gas-importing countries would do well to focus on this point.

One more item needs to be mentioned before closing this chapter. This concerns the real price of oil — which takes into account inflation (and perhaps also exchange rates) — as compared to the money (or 'nominal') price. The real price has to do with how much 'real goods' a certain amount of money will buy. It is often claimed that although the money price of oil may have touched, e.g., $100/b and continued to increase in, e.g., 2008, the real price was much lower on that occasion.

The excellent Josh in the TV show "The West Wing" informed that program's faithful viewers that the highest real price of oil after October 1973 (which was the date of the first oil price shock) was registered in 1979–1980, when a change in the political situation in Iran led to the nominal oil price spiking to $40/b. His contention might be correct if the base date for the calculation of the real price was 1973.

It might be argued though that the base date for calculating the present real price should be in the mid-1980s, or perhaps the mid-1990s, after adjustments had been made to the earlier price shocks, and large industries — including oil-producing firms — as well as consumers were making plans and investments to deal with a future in which there was talk in the corridors and restaurants of power that someday the oil price would stabilize at $28/b (which was OPEC's goal just after the turn of the century), or even in the low 20s range (as predicted by Lord Browne of BP), which some oil firms said they were using as a benchmark for their investment plans.

For the persons and firms who accepted those forecast prices as realistic, and who adjusted their investment and consumption to deal with these expectations, an oil price in the range that we have experienced in the last year or two is capable of bringing about real sacrifices. For instance, at the present time the economy of, e.g., the U.S. has not yet recovered its strength and there is still serious unemployment, while at the same time in the Gulf, economies have been strengthened to a point where until recently there was talk of constructing a new Olympus. The rising oil price might also have a major influence on share prices, and what we may experience at any time is the start of a cycle in which falling share prices will impact GDP via the wealth effect, which in turn will put a further downward pressure on shares and reduce aggregate demand. As they might say on Wall Street: "No, Ingrid, the declining *real* price of oil that ill-informed observers often discuss in those unread journals gathering dust in our academic libraries will not compensate for the damage that could be done by a sustained rise in the *money* price."

And finally, in light of all that has been said above, it might be nice to get the peak oil issue off the table (again) by repeating something worth saying and repeating whenever the opportunity appears. *Peak oil is not about the future — it is about the past!* It is about a (generally unspoken) strategy formulated many years ago by the most important countries in OPEC, featuring a stagnation or a decrease in the output of their invaluable oil (and perhaps also gas) when they get the opportunity. *High oil prices gave them such an opportunity in 2008, and to a certain extent in 2009, and it makes sense to believe that such an opportunity will arise again!* Anybody who does not believe that should be given the opportunity to explain what happened to the oil price when the senseless military venture in Libya was initiated, instead of doing what Pope Benedict suggested, which was to resort to diplomacy and discourse. The Libyan venture provided oil producers with a beautiful increase in income, *and* a future that many oil importers thought that they alone deserved.

Appendix 1. More Comments about Oil Futures Markets

With so much more that could be said about oil prices and their likely development, why turn to some mechanics of futures markets? The answer is that there are many very wrong beliefs about these markets in circulation. For instance, in the August/September 2010 issue of *The Middle East*, the OPEC Secretary-General said that "the emergence of oil as a financial asset, traded through a diversity of instruments in futures exchanges and over-the-counter markets, may have helped fuel excessive speculation to drive price movements and stir up volatility." The key word in this quote is "may", because in reality the force driving price movements during the crucial period 2003–2008 was an increasing demand from Asia, together with the decision of most OPEC countries to concern themselves with their own welfare — future as well as present — instead of that of motorists in the oil-importing countries.

What is needed is an applicable introduction to this topic, and here I can recommend my previous textbooks, as well as my finance book (2001). The plain truth is that students and others often refuse to understand that futures and options are very simple subjects as long as the advanced mathematics are ignored. And despite what your favorite finance teacher might have told you, almost all of the advanced mathematics is completely and totally superfluous. Furthermore, Carol Loomis, in an article in *Fortune* called "The risk that won't go away" (March 7, 1994), claims that few people have more than a sketchy understanding of these assets (e.g., futures) anyway, and that includes what she calls "top brass" in the financial and corporate worlds.

I say "thank you" to that, and as a result I do not feel it necessary to be overly concerned if you have seen some of the materials below earlier in this book. The more you see these materials, the better. One of the difficulties for many observers is comprehending the relation between the oil market and the financial market; however, to deal with energy matters it is necessary to have a reasonable insight into both, and even a 2013 Nobel Prize winner in economics — Professor Robert Shiller — is somewhat vague on the history and mechanics of the oil market, as he demonstrates in a 2007 article in *Forbes*.

Like options and swaps, an oil futures contract is a *derivative* asset, which means that its payoff is tied to the value of some other variable, in this case *physical oil*, which is also referred to as the *underlying* or *actuals*. (The barrels mentioned on a futures contract are often called *paper oil*.)

Because delivery is generally an alternative (but not a necessity), a futures contract is not a fully-fledged *forward contract*, which is a contract obliging one party to buy and receive a specific commodity (or asset) for the price that is quoted at the maturity date, and another party to sell and deliver the asset. (Please note that a conventional forward market typically involves private buying and selling arrangements between identifiable buyers and sellers that call for the future delivery of a commodity.) In the classroom, a futures contract is sometimes called a *standardized* forward contract, because it is traded on an *exchange* (i.e., an *auction market*) where prices are 'transparent' (i.e., visible), and where transactions are impersonal in that buyers and sellers are generally unknown to each other. The genius in futures markets is the mechanism for avoiding delivery.

Futures markets operate as follows. Against a background of speculators betting on the direction and size of oil price movements by buying and selling futures contracts, an impersonal agency can be created which permits producers, consumers, inventory holders and various transactors in physical products to reduce (i.e., hedge) undesired price risk by also buying and selling these contracts. As uncomplicated as this happens to be, there are a great many misunderstandings about this process.

One of these is the failure to realize that there is a social gain from futures trading that derives from the voluntary redistribution of risk between speculators and risk-averse dealers in physical products! In addition, despite what you may have heard or have decided to believe, futures trading usually (but perhaps not always) decreases the volatility and level of the oil price, because by facilitating the reduction of price risk, this trading encourages producers and others to carry larger inventories. By selling from or adding to these inventories, price swings can be dampened.

The success of a futures market tends to be dependent on the satisfaction of several well-defined criteria. For instance, it is essential that the commodity in question (e.g., oil) can be traded in bulk, and that it is bought and sold in circumstances that cause its price to fluctuate in a random or non-systematic manner. Without the latter provision, speculators may not be attracted to the commodity, *and without fairly large-scale speculation, futures markets will not function properly.* Here it should be noted that there are many *maturities* (i.e., time to expiry) of futures contracts in an individual market (e.g., 1 month, 2 months, . . . , maturities for oil contracts), but market liquidity usually declines rapidly for contracts with a maturity of longer than 6 months, and sometimes less. Thus, it was senseless to refer to oil futures contracts with a maturity of several years, as the governor of the U.S. central bank (i.e., the Federal Reserve System) once did when asked about the future supply of oil (and thus its price). *The lack of liquidity of futures with long maturities should be carefully noted by all readers.* Some very important people do not understand this situation!

These days the *modus operandi* of speculators is known to almost everyone with access to a television set; however, a few comments might still be useful. If a speculator believes that the price of oil is going to rise, then he might buy futures contracts — i.e., he goes *long*. This can be done by simply picking up a telephone and calling his broker (who in turn makes the purchase through the futures exchange). Conversely, if he believes that the price of oil is going to fall, he can call his broker and sell futures contracts (i.e., he goes *short*). Please observe that in both cases, at first remove, he is NOT dealing in physical oil. He does NOT have to be in possession of physical oil in order to sell paper oil! True, he may be in the habit of keeping a few barrels of oil in his bedroom for speculative purposes, but that is irrelevant to what we are talking about.

There are many more transactions in paper oil than in physical oil on any given day. To understand this phenomenon, the reader needs to remember that futures contracts are also forward contracts, in that delivery conditions are stipulated on them relating to the movement of a specified amount of physical oil, on or perhaps slightly after the maturity (or expiry) date of the contract, during what is called the *delivery month*. However, in a viable futures market, it is always possible to avoid making or taking delivery on a contract! For example, with a long contract, at any time before the contract *matures* (i.e., before the contract's expiry date), an offsetting (short) sale is made for the same amount of oil, referred to the same delivery month given on the long contract. *If, e.g., Mr. X opened a position by going long, he can close it by simply calling his broker and going short!* Obviously, market liquidity is the most important factor for this operation, as the reader knows from the ease with which shares (or stocks) can be purchased or sold, and which is due to the considerable liquidity in most share markets.

The evidence indicates that delivery takes place on less than 10 percent of futures contracts. This is not just because of the ease of offsetting a contract, but because delivery on futures contracts is made to or from locations that are inconvenient to most transactors. If you live in Chicago, and delivery on your long contract is made to West Texas, this takes some of the joy out of opening a position for the purpose of having immediate access to the physical commodity. In addition, as noted earlier, delivery can sometimes be avoided by resorting to cash settlement. For example, a contract is held until the delivery date, at which time, or shortly after, both long and short positions are closed by 'losers' making a payment to the exchange, and 'winners' receiving a payment. The important thing here is the specification, by the exchange (or its Clearing House) or 'the market', of a settlement (or 'reference') price.

Before making a few comments about the hedging of price risk, there are several extremely important topics that need to be perused. These have to do with margin, 'marking to the market', and the Clearing House.

The Clearing House is a non-profit operation belonging to an exchange. It acts as an intermediary (or 'middleman') in transactions, while at the same time making sure that monies are routed from losers to winners. For instance, if Mr. X opens a long position and the price falls instead of rises, then he owes somebody money. On the other hand, if Ms. Y opens a position by going short and the price falls, she has made a profit. Why is this? She starts by selling a contract for F_1, and the price falls to F_2. Her gain is then $F_1 - F_2$ (minus the broker's fee).

Perhaps the main function of the Clearing House is to guarantee transactions. In order to carry out this function they are involved in *marking-to-the-market*, which means that every night after the exchange closes, Clearing House employees examine the transactions that took place during the day, and inform brokers (who are certified members of the exchange) of winners and losers among their clients. These brokers in turn adjust the accounts of their clients, and perhaps inform them.

Let us take the case of Mr. X. Suppose that the oil price when Mr. X went to bed, and after the exchange closed, was $40/b, and he dreams that it will increase. As a result he calls his broker the next morning and instructs him to buy one contract, which is always for 1,000 barrels, and therefore the cost of the contract is $40,000. But instead of paying this $40,000 he pays his broker a *margin*, which is a security deposit and is usually between 5 to 10 percent of the value of the transaction. Suppose that it is 10 percent, which means that he must make $4,000 available for his broker. Essentially someone is lending him $36,000, and we use the expression *leverage* to describe this state of affairs. (A futures market offers its participants considerable leverage.) Often this margin is already in Mr. X's margin account, which is held by his broker. Suppose that at the time of the transaction the margin account of Mr. X was $5,000, of which $3,500 is specified by the broker as *maintenance margin*, which is a kind of lower limit for Mr. X's margin account.

Now we can examine the situation at the end of that day. Suppose the price of oil futures increases to $41/b. Mr. X's contract is marked to this amount by the Clearing House, which means that his margin account (with his broker) now contains $6,000. Of course, the $1,000 profit realized that day — minus the broker's fee — can immediately be removed, which brings the margin account back to $5,000. (*Note that just as Mr. X gained $1,000 because he was long in the oil market, someone else lost the same amount because they were short. One of the beauties of this arrangement is that accounts always balance!*)

But suppose that during that day the price fell to $38/b instead of increasing. His contract is marked to the market at $38/b, which means that Mr. X is a loser, and his broker owes the Clearing House $2,000 (which will be passed to a person holding a short position). This money is in Mr. X's margin account and can be transferred to the Clearing House, but now Mr. X's margin account is $500 below maintenance margin. A *margin call* then goes from the broker to Mr. X for $500, and if this money is not forthcoming during the day, the broker will usually close Mr. X's position in that contract by immediately selling it at the prevailing price.

Notice that the issue here is *maintenance* margin as compared to positive margin. What the broker wants to do is to make sure that if the oil price suddenly fell from $40/b to, e.g., $34/b, and Mr. X was in his favorite jazz club in Paris and unreachable, his firm would not have to pay for the total decline ($6,000) of this particular contract. Instead, the brokerage firm would have to account for $1,000 of this decline, following which they would curse themselves for not requiring more maintenance margin. It might happen though that they had an agreement with Mr. X to transfer excessive margin from other contracts the brokerage might be holding to this contract if a price decline caused margin to move below the maintenance amount (= $3,500).

Once we understand the above, and the convergence of 'paper' and physical (or 'actuals') prices, the explanation of risk avoidance (or price insurance) becomes a detail. Convergence comes about because in its absence there is arbitrage (which means the ability to realize a riskless profit). If the price on the physical market is greater than the price on the paper market, then holders of long contracts take delivery and immediately sell on the physical (or 'spot') market. This reduces the spot price. On the other hand, if the price on the physical (spot) market is less than the price on the futures market, then holders of short contracts buy spot and deliver oil. This raises the spot price.

Now for hedging (i.e., price insurance). Suppose that Mr. X must buy some oil in 30 days, and he is afraid that the price will escalate. He then buys a futures contract (i.e., goes long). If the price of physical oil goes up, and there is a convergence of the physical and paper prices, then what Mr. X loses on physical oil he gains on paper oil. As the reader can easily show, he has 'locked in' the price of oil. Suppose that Ms. Y is producing oil but is afraid that the price will fall. She might then sell futures contracts: if the price of physical oil fell, so would the price of paper oil, and what she loses in the physical market she would gain in the paper market. She too has locked in a price.

That brings us to a short mathematical exercise touching on the famous case of MGRM, a U.S. subsidiary of one of the largest firms in Germany, Metallgesellschaft, which lost about $1.3 billion in a flawed hedging project.

What MGRM did was to offer U.S. firms fixed-price *forward contracts* for (physical) oil products. These forward contracts had maturities of up to 10 years, which means that MGRM was accepting a considerable price risk; however, it was the theory of their management group, which included a former professor of economics, that all would be well if a hedging program was employed that involved 'rolling over' short-term contracts. This is sometimes called a 'stack hedge', or 'stack and roll', and the magic in the scheme was supposed to be injected by what is defined as *backwardation*, with current futures contracts selling at a premium to far-dated futures contracts.

Here I go to some algebra of the kind in my international finance book (2001), and which many readers might choose to ignore. At time $t = 0$, for example, a 3-month contract is *purchased* for a certain amount of oil or oil product, and I designate this operation $-F^3(0)$. This contract was then *sold* at, for example, $t = 2$, at which time there is still one month to go to its maturity. I designate this selling operation $+F^1(2)$. Moreover, at this time another 3-month contract was bought, which can be designated as $-F^3(2)$, which was sold in two months (at time $t = 4$), and so on. If the physical item was sold forward for $C(0, T)$ at $t = 0$ for delivery at time T, then total *undiscounted* profit V over the period T takes the following form:

$$V = -F^3(0) + [F^1(2) - F^3(2)] + [F^1(4) - F^3(4)]$$
$$+ \cdots + [F^1(T-1) - F^3(T-1)] + C(0, T). \qquad (A1)$$

This can immediately be written as:

$$V = -F^3(0) + \Sigma[F^1(2t) - F^3(2t)] + C(0, T). \qquad (A2)$$

To make this work, the summation is from $t = 1$ to $t = (T - 1)/2$.

The thing to notice here is that if the majority of expressions in the brackets are positive, then the profit (V) might also be positive. For a typical parenthesis to be positive, we must have $[F^1(\ ) - F^3(\ )] > 0$, which means that a near-term futures contract has a higher price than a distant contract. As noted above, this is backwardation (or inversion), and MGRM's hedging team thought that this was almost always true for oil. They were essentially correct; however, 'contango' (when the opposite happens) is always possible, and in the case of MGRM it happened, and kept on happening. At the present time it is often claimed that the long spell of contango that we have experienced over the past year or two is responsible for the enormous amount of oil that is not stored all over the world.

Something that is often overlooked in the populist crusade against futures markets is that speculation offers hedgers some extremely important insurance against unpleasant price arrangements. In an efficient market, speculators should expect to be rewarded for providing this service. If we consider only short hedgers (who are afraid of a price decline), then we must have $E(P_{t+n}|P_t) > F_{t+n}$. E is the expectation at time t of the price at time $t + n$ of the oil to be delivered at that time. F_{t+n} is the relevant futures contract.

Arguments have occasionally been presented that there is a declining liquidity in the futures market for crude oil, and so another arrangement should be introduced in order to derive some kind of reference price for crude.

I touched on this matter in my earlier work, and especially in my book on copper (1974). My conclusions about eliminating existing arrangements and introducing new arrangements — derived during the three years I spent in Geneva (Switzerland) doing research on commodity markets at the United Nations Commission on Trade and Development (UNCTAD) — basically had to do with the futility of fixing or planning to fix something that was not broken. This issue was discussed extensively on a later occasion at a workshop in Paris that was organized by the most brilliant analytical development economist of the 20th century, the late Professor Hollis Chenery.

A problem inevitably arises though whenever there is a very great deal of money in play, in that some very intelligent people might be tempted to manipulate prices. Can they do this? Frankly I don't know, although I suspect that for a commodity like oil, where the amount traded — both physically and in money terms — is enormous, it would take a cartel of investment banks and/or hedge funds to influence the price. I happen to believe that a cartel of this sort would be difficult or impossible to form in the industrial world, and if it was possible, concealing it from law enforcement would be difficult.

One more comment on this subject might be useful. In his article Professor Solow (1974) says that he wonders whether public policy can contribute to stability and efficiency where reserves, technology and demand in the fairly far future are concerned. This leads him to encourage "organized futures trading in natural resource products. To be useful, futures contracts would have to be much longer-term than is usual in the futures markets that now exist." Well, readers, that is more rain on our parade, because where oil is concerned, contracts tend to be illiquid after six months and sometimes less. Of course, some people do not know this, while others do and pretend otherwise.

Appendix 2. Deriving a Logistic Curve for an Exhaustible Resource

It does not take a background in mathematical economics to suggest that if we start out with an amount R^* for the supply of an exhaustible resource, the law of motion for the depletion of the resource might be $dR/dt = \lambda(R^* - R)$, where R is reserves extracted and dR/dt is the change in reserves *being* extracted per unit of time, as the deposit goes toward exhaustion. As for λ, on this occasion it is a constant (although conceptually it might be something else), and in the derivation below it has to do with the initial depreciation rate of reserves. dR/dt is production from the deposit, or q units per time period, but this can be overlooked at present because while the above equation might be interesting in an unsophisticated context, it does not lead to the 'bell' (or 'normal-like') curve associated with output from a typical oil deposit over a long time period. The beauty of the logistic equation is in justifying its use on the basis of economics, and not just because it generates a recognizable piece of geometry, but it is not so beautiful that readers uninterested in mathematics should spend time deciphering its charms.

As with the theory of economic growth, what has to be done is to work with *rates* rather than the derivative. Using the same symbols as above, the expression that will be employed as the first step in the derivation of the logistic equation is given in (A1'):

$$\lambda' = \frac{1}{R}\frac{dR}{dt} = \lambda\left(\frac{R^* - R}{R^*}\right). \tag{A1'}$$

Now for a key recognition. First of all, when $R \to 0$, then $\lambda' \to \lambda$, but when $R \to R^*$ — or when the amount of the deposit that has been extracted is approximately equal to the amount available (R^*) — then $\lambda' \to 0$, because there is no more to extract. Clearly, as R increases, then λ' — the rate at which exhaustion is taking place — decreases. At this point it might be useful to recognize that λ' is also analogous to a growth rate: it is the rate at which the deposit is being depleted, and it declines as reserves are exhausted. Observe that instead of using ($R^* - R$) on the right-hand side of (A1'), I used $[(R^* - R)/R^*]$. This was necessary in order for the units on both sides of the equation to match. As noted, λ is analogous to a growth rate, and *ceteris paribus* can be taken as constant, but $[(R^* - R)/R^*] = [1 - (R/R^*)]$ is a 'damping factor' that is directly related to the limit put on production by the availability of the resource: it reduces λ'.

A problem with (A1') is that it does not look like the logistic equations you see in your favorite mathematics book, and so it will have to be adjusted.

Doing this is uncomplicated and involves no more than treating (A1′) as a differential equation. Then we get:

$$\frac{R^* dR}{R(R^* - R)} = \lambda dt \Rightarrow \int \frac{R^* dR}{R(R^* - R)} = \lambda t + \text{constant}(c). \qquad (A2')$$

Dealing with this apparently complicated integral is straightforward:

$$\frac{R^* dR}{R(R^* - R)} = \int \left(\frac{1}{R} + \frac{1}{R^* - R} \right) dR$$

$$= \ln R - \ln(R^* - R) = -\ln \left(\frac{R^* - R}{R} \right).$$

Thus we get with the expression just above and (A2′):

$$\ln \frac{R^* - R}{R} = -\lambda t - c \quad \text{or} \quad \frac{R^* - R}{R} = e^{-\lambda t} e^{-c} = a e^{-\lambda t}. \qquad (A3')$$

Observe that the constant e^{-c} is now written as a. From this we get the logistic equation:

$$R = \frac{R^*}{1 + a e^{-\lambda t}}. \qquad (A4')$$

Since $dR/dt = q$, it is a simple matter to derive a Bell-like curve from (A4′), particularly when a few more manipulations yield the curve's inflection point, which is $R' = R^*/2$ and $t' = 2 \ln a$. This inflection point is the maximum for the Bell-like curve, and returning to the previous discussion, Hubbert's t' was 1970. A great deal of the subsequent discussion about applying Hubbert's work to the entire world had to do with the value of R^*. Persons who want to believe that there is plenty of oil in the crust of the earth claim that t' will not be soon, because as far as they are concerned, R^* is extremely large.

There is an important question remaining here, which is how can we get a curve with a plateau for a peak rather than a bell-like arrangement? Well, I do not know, and when M. King Hubbert was doing his work, I doubt whether he bothered with an issue of this sort, since regardless of what it looked like, it was crystal clear to him and his colleagues that in the great world of oil, what went up had to come down.

Key Concepts and Issues

Carter Doctrine	options
Cantarell and other large oil fields	oil products
critical R/q ratio	peak oil output: reality or fantasy
logistic curve	plateau production
M. King Hubbert	Russian oil
natural depletion	stock-flow model
oil refinery	swaps
OPEC	T. Boone Pickens

Questions for Discussion

1. What do you think about the work of M. King Hubbert?
2. Where do you think oil refineries should be constructed?
3. Suppose that there are excess inventories. Using a stock-flow model, explain how this matter is handled.
4. Do you think that there will be a global peaking of oil production?
5. The disadvantages of oil from tar sands, and heavy oil, are discussed in this book and elsewhere. What do you know about this topic? What about shale oil?
6. Do you think that it is easier or more difficult for OPEC countries to agree now than it was 10 or 15 years ago?
7. How worried should OPEC countries be about the introduction of synthetic motor fuel?
8. Explain in detail the use of the critical R/q ratio.
9. How do you take advantage of expected oil price rises with futures? How do you hedge against expected oil price falls with futures?
10. Do you think that the availability of futures markets has increased or decreased the volatility of the price of physical oil (i.e., *wet* barrels)?
11. In my lectures I have always argued that there is a clear relation between inventories and the (short-run) price of oil, but at the present time (March 2011) oil prices are increasing although there appear to be excessive 'private' inventories. What might be the cause of this?
12. The former director of the United States central bank — i.e., the Federal Reserve System — has stated that the war in Iraq that commenced

in 2003 was about oil. What is your opinion of this? If you have no opinion, turn to Google or some other 'search engine' and try to learn a few things.

Bibliography

Albouy, Michel (1986). "Nouveau Instruments Financiers et Gestion du Couple Rentabilité Risque sur le Marché du Petrole." Stencil, Grenoble Université.

Allen, R.G.D. (1960). *Mathematical Economics*. New York: St. Martin's Press.

Banks, Ferdinand E. (2008). "Economic theory and OPEC." *PetroMin* (forthcoming).

_____(2007). *The Political Economy of World Energy: An Introductory Textbook*. London and Singapore: World Scientific.

_____(2006). "Logic and the oil future." *Energy Sources, Part B*, 1(1): 97–114.

_____(2004). "A new world oil market." *Geopolitics of Energy* (December).

_____(2001). *Global Finance and Financial Markets*. Singapore: World Scientific.

_____(2000). *Energy Economics: A Modern Introduction*. Dordrecht and Boston: Kluwer Academic.

_____(1994). "Oil stocks and oil prices." *The OPEC Review*, 18(2): 173–184.

_____(1991). "Paper oil, real oil, and the price of oil." *Energy Policy* (July/August).

_____(1980). *The Political Economy of Oil*. Lexington and Toronto: D.C. Heath.

_____(1977). *Scarcity, Energy, and Economic Progress*. Lexington (Massachusetts) and Toronto: Lexington Books.

Cinti, Veronica (2011). "The development of clean energy in the Gulf: opportunity or risk?" *Geopolitics of Energy* (Vol. 33).

Dahl, Carol A. (2004). *International Energy Markets*. Tulsa, Oklahoma: Pennwell.

Deffeyes, Kenneth S. (2003). *Hubbert's Peak: The Impending World Oil Crisis*. Princeton: Princeton University Press.

Flower, Andrew (1978). "World oil production." *Scientific American*, 283(3): 41–49.

Fox, Justin (2002). "The truth about oil security." *Fortune* (September 16).

Helman, Christopher (2006). "Really, really cheap oil." *Forbes* (October).

Hotelling, Harold (1931). "The economics of exhaustible resources." *Journal of Political Economy*, 39(2): 137–175.

Lindahl, Björn (2007). "Opec ökar inte produktionen." *Svenska Dagbladet* (December).

Salameh, Mamdouh G. (2013). "Impact of US shale oil revolution on the global oil market, the price of oil and peak oil." Conference Paper; 13th European IAEE Conference (August).

———— (2007). "Peak oil: myth or reality." A paper presented at the World Bank in Washington, DC (April 10).

Solow, Robert M. (1974). "The economics of resources or the resources of economics." *American Economic Review*, 64(2): 1–14.

Stanislaw, Joseph and Daniel Yergin (1993). "Oil: Reopening the door." *Foreign Affairs* (September–October).

Chapter 5

An Introduction to Natural Gas Economics

In the short course on energy economics that I was asked to present to a research organ of the European Union, the topics I decided to cover were oil, nuclear energy, natural gas and electricity, and later I concluded that the main topic for my lectures was logically natural gas. Moreover, when I examined an issue of the *Energy Journal* dedicated to gas, and realized that of the 14 papers it contained, only a few would have received passing grades in a remedial course at Boston Public, I knew that it was time for me to examine and upgrade my previous work.

This long chapter has its origin in a very long lecture that I once presented at the ENI Corporate University in Milan (Italy), and which I later extended and revised for the natural gas portion of the course on oil and gas economics that I gave at the Asian Institute of Technology, Bangkok (Thailand). The goal in what follows is to point out certain basic concepts and conditions that characterize natural gas markets at the present time, and to do so in a manner that will enable readers to obtain some important information about these markets without having to deal with an overload of abstraction.

A considerable amount of attention is now being directed toward gas markets because of what appears to be a major technological breakthrough in the United States. In that country a large amount of 'shale gas' appears to have become available, and an important question often raised is how much of this resource can be found in other countries? According to a study published by the U.S. Energy Information Agency (EIA) in April 2011, global reserves of shale gas are at 6,620 trillion cubic feet, with China possessing 1,275 Tcf, the U.S. 862 Tcf, Argentina 774 Tcf, Mexico 681 Tcf, South Africa 485 Tcf, and Australia 396 Tcf. Where production is concerned, however, the only serious production is taking place in the U.S. and Canada, although there are ambitions in China to exploit their large reserves.

Where conventional gas is concerned, the latest *British Petroleum Statistical Review of World Energy* lists global reserves at 6,608 trillion cubic feet, with the top three countries being Russia with 1,580 Tcf, Iran with 1,045 Tcf, and Qatar with 894 Tcf. Saudi Arabia and Turkmenistan are next with about 283 Tcf each. As it happens, I have certain doubts about the estimate of shale gas resources, because qualitatively I do not regard them as being equivalent to conventional gas. Qualitatively they may — because of 'decline' and environmental issues — belong in a different category, although this could be a case in which quantity is a good surrogate for quality.

Despite the boisterous talk about shale gas revolutions, there are many observers who are not bursting with enthusiasm over this resource. Some very smart people claim that it is not a 'cure-all', and one of them, Murray Duffin in the forum *EnergyPulse*, pointed out that the average 'sweet spot' in natural gas properties — the area in which the concentration of the resource is relatively large — amounts to no more than a modest 30% of the total area. This may or may not be true, but Duffin also noted that the physical depreciation/decline taking place in those deposits tends to be very large, which is something one hears all the time. Moreover, it is possible that only the presence of high-value liquids in shale gas deposits assures that those deposits can be profitably exploited, especially with the low gas prices now prevailing. On this score, much effort has gone into boosting the selling price for shale gas properties to a level that some observers feel is manifestly incorrect, given the present and predicted price of natural gas.

Environmental doubts have kept France from developing (or attempting to develop) what may be the largest shale deposits in Europe, with the possible exception of Poland. At the same time, the French firm Total and the Chinese firm Sinopec have each invested about 2 billion dollars in various gas enterprises in the U.S. If I had to make a guess, I would say that these investments have a lot to do with learning how to produce, sell and advertise shale gas in an optimal manner, because it is very doubtful if the French government will restrict the production of shale gas over the indefinite future. It has also been suggested that the properties acquired by Total are rich in natural gas fluids. In any event, China is the world's most enthusiastic searcher for energy resources, and does not lack the billions of dollars needed to finance what it regards as attractive purchases.

I am willing to hope that shale gas will live up to expectations in the energy future of North America and elsewhere, but at the same time I feel it necessary to ignore some of the rosier estimates that one encounters almost daily about the availability of various non-conventional energy resources and alternatives, such as shale gas, wind and solar energy. However, I recognize that because turbines burning natural gas can rapidly vary their output, and plentiful gas may now be available, they are ideal for complementing wind farms and solar panels that might be featured in various energy undertakings, but whose output of electricity tends to be irregular because of occasional shortages of wind and sun. Fast-start gas turbines have a prominent role to play in districts where gusts of wind not only lose their force suddenly, but occasionally become so strong that wind turbines must be shut down for safety. Of course, when dealing with this subject, attention should be paid to the additional expenses associated with installing these gas turbines, as well as investing in infrastructure to move gas to relevant facilities.

1. Introduction

What I was particularly concerned with when planning my lectures in Bangkok and Spain was an unambiguous description of natural gas markets and their mechanics. This is where courses in energy economics often disappoint their participants, because many teachers of these courses are unaware of some of the most useful elements of the subject. As a result, even interested students could be in danger of experiencing unpleasant surprises in routine discussions with other students, future or potential employers, and genuine or bogus energy professionals. Natural gas is a subject where we do not want this to happen, because for various reasons that resource may be more important than ever, particularly for countries that have a large and perhaps growing natural gas deficit, but whose gas transportation and distribution network is intact. The UK obviously fits this description.

Before commencing a systematic examination of this topic, something should be made clear. According to Robert Bryce, managing editor of the *Energy Tribune Magazine*, coal dominated the energy picture in the 19th century, oil dominated in the 20th century, and — in his opinion — natural gas will be the dominant "fuel" of the 21st century. A statement like this gives the impression that gas is capable of outshining coal and nuclear as, e.g., a source of base load electricity throughout the entire 21st century, which I doubt. As pointed out by Len Gould in the forum *EnergyPulse* (www.energypulse.net), it is now possible to construct (in China) a 1,000 Megawatt (=1,000 MW) nuclear facility in less than 5 years — from 'ground break' to grid power — and this reactor should have a 'life' of at least 60 years, while the known supply of its fuel may be much larger.

Anne Lauvergeon, the former director of Areva, was also aware of this situation, and regarded it as "worrying" — by which she meant worrying for the management and shareholders of her firm. Given that nuclear reactors do not release any carbon, and considering the likely supply of uranium and thorium, it is difficult to believe that nuclear-based power will be dominated by a fossil fuel (e.g., natural gas) whose global availability was once questioned by people like Alan Greenspan, the long-serving director of the U.S. Central Bank (i.e., the Federal Reserve System). On considering shale gas, Professor David Victor of the University of California ostensibly stated that "We don't know if it will be truly awesome or only theoretical in its impact."

As bad luck unfortunately has it, some of us know enough about energy issues to become unreservedly suspicious. Not only shale gas, but coal deserves some scrutiny. Coal is considered a near-toxic resource by a number of politicians and environmentalists, and so daily we hear about the strenuous efforts that will be made to replace it with renewables and/or natural gas (since natural gas has about 50% of the carbon dioxide (CO_2) emissions of coal). I have been informed about some of these goals for decades, and it is clear that in every part of the world there are politicians and civil servants who are serious about putting some sort of 'cap' on carbon emissions. Even so, I am afraid, however, that a large fraction of these intentions might best be described as hypocrisy or bunkum. There is too much energy in coal for it to be dismissed.

For example, decision makers in China and India have not provided any proof that they are going to augment their high-minded rhetoric with tangible efforts to reduce their huge dependency on coal, and the scientific elite in Europe and North America are not making the efforts that they should make to provide us with the information we need to choose — or merely to think about — what might be called the optimal energy future, given the resources that are or will become available. Instead, many of them are running scared that they will encounter persons like Jeremy Leggett, an environmental celebrity, who mistakenly thought that he had something useful to tell me about energy issues after hearing my paper "The Real Nuclear Deal" at the Singapore Energy Week.

2. Energy Units and Heat Equivalents

In the most elementary yet most comprehensive sense, energy can be defined as anything that makes it possible to do work — i.e., directly or indirectly bring about movement against resistance. Energy takes many forms, and one of its most interesting characteristics is that all aspects of motion, all physical processes, involve to one degree or another the conversion of energy from one state to another. For example, the chemical energy that is found in natural gas can be converted to active heat, which in combination with water will generate steam in a boiler. This steam can then be used to drive a turbine which, in turn, rotates the shaft of an electric generator, and thus produces electricity. Note that the rotating shaft implies the ability to do physical work.

All this is perfectly straightforward, but unfortunately heat cannot be converted into work without loss, and that loss takes the form of heat transposing to a temperature closer to that of the surroundings, and away from that of the heat source that made the work possible. Once heat has descended to the *ambient* (i.e., surrounding) temperature, it is no longer available to do useful work. What we are dealing with here is a highly abstract concept from thermodynamics known as *entropy*, sometimes called "time's arrow", which signifies energy going down the thermal hill and being diffused into space — lost forever we might say — which implies that the universe itself is in danger of 'running down' (in, e.g., a few million years or so).

John von Neumann has sometimes been called "the best brain of the 20th century", and one of his advantages was to have virtually every physical constant known to mankind stored in his brain and ready for instant recall. That sort of achievement is not normally required to convince friends and neighbors of your acumen, but it is always useful to have a few numbers on hand when studying the present topic. First of all, I suggest knowing that one metric ton (= 1 *tonne* = 1 t) equals 2,205 pounds, and that 2.2 pounds = 1 kilogram (=1,000 grams). Similarly, 1 inch = 2.54

centimeters, 12 inches = 1 foot, 100 centimeters = 1 meter, and thus 1 meter is approximately equal to 3.28 feet = 39.37 inches. One cubic meter ($=1\,m^3$) is therefore equal to $3.28^3 = 35.3\,ft^3$. In everyday life the usual ton is the *short ton*, or simply *ton*, which equals 2,000 pounds. Thus, 1 t = 1.103 tons.

When dealing with energy we are often interested in heat equivalents, and when the topic is gas, the most favored unit is the British Thermal Unit (or Btu), which is the amount of heat required to increase the temperature of one pound of water by 1 degree Fahrenheit. (One pound of water is approximately equal to one *pint*.) Here it might also be useful to remind readers that with F Fahrenheit, and C Centigrade (or Celsius), we convert C to F with the equation F = (9/5)C + 32. In scientific work, and in certain countries, *joules* might be preferred to the Btu as a unit of heat energy; however, since the price of natural gas is commonly given in dollars per million Btu (=$/MBtu), there is no reason in energy economics to spend a great deal of time pondering the utility of the joule or for that matter the calorie or kilocalorie (=1,000 calories = 3.968 Btu), which are other heat units.

That brings us to a key observation, which is that 1,000 cubic feet (=1,000 cf = 1,000 ft^3) of natural gas has an approximate energy content of 1,000,000 Btu. (To be exact, 1 ft^3 of natural gas has an average heating value of 1,035 Btu, but 1,000 Btu is almost always used.) Moreover, the *average* energy content of natural gas varies from a low of 845 British Thermal Units per cubic foot (845 Btu/cf = 845 Btu/ft^3) in Holland to 1,300 Btu/ft^3 in Ecuador. Now let us make a calculation involving gas and crude oil, where one barrel (=1 b) of oil has an average energy content of 5,686,470 Btu ($\approx$5.686 MBtu). If we assume the price of oil to be $100/b, and the price of natural gas to be $5 per million Btu (= $5/MBtu $\equiv$ $5/mBtu $\equiv$ $5/MMBtu), then it is easy to compare Btu prices of these two energy resources. The cost of a *million* Btu of oil is thus 100/5.686 = $17.587/MBtu (as compared to $5/MBtu for natural gas). There is a large difference between these two prices, and it has been suggested that this difference will result in the (dollar) price of oil falling by a great deal. I do not share this belief.

Persons who find this approach interesting or important can search the academic literature for information about the 'burner tip parity rule', which has to do with an inevitable convergence of oil and gas prices (in Btu terms), and has virtually the same authority as Albert Einstein's equivalence theorem. The opinion here is that a substantial decline in the oil price is out of the question as long as OPEC retains its present unity. In these circumstances the 'quasi-invisible' hand of OPEC determines the oil price, while for the time being the natural gas price is likely to follow a conventional supply-demand relationship, but with a stochastic component.

It is possible that the economics of natural gas markets will eventually be transformed by technological advances that have made the exploitation of large amounts of shale gas possible, although the extent and character of this transformation is not certain at the present time. More importantly, in the short run, the availability of large amounts of shale gas will tend to depress the price of natural gas, while OPEC will act to keep the price of oil on an upward trend (although hopefully retaining full awareness of the importance of the oil price for the global macroeconomy). It is also pertinent to recognize that oil is generally rated as a more 'efficient' resource than gas, mostly because it is more economical to deliver and to process into oil products (e.g., motor fuel) or ultimately into petrochemicals (e.g., fertilizers).

Now let us consider an example that involves two light bulbs. One of these produces a great deal of illumination and has a power rating of 500 watts, while the other is considerably weaker and has a rating of only 50 watts. If the heat energy in coal (or oil or natural gas) is totally and perfectly transformed into electrical energy (i.e., with 100 percent efficiency), then 3,412 Btu are required to generate a kilowatt-hour (kWh) of electrical energy (where the kWh is the unit in which *electrical energy* — as distinguished from *power* (in, e.g., watts or kilowatts) — is measured). Note that a kilowatt is 1,000 watts, a megawatt is 1,000,000 watts or a thousand kilowatts, a gigawatt is a billion watts, and a terawatt is a trillion watts.

The power rating of the bulbs — 500 watts and 50 watts — informs us of the rate at which the energy potential of the coal is consumed, and so if the 27,700,000 Btu in an average tonne of coal is transformed into electricity in a *perfect* system, then it could provide exactly $27,700,000/3,412 = 8,118$ kilowatt-hours of electrical energy. In other words, in a *perfect* system, the stronger of the two bulbs — which consumes power at the rate of 500 watts (=0.5 kW) — could function for $8,118/0.5 = 16,236$ hours. The other bulb would require $8,118/0.05 = 162,360$ hours to consume a tonne of coal.

In reality, the efficiency with which fossil fuel can be converted to electrical energy is well under 100 percent. An efficiency of about 32 percent seems typical for much of the industrial world, and so, on average, it would require $3,412/0.32 = 10,662$ Btu to obtain a kWh of electrical energy. A number of this type is conventionally referred to as a *heat rate*, and is sometimes defined as the energy content of a fuel! Using the above numbers, this can be put another way: $1 \text{ kWh(e)} = 3.12 \text{ kWh}$ (fossil fuel). Some time ago the UN and OECD calculated that $1 \text{ kWh(e)} = 2.6 \text{ kWh}$ (oil). Naturally, we are dealing in averages here.

As simple as all this seems, many readers may feel that something is missing. While electrical power is defined as a 'rate', it is not always explicitly associated with a time dimension; for instance, the 'rating' of a power station is likely to be in megawatts. However, in the example above with the bulbs, we saw that a large bulb exhausted the energy potential of a tonne of coal more rapidly than a small bulb, which trenchantly suggests that the dimension for power is energy per unit of time. Furthermore, a watt is one joule per second (which is immediately recognized as a rate) or 3,600 joules per hour; and since 1,055 joules is 1 Btu, one watt is 3.412 Btu/hour (which is more easily recognized as a (time) 'rate' by those of us accustomed to working with the Btu). Observe that $1 \text{ kW} = 1,000 \text{ J/second}$, where J signifies joules.

Finally, there is the very small unit called a calorie, and here we have 1,000 calories equal one kilocalorie (kcal), and $1 \text{ kcal} = 3.968 \text{ Btu}$. Where equivalencies of this nature are concerned, we are talking about the outcome of perfect experiments in a perfect laboratory. This kind of perfection is not easy to achieve in the real world, however, which is why the term "heat rate" had to be introduced into these proceedings.

Let us conclude this discussion with two simple examples. For the first, the fuel in the tank of a vehicle may be depleted of 10 million Btu ($=10 \text{ MBtu}$) during an hour of driving. A portion of this energy — for example, 3.5 MBtu — might be transformed into work in the form of rotating a shaft that turns the wheels of a vehicle. The rest of the energy is discharged as heat into the air (or perhaps into cooling water). Fuel efficiency in this case is only 35%, which is the percentage of the fuel that is actually transformed into useful work. Just as unfortunate, as the temperature of the 'non-useful' work falls, we are losing forever its availability to do work; in other words, its unavailability is increasing. As alluded to earlier, this is what *entropy* is all about: the permanent degradation of energy.

To continue, once we have the heat rate, obtaining an estimate of the fuel cost is elementary. For instance, if we have a natural gas turbine with a heat rate of, e.g., 10,000 Btu per kilowatt-hour ($=$10,000 Btu/kWh), and in addition a fuel (i.e., natural gas) cost that at the present time is about \$3/MBtu, the *fuel* cost of the electric output is clearly:

$$\left(\frac{\$3.00}{1,000,000\,\text{Btu}} \right) \left(\frac{10,000\,\text{Btu}}{1\,\text{kWh}} \right) = \frac{\$0.03}{\text{kWh}}.$$

In the U.S. this would be called 3 cents per kilowatt-hour. This is not a very sophisticated estimate of the fuel cost of electricity generated with a natural gas turbine, but it can be regarded as a satisfactory first step for the present. Later in the exposition a calculation will be presented dealing with the capital cost.

3. Some General Remarks about Natural Gas

Before turning to more abstract materials, a few general observations about energy are perhaps in order. Despite the importance of natural gas and various other energy topics, the big issue at the present time is still the availability of oil, and its price. Shale gas may change a great deal on the energy front, and not only in the U.S., but perhaps not in the short run. I think it best to wait for a final judgment. In plain language, some of the things I hear about that resource sound like hype rather than geological and engineering truths.

It has also been suggested that global natural gas exports (totalling perhaps 79 billion cubic meters in 2011) could be negatively impacted if the recent disruptions or unrest in North Africa and the Middle East were to intensify. Some observers are concerned about Algeria, which produces almost 3 percent of the global output of natural gas, with 60 percent of this output exported via pipelines under the Mediterranean Sea to Italy and Spain. Algeria also has a substantial export capacity in liquefied natural gas.

The serious problem at the nuclear reactors in northern Japan has raised new doubts about the safety of nuclear energy, which is somewhat amazing, because the principal cause of the damage in Japan was the horrendous tsunami. Put another way, there was an inexplicable display of carelessness in locating the nuclear reactors at the water's edge in Fukushima, although as far as I am concerned, a much larger amount of economic damage could result if assorted influential observers in Japan and elsewhere accept the position of Nick Clegg, the Deputy Prime Minister in the present government of Great Britain. He believes that nuclear energy will be too expensive in the future, which on the basis of conventional economic theory is completely wrong. Another illogical departure from an economic point of view is the comprehensive nuclear retreat in Germany that has been proposed by the government of Angela Merkel, although it may result in her and her foot soldiers temporarily enjoying a political profit.

Oil has lost a great deal of its charm where the generation of electricity is concerned, while natural gas has been put in a relatively favorable light because whenever coal is mentioned, reference to global warming tends to follow. It has also been claimed that the price of natural gas is now both lower and less volatile, and that gas supplies in the U.S. are less questionable. According to Lawrence J. Goldstein of the Energy Policy Research Foundation, "gas comes out a winner. It's a technical knockout." Unfortunately he chose the wrong sport to use in his praise of natural gas, because as I make it my business to suggest, what we have in the energy field is a team effort, which as far as I am concerned will eventually emphasize nuclear energy and various renewables and alternatives.

Let us put it as follows. At the present time, natural gas is the resource attracting the largest amount of attention, largely because of the technological upgrading of shale gas. Instead of trying to work out the magnitude of this upgrading though, I intend to spend what free time I have trying to work out the likelihood of a GAS-PEC, or producers' organization for gas. This ties in with shale gas, because if it is as plentiful as often pictured, it could seriously press down the price of all gas as increasing amounts of it come to market.

That immediately leads some of us to ask what countries would be interested in a GAS-PEC. First and foremost, the countries of the Middle East, who seem to have mastered the logic of cartel management, and in addition have about 40 percent of the world's proven gas reserves. Iran deserves particular attention, since although they possess the world's second largest gas reserves, and a great deal of oil, they have not been able to use these resources to accelerate their macroeconomic growth. In fact, the three largest natural gas producers — Russia, Iran and Qatar — possess a larger share of gas reserves than the three largest producers of crude oil.

There is also the matter of continuing to index the price of natural gas to the price of oil. In an abstract or classroom sense, this activity means a great deal less today than a few years ago, although it is likely that most suppliers of natural gas will be able to force some kind of indexing relationship into the contracts they sign with their clients.

It also needs to be appreciated that 40 percent of the world's gas cannot give the Middle East the 'juice' they obtain from possession of 60 percent of the world's proven oil reserves. However, it is not unthinkable that the world's largest holder of gas reserves, Russia, might someday be interested in becoming part of a cartel, and if not becoming a part, supporting a cartel by moderating its output of gas. Unless the global economy collapses, the price of natural gas in Europe and Asia should be satisfactory for sellers into those markets. But regardless of what happens, formal or informal membership in an efficient gas cartel might eventually prove extremely attractive — or irresistible — to many gas producers.

Such a cartel would be easier to maintain than many researchers think. Readers should remember that one of the most celebrated economists of the 20th century, Milton Friedman, took what was virtually a sacred oath that OPEC would collapse, and in the March 15th (1974) edition of the widely read American magazine *Newsweek*, he predicted that the price of crude oil would decline to about $5 a barrel. For this to have happened would have required OPEC members to believe that the kind of economics that Mr. Friedman taught at the University of Chicago made more sense than the kind of mathematics I was taught in my engineering school on the South Side of Chicago, and as it happens, this was not the case. A new sophistication has moved across much of the Middle East in the crucial matter of oil and economic growth, and perhaps it also applies to gas: the basic strategy is to produce less in order to earn more. "More" in this context specifically means income, *money*, while producing (i.e., supplying) less has to do with obtaining that income over a longer period of time. Oil products and petrochemicals also have a big part to play in this drama.

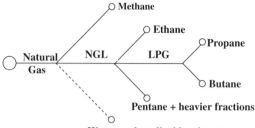

	Liquefied at	Definitions
Butane (normal)	– 0.5°C	**NGL: Natural gas liquids**
Butane (iso)	– 12.0°C	**LPG: Liquefied petroleum gas**
Propane	– 42.0°C	**LNG: Liquefied natural gas**
Ethane	– 88.0°C	**SNG: Synthetic (substitute) natural gas**
Methane	–161.0°C	

Figure 1.   Natural Gas Components

Notes: Natural gas = methane + NGL + (water, nitrogen, CO_2).
NGL = ethane + LPG + (pentane and heavier fractions).
LPG = propane + butane + mixtures of propane and butane.

Before looking at Figure 1, which in a certain sense is the beginning of the more analytical part of this contribution, readers should make sure that they have the following definitions at their disposal. Renewable energy sources include solar radiation, water power, wind power, tidal energy, petrochemical energy stored in plants and animals (food, wood, vegetable refuse), animal (and human) power, and geothermal energy in the form of a heat flow.

Having mentioned these items, I see no reason to conceal my belief that much later in the present century, probably after 2050, nuclear energy and renewables will have to carry the largest portion of the energy load; but precisely when, or exactly or approximately how this will work, is a complete mystery to this teacher of energy economics. I am also completely ignorant in the matter of fusion and a few of the items mentioned in the previous paragraph, and have made plans to stay that way. There is far too much hype out there about renewable energy for me to sacrifice valuable time trying to find out about energy sources that might not have a future.

As illustrated in Figure 1, natural gas from a well consists of methane (on average 85%), heavier hydrocarbons collectively known as natural gas liquids (and composed of ethane, propane, butane, pentane and some heavier fractions), water, carbon dioxide (CO_2), nitrogen, and some other hydrocarbons. (Note here that gas is *not* free of carbon dioxide. It merely has less than oil and coal.) Before dry natural gas can be distributed to consumers, some undesirable components must be removed and, by decreasing the share of heavier hydrocarbons, a uniform quality attained.

The last-mentioned operation takes place either at or near the gas well itself or in special installations farther away. It is at this point that the natural gas liquids (NGL) can be separated out. (NGL should not be confused with liquefied natural gas (LNG), which mostly consists of methane and ethane.) The most important constituents of NGL are butane and propane, which I have heard called 'wet gases', and in liquid form these are called liquefied petroleum gas (LPG). In many countries, LPG is sold under the name "gasol" or bottled gas, and at one time in Australia the government wanted a greatly increased use of LPG, although I never heard anyone say how this could be brought about in that energy-rich land unless it had the 'right (i.e., low) price'.

As is made clear in most of my work, the relation between the very important (levelized or annualized) capital cost A — i.e., *annual* payment at the end of each year for T years for an asset costing I (which is the *investment cost*) — can be calculated using the *annuity* formula. That simple expression is:

$$A = \left[\frac{r(1+r)^T}{(1+r)^T - 1} \right] I. \tag{1}$$

This equation is derived elsewhere in this book, but it is just as well to mention how we look at it from the standpoint of economic theory, beginning with a fundamental (though theoretical) neo-classical economic concept: the capital cost of an investment is the uniform return per period that an asset must earn in order to achieve a net present value of zero. In other words, the asset price is the present value of future net yields (i.e., revenues minus costs). The usual notation in this derivation is changed somewhat in order to correspond to standard usage. Taking I as the asset price (i.e., the investment cost or present value), A as the levelized capital cost per period, and r as the market discount rate, we write for T periods:

$$I = \int_0^T Pe^{-rt}dt = \frac{A}{r}\left(1 - \frac{1}{e^{rT}}\right). \tag{2}$$

It takes a little manipulation to obtain $A = re^{rT}I/(e^{rT}-1)$, and remembering that we can approximate e^{rT} by $(1 + r)^T$ for small values of r, we get Equation (1), though ordinarily in my books and lectures, instead of I (for the value of capital investment), readers will find PV (for present value). The discount rate here is usually the market interest rate, which assumes no risk/uncertainty on the part of lenders and borrowers. This is not the kind of stipulation to be taken seriously outside a seminar room. And note: A intrinsically has to do with capital and *not* operating costs. The bank that financed my house would hardly agree to finance any *ex-ante* (before the fact) mistakes I might make during my occupancy!

The reason this relationship is repeated so many times in my work is because of its extreme importance. For instance, if an asset costs $1,000 ($=I$), $T = 2$ (years), and we have a discount (e.g., interest) rate of $r = 10\%$, then a simple calculation gives annual payments A. From Equation (1) these are $\{[r(1 + r)^T]/[(1 + r)^T - 1]\}I = \{[0.10(1 + 0.10)^2]/[(1 + 0.10)^2 - 1]\}1,000 = \$576/\text{y}$. Moreover, for theoretical purposes, there is no difference between paying $1,000 for the asset now, or $1,000(1 + 0.1)^2 = \$1,210$ after two years, or $576 at the end of each of the next two years, taking care to remember that theory is one thing and practice another, because payment schedules depend on more than theoretical considerations. T is sometimes called the amortization period, and readers can easily show — formally or otherwise — that an increase in T will bring about a decrease in A.

Now for what might be a useful simplification. When T is very large, then A approaches rI, and this expression can be used as an approximation for A — though in some cases not a very good approximation. Suppose that we have a very large natural gas facility of the size and cost that TEPCO (The Tokyo Electric Power Company) installed in Tokyo many years ago, and for this exercise assume that the investment cost was 10 billion dollars ($=\$10G$) and the amortization period was 30 years. With an interest (i.e., *discount*) rate of 6%, and using the above approximation, the yearly (i.e., annual) cost is $0.06 \times 10,000,000,000 = \$600,000,000$ (=\$600,000,000/y). Now let us make a calculation of the annual cost using Equation (1), with an amortization period of 40 years and the same discount rate. The value now is $\$660,000,000/y = 0.66 \times 10^9$ dollars/year.

There is a substantial difference between the exact and the approximate values in the above example; however, I am aware of reputable teachers of engineering making the same mistake in this regard. (A perhaps more serious shortcoming is my inability to obtain the exact cost of the installation, and my uncertainty about the discount rate. 6% though might be close to the 'prime' rate in Japan at that time, plus one or two percent to account for uncertainty.) More important though, is that if we have an estimate of the annual payment (= levelized capital cost) for the investment, we can obtain the capital/investment component of the cost of generated electricity.

The TEPCO facility being discussed was composed of 12 gas-based generators, inclusive of the (steam) boilers available for each, and the rated capacity/power was 2,000 Megawatts. With 8,760 hours in a year, and a capacity factor of 0.90 percent, the annual energy output is 2,000 (MW) $\times$ 8,760 (hours) $\times$ 0.90 $= 15 \times 10^9$ kilowatt-hours. (The capacity factor is explained below!)

The capital component of the hourly cost of electricity can now be determined. As with most commodities, this would be expenditure divided by quantity, or $(0.66 \times 10^9)/(15 \times 10^9) = \$0.044/\text{kWh}$. There may be some objections to this number, but a year or so ago I calculated a kWh cost for nuclear-based electricity in the U.S., and since it came to $0.09/kWh, the estimate above that I obtained for a Japanese project carried out many years ago should not be classified as outrageous. In any event, the main issue here is knowing the technique for calculating the capital/investment component of the cost per kWh of electricity. Moreover, a very similar calculation would be used for nuclear energy.

A comment might be useful here. Where the amortization period (=40 years) is concerned, until a few years ago 40 years was a quite common amortization period for nuclear reactors in the United States. I can also mention that the amortization period for my house in Sweden is 40 years. This is the period that I am given to repay the loan that I received from a bank in order to purchase my humble home, and in the course of repaying the loan I also pay interest charges. After 40 years my expenses will be the usual upkeep and taxes, and perhaps a few other things, but the largest financial burden will be terminated.

As for the *capacity factor* mentioned above, gas-based turbine generators, nuclear reactors, etc., occasionally break down and must be repaired, or for that matter there are routinely scheduled periods for maintenance, during which no electricity is being delivered. Obviously, in the TEPCO installation mentioned above, the 12 generators and collateral equipment would not break down or be serviced all at once, and so a capacity factor of 0.90 is perhaps too low. Having 12 distinct pieces of equipment delivering 2,000 MW perhaps makes more economic sense than a single generator servicing the same load. Perhaps?

4. Natural Gas Reserves and Locations

In many respects, natural gas is an ideal fuel. Its environmental qualities (in terms of its 'comparatively' modest emissions of greenhouse gases) have occasionally caused it to be labelled a *premium* fossil fuel (as compared to oil and coal), and there is a great deal of it in the crust of the earth — though it is uncertain if there is as much as many observers now choose to believe.

As pointed out by Ken Chew (2003), the global amount of gas resources discovered annually peaked in the beginning of the 1970s, and the *number* of discoveries peaked early in the 1980s. (This lends weight to a contention by Julian Darley (2004) that gas production could 'plateau' before 2025, but the developing situation with shale gas very likely nullifies these estimates.) As already noted, natural gas is found in very many countries, with Russia having the world's largest reserves of conventional gas and in addition being the biggest producer of that commodity. The United States is the largest consumer of natural gas, and because of shale gas, its total production may be as much as Russia's. Europe (excluding Russia and Eurasia) consumes roughly 20% of global output.

A considerable amount of gas has attained the status of *stranded gas*, which means gas that has been discovered but is not economically viable for extraction for physical or economic reasons. For instance, it is too far from main pipeline routes, or from localities where gas consumption is high, to make its exploitation attractive. Perhaps the best example here is the natural gas in Alaska. Sarah Palin, the vice-presidential candidate in 2008, often spoke of liberating this stranded resource with a pipeline running from Alaska to the vicinity of Chicago, which several representatives of 'Big Gas' called politically wise but economically dumb. However, I suspect that before talk surfaced about the availability of large amounts of shale gas in the 'Lower 48', that pipeline might have finally been constructed.

A belief has started to circulate that a large-scale exploitation of shale gas deposits in many parts of the world outside the U.S. may soon be possible, although I am surprised by the inactivity in some countries that are ostensibly rich in this resource — particularly China. If shale gas lives up to its promise, though, then some portions of the discussion of natural gas being presented in this chapter will have to be modified. However, until more evidence is available, it may not be wise to expect that shale gas will drastically alter the global gas picture.

Since 1980 natural gas has exhibited the fastest consumption growth of all fossil fuels — averaging somewhere between 2% per year (=2%/y) and 3%/y — and occasionally there is talk of a more rapid growth for some regions. Gas is mainly used for heating and cooking, power generation, providing energy for industrial processes, and as a *feedstock* for the manufacture of petrochemical products such as fertilizers, chemicals and plastics. Where, e.g., power generation is concerned, gas functions as a *primary* energy input, while the electricity being generated is a *secondary* energy resource. Primary energy is energy obtained from the *direct* burning of coal, gas, and oil, as well as electricity having a hydro or nuclear origin. To these it is possible to add wood and waste, geothermal energy, solar thermal energy, wind, etc., although they provide only a small fraction of electricity in modern societies and are classified as renewable resources.

Here again you should also be aware of the difference between *energy* (which might be measured in kilowatt-hours) and *power* (which would be, e.g., in kilowatts). For instance, the size of a bulb often indicates its power, and this size multiplied by the number of hours it 'burns' is energy in, e.g., watt-hours (Wh) or, more commonly, kilowatt-hours (kWh). This observation is highly relevant later in the discussion of compressors for gas pipelines, only there the most relevant units are *horsepower* (for power) and *horsepower-hours* (for energy).

In considering fertilizers in the U.S., the escalating price of natural gas — from $2 per thousand Btu in 1999 to a *spike* of $13 per thousand Btu in

July 2008 — was the worst possible news for the fertilizer industry, and resulted in the closing or downsizing of considerable fertilizer production capacity. This outcome was predicted years ago by a Nobel Prize winner in chemistry, Sir Harry Kroto, who also implied that food prices would soar as a result of price increases for oil and gas. That natural gas spike soon declined, but it provided an unwelcome wake-up call for many inveterate gas enthusiasts, especially since it accompanied a record spike in crude oil prices (to \$147/barrel), and it had been preceded by two other gas spikes (or bubbles) earlier in the century.

I can also mention that the subsequent decline in natural gas prices following the last spike was even more dramatic than the rise, and many gas producers closed down operations. It is interesting to note that the decline in oil prices was just as dramatic as that for natural gas, only in the case of oil, OPEC demonstrated their command of the market, and fairly soon the oil price was well above \$75 a barrel again. Moreover, the gas-on-gas arbitrage that certain investigators believe will happen, and which involves an expanding transport over long distances of gas from regions of low prices to the heavy consumers of gas in, e.g., Asia, has given no indication as yet of taking place in reality as well as theory.

Using the *British Petroleum Review of Energy* to ascertain gas reserves, share of total production, reserve-production (R/q) ratios, and consumption — all slightly modified in an effort to bring them up-to-date — seven vectors are presented below. In each we have (Reserves in Tcf, share of total in percentage, R/q ratio in years, and consumption in million tonnes of oil-equivalent). At face value 'aggregate' R/q ratios are interesting, but they are often misinterpreted, although, as will be indicated later in this chapter, the R/q ratio for individual gas fields can have a great deal of significance. (The world (or global) R/q ratio for gas is at least 60 years; however, this has little meaning as compared to the (unknown) date on which the output of gas will peak.) As an example of the conversion from cubic feet to cubic meters, it can be noted that for the world vector, using the 2010 BP statistics, we obtain 6,263 Tcf $=177$ Tcm.

One large gas producer in the Middle East announced last year his intention to at least double the time horizon over which sizable quantities of (conventional) natural gas are produced in his country. This intention to restrain from producing very large amounts of natural gas will almost certainly increase in the future as the oil reserves of some large gas producers decrease, and they think more about maximizing the price at which they sell their gas. (Incidentally, about 62% of both oil and gas are found in just five countries.)

Now for the aforementioned vectors, which are slightly out of date, and have not been adjusted for the large exploitation of shale gas (which has just begun):

- North America: (281, 4.5, 10.3; 729)
- South and Central America: (273, 7.73, 4.4; 121)
- Europe and Asia: (2,097, 33.5, 55.2; 1,040)
- Middle East: (2,585, 41.3, ? ; 269)
- Africa: (515, 8.2, 76.6; 75)
- Asia-Pacific: (511, 8.2, 36.9; 403)
- World: (6,388, 100, 60.3; 752)

So much for discovered reserves. What about undiscovered reserves? A year or so ago this author would have placed these at about equal to discovered reserves, which would have meant that ultimate resource recovery (URR) — or past, present and future — is about 12,000 Tcf. It must be admitted that this estimate would not satisfy all investigators. For example, Professor William Fisher of the University of Texas puts URR at 25,000 Tcf, to which he adds exotics such as hydrates, and I will not insist that he is incorrect.

As things have turned out, shale trapped in huge shale beds, largely exploited with horizontal rather than vertical drilling, together with a large input of liquids (especially water), may drastically alter the U.S. energy picture. In contrast to estimates cited earlier in this chapter, one consultancy has claimed that there could be as much as 840 Tcf of readily 'accessible' reserves in various U.S. shale deposits. The U.S. Department of Energy (USDOE), however, estimates that the correct figure is 125 Tcf, and it may be true that there are large shale deposits in some European countries. (Russia and Norway are frequently mentioned, but also the 'Baltic countries' and Poland.) Moreover, recently the Energy Information Agency (EIA) of the USDOE greatly downgraded its estimate of recoverable reserves in the Marcellus Shale region.

Unfortunately, the circumstances here are not as unequivocal as they may seem to shale gas enthusiasts. Many researchers are aware that much talk about 'shale oil' was wrong because it ignored the enormous amount of water required for an economical exploitation, but now it appears that there are a number of disturbing environmental issues associated with exploiting shale gas, some of which also involve water and chemicals, and these issues appear more acute today than ever. Furthermore, the reported depreciation of some shale deposits is so high as to suggest the need for extensive investigation rather than immediate acceptance. Employing economic logic, some of the listed depreciation rates translate into soaring economic costs. Just as serious, widely accepted but false expectations could lead to a rash of sub-optimal investments. This may already have happened in some places with both conventional and shale gas.

It can also be mentioned that, surprisingly, the Netherlands has more proven gas reserves than the UK ($44\,\mathrm{Tft}^3$ versus $15\,\mathrm{Tft}^3$), but much less than Norway ($105\,\mathrm{Tft}^3$). Because of its location and gas storage facilities, the Netherlands is regarded as the *swing producer* for the Western European gas market, which implies that if there were a sudden decline in the physical availability of gas, that country could (and ostensibly would) fill the gap — which is very definitely not certain. Saudi Arabia has often been pictured as playing this part for the world oil market; however, with the rapid increase in the global consumption of oil, it may not be able or want to perform this function anymore, and in truth may never have performed it to the extent that some people believe.

An important observation that can be offered here is that the extensive deregulation of gas that for one reason or another may take place in the European Union (EU) could result in greatly increasing the *market power* — i.e., the *monopoly* or oligopoly power — of Russian gas exporters. The sponsors and supporters of this deregulation are apparently unfamiliar with the economic logic supporting this contention, and so it might be suitable to point out that collusion on the buying side of the *wholesale* gas market (which is where large buyers obtain gas from external producers) may optimize the welfare of Western European firms and households who are becoming increasingly dependent on foreign resources. This is obvious because the *bargaining power* of an aggregate of large buyers is greater than that of a 'patchwork' of small, competitive buyers. Several governments apparently are aware of this dilemma.

As is the case with other energy resources, the Chinese government seems to be increasingly interested in obtaining a very large amount of gas from the large exporting countries, and perhaps a great deal more than is or will become available from its present main suppliers, Australia and Indonesia. Here it should be kept in mind that we hear very little about the exploitation of the huge amount of shale gas that, e.g., the EIA claims is within the present boundaries of China. This situation gives Russian gas sellers a market opportunity that they have every intention of exploiting, because, among other things, it changes their position with respect to European buyers. The U.S. remains the world's largest gas market, but, e.g., Japan is the largest importer of Liquefied Natural Gas (LNG), and may be able to exercise some market power — in this case *monopsony* — when confronting its usual suppliers.

As has become evident, a region that might be moving into an unfavorable gas position is the UK. Oil production has peaked there, and domestic gas reserves are insufficient to support expected consumption increases. As in California, a very large inventory of gas-based facilities came into existence because of technical improvements in gas-burning equipment and low gas prices; while later (for a while) it appeared that it was more economical to pay higher prices for gas than to invest in alternative sources of energy. This outcome should be carefully noted, because countries that have very large investments in pipelines and other facilities for receiving and using gas will be more inclined to pay higher prices for this commodity than, e.g., importers with access to other energy resources. The Swedish parliamentarian Carl Hamilton has stated that this explains his preference for nuclear energy.

The final topic in this section has to do with the concept of a GAS-PEC. It seems clear, I think, that large suppliers of natural gas are unlikely to view with joy the possibility of a sustained decline in the price of natural gas, and this is especially true since the bad news has to do with a non-renewable (or exhaustible) asset. Moreover, it must be clear at this point in history that any firm that can legally participate in a strong oligopoly — the extreme form of which is banding together with other sellers to form a cartel — would be irrational to forsake the opportunity.

This does not mean that countries like Russia or Norway were irrational to stay out of OPEC, because obviously they were able to profit by the presence of OPEC and its ability over the last decade to play a decisive role in elevating the price of oil. But some questions must be raised as to whether the sale of natural gas for the prices now being paid in the U.S. is something that suppliers are willing to accept without thinking in terms of a radical adjustment of the OPEC type.

In any event, the formal argument is roughly as follows. If one firm in an oligopoly decides to tacitly ignore the agreed upon arrangements, and to compete aggressively with its 'teammates', and there is no effective response from other members, then (in theory) that firm might be better off, while the remainder of the oligopoly might be worse off. Virtually every economics textbook that was written before OPEC was formed (and some after it was formed) insisted on this outcome, and cited this as the reason that strong oligopolies of the OPEC type had no future. Put another way, ordinary human selfishness — including the desire of the rich to become richer — would ensure a return to the 'normal' state of affairs.

The normal state of affairs might mean all-out competition, with all suppliers ending up worse off. Here we might have the conventional "prisoner's dilemma" outcome, where so-called rational 'individual' behavior can work against the greater good. Moreover, rational individual behavior often means "cheating". On a higher intellectual level, however, another approach might be useful.

The basic item that belongs here is time. In conventional textbooks of the kind Milton Friedman read and authored, the passage of time leads to the collapse of cartels, because early success leads to some cartel participants misjudging their strength and concluding that they can 'go it alone'. But what happened with OPEC was that as time passed and the oil price weakened, it finally became clear to OPEC participants that cartel discipline was the key to maximizing income, and that income in turn was the key to a priceless social stability in the major oil-exporting countries. The relevant folk theorem here does not have to do with the greed of the rich, but a remembrance of things past, by which I mean the fate of colleagues and kingdoms whose appetites subverted their common sense. Also, in case you have forgotten, a folk theorem is a simple idea that has been in circulation throughout the length of recorded history, perhaps all the way back to the Stone Age, but awarded a fancy name by game theorists.

Let us also remember that in a repeated prisoner's dilemma type game, one of the basic textbook prescriptions is to reward cooperation with cooperation, and to repay cheating or aggression with the same coin. While this kind of behavior leads to what economists call an "equilibrium" (of sorts), the success of OPEC suggests that it is an inferior strategy when dealing with exhaustible resources. Instead, where an OPEC or a GAS-PEC is concerned, it seems to be optimal to tolerate a certain amount of cheating, while patiently trying to educate the 'cheaters'.

5. Some Aspects of Natural Gas Production Economics

When we examine natural gas production profiles in major oil or gas regions, e.g., in the United States, what we see is a rising output (or build-up) for oil and gas that eventually peaks, perhaps forms a plateau, and eventually begins to decline, even though there may still be a huge amount of the resource in the ground. Obviously, the same kind of economic logic (i.e., economic theory) that was in the background and foreground for oil is relevant here.

As shown in Figure 2, there is a decline with or without additional investment in measures to maintain the flow from gas wells, and thus without raising the rate of production to extend the plateau. *The reason (and perhaps the only reason) for this is that on the basis of reserves that have been identified in a particular deposit or field, it is uneconomical to attempt to*

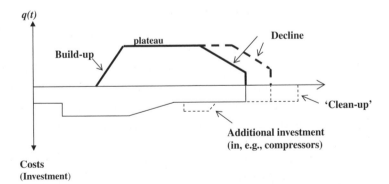

Figure 2.   Production-Investment Profile for a Natural Gas Deposit

prolong the plateau indefinitely! This is something that was often forgotten by economic theorists who attempted to deal with the natural gas scene, but who forgot that there is a difference between an oil or gas deposit and an automobile factory.

Accordingly, the explanation for this configuration turns on economics and not geology: geology is a *constraint*. Ideally, as in Economics 201 or 301, we can write an equation for profits in every period of some time horizon, and proceed with some mathematics whose purpose is to determine the intertemporal values of output that would permit the maximization of (discounted) profits over the relevant time horizon. But assuming that the marginal (and average) cost of production is constant over time, and S is available supply, we can start with the familiar relationship $V = \sum_{t=0}^{n}(p_t q_t - c(q_t)q_t)(1+r)^{-t} + \tau(S - \sum_{t=0}^{n} q_t)$, and by differentiating partially with respect to the values of q, end with the famous Hotelling (1931) expression $\Delta p/p = r$, where p here is defined as the 'net' price — or price minus the marginal cost of the *last* unit of output — and this net price increases at the rate r. V is profit, q output, and τ a Lagrangian multiplier.

If you have studied energy economics from virtually any book, except mine of course, you have probably encountered this Hotellian outcome, often accompanied by an assurance of its importance in the great world of energy economics. In reality it has a very limited scientific value, although if cost is made an explicit function of output, it is capable of revealing an important truth. The truth being referred to is that production from conventional natural gas deposits will take on the form shown in Figure 2, and this is because of a decline in deposit pressure. The key problem here, as you have probably guessed, is finding the right kind of function for $c(q_t)$.

Moreover, in terms of the real world, where the ex-post (i.e., after the fact) production curves of gas take on the appearance of the curve in Figure 2 (or even a distinct 'bell' or 'normal' appearance), the Hotelling result barely merits mention in a footnote, although it has been a godsend for many academic careers. In order to obtain something approximating realistic production curves, it is necessary to comprehend that the most important variable for an individual deposit is NOT the rate of interest (r) — which your favorite economics teacher might have claimed — but, as with oil, *deposit pressure* and its significance for the cost of production. As gas is removed and deposit pressure falls, it may be necessary to introduce additional wells or pressure-augmenting activities in order to maintain output. Genuine oil and gas people know this, but it cannot be repeated often enough in an academic setting.

If you saw the film *Five Easy Pieces*, the good Jack Nicholson was apparently occupied with work that was intended to compensate for the decline of pressure in an oil deposit. The same observation would apply to natural gas, and Professor Eric N. Smith (of Tulane University) has said that in the U.S., cost increases in marginal deposits are especially large.

In considering the present topic, it is also essential to be aware of something called the 'natural decline rate', which involves the deterioration (= depreciation) taking place in a deposit that is *independent* of previous production (where previous production tends to 'wear out' a deposit, and also contributes to depreciation, but like deposit pressure does not explicitly enter into mathematical presentations). Readers should never forget that the same kind of production pattern illustrated above might someday be duplicated on a global scale, and perhaps sooner than many readers expect. On the basis of present supply-demand trends, it is possible that in 25–35 years the output of gas will peak, and after a short or long *plateau*, begin to decrease. Had it not been for the appearance of large quantities of shale gas, the bad news about natural gas might not have taken 25 years to appear, as the former Chairman of the Federal Reserve System, Alan Greenspan, flatly stated in his testimony before the Committee on Energy and Commerce of the U.S. House of Representatives in June 2003. On that occasion the Chairman was not thinking in terms of depreciation, but of price, and he had good reason for his concern.

Clearly, everything else remaining the same, the presence of natural decline reduces the value of a deposit. More importantly, the mere fact of depreciation means that output can only be maintained as a result of investment, and in the long run, investment in a natural gas deposit or field might become too expensive to contemplate if there are alternative uses for money. Some mathematics here might be useful, but not as useful as common sense. Over the last few years, billions of dollars have probably been paid for properties worth considerably less. The reason they are worth less is that while the initial production might be impressive, depreciation due to production or, for that matter, natural depreciation means that millions must eventually be spent on investments of one type or another in order to maintain a given level of production, and in the very long run, even this might not suffice.

In any event, since I was once reproached by a self-appointed expert for not using calculus in a discussion of natural depletion, interested parties can add the following trivial integral to the lectures they might give on this subject. Natural decline (or depreciation) is analogous to what we call in capital theory 'depreciation by evaporation', and in which an asset with the initial value of $V(0)$ is subject to a constant force of mortality θ. The relevant expression has the following appearance:

$$V = V(0) \int_{t}^{t+T} e^{-(\theta+r)(\tau-t)} d\tau = \frac{V(0)}{\theta+r} \left[1 - e^{-(\theta+r)T} \right]. \qquad (3)$$

Needless to say, if we are thinking of something like shale gas, θ could be made a function of the deterioration of the asset as a result of production. I have elaborated on this in many lectures, but it seems too obvious to pretend that a full-fledged mathematical exercise has a great deal to offer. Moreover, although the importance of natural gas is undoubtedly increasing, readers should remember that oil remains the most valuable energy resource.

For instance, as energy resources must be moved over longer and longer distances from large suppliers to large buyers, gas's relative inferiority to oil increases. Whether by pipeline or by tanker, the unit transport costs of oil are lower than those of gas. If we consider a given volume of pipe, oil contains (on average) 15 times as much energy as gas, which immediately reflects negatively on pipeline investment costs for gas. Furthermore, when considering intercontinental trade, transporting a given quantity of gas by ship over all except very long distances is more expensive than by pipeline, while transporting oil by tanker over the same distances tends to be less expensive than by pipeline. *This is one of the reasons why, quantitatively, the kind of global competitive market that various observers hope will come into existence after enormously expensive LNG investments are carried out, may prove to be illusory.*

As with oil, there are plenty of energy professionals ready to claim that technological advances will ensure that we will always be able to obtain the gas and other energy resources we need at prices we can afford. The technology booster club is now turning its attention toward innovations that might make it possible to exploit vast deposits of crystallized natural gas suspended in Arctic ice, or buried just below the ocean floor, and which are known as methane hydrate. Optimists even claim that it is now possible to obtain controlled volumes of methane from hydrate-rich areas in North Canada, and apparently some of this hydrate-based gas has been officially classified as a viable energy reserve. One hopes that they are correct, because the productivity of conventional gas wells in, e.g., North America has been falling, and the increased difficulty in finding and bringing into production new conventional wells is widely acknowledged.

As was alluded to in many places in this book, a new drilling boom may have started in North America in which the object is to use advanced technology to obtain gas found in large shale beds. This advanced technology is nothing more than horizontal (instead of vertical) wells into which water, chemicals and sand are pumped in order to *fracture* sedimentary rock. I can recall conferences in the United States and elsewhere in which horizontal drilling was advertised as an approach that would revolutionize the access to oil; however, this did not happen because the oil — unlike perhaps the gas — was not there.

In these circumstances I think it best to wait before accepting that the basic gas situation in the U.S. (or in Canada) is in the process of a *major* change, and also wait before accepting the estimates of some natural gas producers about the richness of the natural gas deposits they own. There are several ways that these deposits can make producers super rich. One is for them to extract and sell large quantities of natural gas, and another is to sell deposits that they have convinced naïve potential buyers to purchase because these deposits ostensibly contain more natural gas than can be found anywhere else on the face of the earth.

6. An Introduction to Natural Gas Pipelines

There is often a problem in a book of this nature concerning the choice of topics to be examined. A few students in my course on oil and gas economics in Bangkok desired a thorough discussion of gas pipelines, while others regarded it as a distraction, given the importance of oil. The attention being given to shale gas may change this situation, and certainly it must make sense to pay some attention to the very large pipelines that bring natural gas from, e.g., Russia to Western Europe, or from Russia to China.

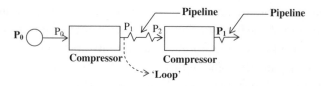

Figure 3.   Pipeline–Compressor System

To be honest, however, the materials below have mostly a pedagogical value. They are intended to indicate to students and others the most important components of a pipeline. The pipelines beginning in Russia and near Asia, and running eastward, are where they are because of politics as well as the ordinary, everyday economics of the kind presented in Economics 101. Schematically we have an arrangement of the type shown in Figure 3.

In any case, it is important that readers who want a thorough exposition of the *economics* of gas pipelines should be aware of some work by the late Hollis Chenery (1949, 1952). Chenery was particularly interested in the relevant technology, and in particular those aspects of pipeline technology in which economies of scale are especially significant. Perhaps the best way to begin the exposition is with a diagram that I have often used, and which, in an elementary manner, shows some of the elements in a typical pipeline (see Figure 3). As we might suspect from this diagram, a natural gas pipeline is a fairly complicated structure. In fact, it is much more complicated than indicated in this introductory presentation.

To begin, readers should understand the expression *sunk costs*. Sunk costs are expenditures that, once made, cannot be recovered; they are associated with decisions that cannot be reversed later. A pipeline that costs a few billion dollars can later be chopped up and sold to scrap dealers for a few thousand or a few hundred thousand dollars, but conceptually it seems appropriate to regard the main gas transmission lines as sunk investments. (On the other hand, a *fixed cost* is a cost that may or may not be fixed in the short run. A 'crack house' can be renovated and turned into a luxurious townhouse — at least in theory.)

The most important costs associated with a natural gas pipeline are planning and design, acquisition and clearing of right-of-way, construction and material costs (e.g., labor costs, the cost of pipes and compressors, etc.), the cost of monitoring the pipeline and performing maintenance, and energy to power the *compressors* (which are analogous to pumps in an oil pipeline, in that they transfer mechanical energy from, e.g., a motor to the gas that is to be transported).

The expected life of a gas pipeline can exceed 30 years, and it could be argued that expensive pipelines should not be constructed if there is not enough gas to keep it operating for about that length of time. Russia has elected to build (at least) two pipelines to China, and their cost is estimated at 10 billion dollars. This sounds like a lot, but it may be less than the estimated cost of a pipeline from northern Canada or Alaska to the U.S. Midwest. In addition, if Europe does not want Russian gas, these pipelines should ensure that Russia could continue to acquire a large income from its gas. In what follows, for pedagogical reasons, no distinction will be made between sunk and fixed costs, since in most economics textbooks the word "sunk" seldom appears in chapters on production theory. On the other hand, the expression "increasing returns to scale" is often used in discussions of the sort presented here, and so concepts are presented which in some measure apply to both oil and gas pipelines.

In designing a gas pipeline, engineers tend to think in terms of varying the pipeline diameter and the number and size of compressors, as well as things like the amount of maintenance that will be required. Increasing the size (and energy output) of a compressor, without changing the diameter, raises the speed at which the commodity goes through the line, and thus increases the *throughput* of a given pipe size. Similarly, increasing the diameter of a pipeline while keeping the compressor size constant might also raise throughput, since there is less resistance to flow (per cubic feet of gas) in a larger pipeline, but please notice the following: *substitution is possible between compressor size (i.e., power or horsepower) and pipe size (i.e., diameter) only within strict limits.* (In other words, remembering Economics 101, isoquants in power–pipe-size space eventually flatten at both ends.)

By way of extending the above observations, remember that the volume of a pipe of length L, with radius r, is $\pi r^2 L$. (For convenience, take $L = 1$ foot or 1 meter.) The inside surface area of the same pipe is $2\pi rL$. If the radius is doubled, the surface area is also doubled, *but the volume of the interior of the pipe is increased by a factor of four!* Furthermore, a small amount of algebra informs us that there is *less surface area per unit of volume* for larger diameter pipes than for pipes with a smaller diameter, and as a result there is less frictional resistance per unit of throughput for a larger than for a smaller pipe. Then why not increase the radius of the pipe indefinitely in order to exploit the returns to scale being described? The answer — as you discovered in Economics 101 — is that at some point it is less expensive to increase throughput by an addition to compression than by increasing the pipe diameter. It can be added that just as (*ceteris paribus*) we have 'returns to scale in the pipe', we might also have returns to scale in compression, at least up to a certain point.

Some elementary algebra might be useful to begin, however. Calling Y the desired 'throughput' (i.e., output) of the above system, H the capacity of the compressor (in, e.g., horsepower), D the diameter of the pipe, p_1 and p_2 as indicated in Figure 3, and N the ratio p_1/p_2, the following arbitrary equations might provide the beginning of an academic economics discussion of a typical pipeline. (How relevant they are for engineers, however, is another matter!)

$$Y = \theta_1 D^\alpha f(p_1/p_2) = \theta_1 D^\alpha f(N).$$
$$H = g(p_1/p_2) = g(N).$$

θ_1 is a constant, and by writing $N = g^{-1}(H)$, and assuming that these expressions have the 'necessary' algebraic and geometric characteristics, a simple substitution brings us to the following relationship that we knew was inevitable when we casually glanced at Figure 3:

$$Y = F(H, D). \tag{4}$$

Note that I called the above the *beginning* of a systematic analysis. The length of the pipeline between compressor stations does not enter into the analysis; however, I remember that in the equations supposedly applicable to the Alaska Highway Pipeline it was possible to solve for this term by using a variant of the first equation for Y that is shown above. There is also the matter of the thickness of the pipe. In your engineering studies this would be a key item, while at the same time I doubt whether your engineering teachers would insist that you make a great deal of the equations shown above. By the same token, in my economics lectures, they serve only a limited purpose.

By that I mean that no equations derived by conventional economists without a relevant engineering background can substitute for the experience

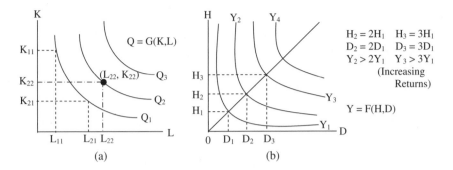

Figure 4.   (a) Neo-Classical Isoquants, and (b) Isoquants with Limited Substitutability

of genuine pipeline engineers, and this is the best reason in the world to avoid an elaborate pretence that what we are dealing with here is a facet of economic theory that should cause engineers to rush to the faculty of economics for additional wisdom.

Moreover, in the modest economics literature on this subject, there is generally a failure to note that there is a limited substitutability between H and D. In my lectures I show this complication by indicating how isoquants for Equation (4) should look. These isoquants are shown in Figure 4(b), while more typical isoquants are shown in Figure 4(a), where the production factors are those you mostly encountered in your microeconomic courses on all levels. Observe in particular the situation for an output of Y_1 in Figure 4(b), where the non-substitutability is very evident.

Since the isoquants in Figure 4(a) are typical of those found in conventional academic literature, it is certainly possible to imagine representing them with a Cobb–Douglas (e.g., $Y = D^{\alpha} H^{\beta}$) production function, but this is not the case with the isoquants in Figure 4(b). It might be possible to write a conventional (neo-classical) equation for the 'stretch' of the isoquant between D_3 and H_3, but obviously there comes a situation where increases in the diameter of the pipe cannot substitute for the absence of compression. Moreover, if the thickness of the pipe being used finds a place in the analysis, as was true with the Alaska Highway Pipeline, then for both structural and economic reasons there is a limit to the ability of compression to substitute for the pipeline diameter, and vice versa. If this substitution cannot take place, representing or trying to represent the isoquants in equation form is not recommended in a serious classroom.

In the work on this topic by the late Professor Hollis B. Chenery, an auxiliary equation was provided to link the pressure in a pipeline with the thickness of the pipe being employed. I seem to remember from my engineering studies that an equation of this nature was a key item in discussing the characteristics of a pipeline. If we consider the pipeline at the bottom of the Baltic Sea transporting gas from, e.g., an outlet in the vicinity of Saint Petersburg to Germany, the sequence of decisions would have begun with the amount to be transported, which would require a minimum diameter, which in turn would determine pipe thickness, and from there the matter of compressors would be taken up and solved by ladies and gentlemen who have been dealing with such matters for many years, and who regard them as routine — which of course they are.

If there were any complications they would involve adjusting the diameter of the pipe so that — as is sometimes customary — it could store a certain amount of gas, and also obtaining clearance for the pipeline, which could involve paying some transit fees. Of course, conduits at the bottom of the Baltic Sea can in theory have any configuration, and tend to be free of harassment by various police or tax inspectors, as long as they are not detected facilitating the movement of Russian Special Forces soldiers who might be tempted to interfere with the lovely parties that take place in the Midnight Sun which shines down on the magnificent Swedish archipelago during the summer months.

It has been suggested that if a pipeline manager is in a position to raise (transmission) prices after gas producers have made their drilling and development investments, it will cause risk-averse gas producers to limit the size of their investments in order to avoid being unpleasantly surprised by an increased price of transmission. This reasoning also works in the other direction. If gas producers have several pipelines through which to transmit their output, it could place individual pipeline managers in a dilemma in that they face the threat of gas producers transferring their affections to another carrier. On a regional level this suggests that an optimum arrangement might prescribe a single owner for gas deposits and pipelines, but 'optimality' does not have a great deal of significance outside seminar rooms. Instead, the 'second best' solution is probably to be recommended in many cases. It involves long-term contracts featuring 'take-or-pay' arrangements. Only in very special cases does it mean a resort to the kind of ineffectual or wasteful short-term provisions that the European Union Energy Directorate and people like Professor Jonathan Stern find so attractive, and are still keen to impose as part of a comprehensive regional energy plan.

One of the things being implied above is that firms (and consumers) do not want to find themselves in possession of a large amount of worthless capital equipment — e.g., equipment that loses its value because the demand or supply for pipeline capacity or gas suddenly and drastically collapses. The algebra for this situation is interesting, but hardly necessary to comprehend the basic issues. *For example, if a firm is risk-averse and wants to avoid the financial dangers associated with excessive investment in fixed or sunk capital, then once again long-term commitments make significant economic sense.* It is difficult to understand how a rational person could come to another conclusion.

As an example relevant to the above discussion, studies for the Polar Gas route in northern Canada resulted in choosing between pipelines having diameters of 30, 36 or 42 inches, with operating pressures respectively of 1,260, 1,449, or 1,680 pounds per square inch. Each of these pressures calls for a certain compressor size. The middle of these was deemed optimal, and provisions were also made to add additional compressors if necessary. Throughputs were to range from $80\,\text{mft}^3/\text{d}$ to $1.6\,\text{Gft}^3/\text{d}$, where $\text{ft}^3 = \text{cubic feet}$.

As for the "loop" in Figure 3, this generally amounts to a parallel section of pipe that is sometimes added in order to increase capacity, since using a loop can often be preferable to increasing the size of the pipe. By supercharging the existing compressors, and/or adding compressors, it can become economical to add parallel sections to the existing pipeline. Note that this does not strictly mean duplication, since the cost-output relationship turns on the amount of supercharging or additional compression, the diameter of the pipe used for looping, and the construction expenses associated with the looped section. As I have mentioned on many occasions, an interesting looping exercise was carried out on the Roma-Brisbane pipeline in Southern Queensland (Australia), where sections of pipe were laid parallel to the main line, with a separation of 4–8 meters, and ostensibly capacity was doubled. The price of the gas being delivered increased, but this was not surprising since a price increase was necessary in order to justify initiating this particular looping exercise.

In theory, output expansion via looping can feature constant, increasing, or decreasing unit costs. It should be emphasized that when there are increasing returns to scale in both compression and transmission, which is likely, then neo-classical economics suggests that optimal behavior calls for initiating (and in some cases completing) projects well ahead of the demand for new capacity, in order to avoid any (per unit) increasing costs associated with, e.g., looping if sizable increases in capacity are necessary at a later date.

7. Some Contemporary Price Issues

The first price issue below is strictly of an academic nature. It is placed in this exposition so that students and researchers who like to manipulate supply and demand curves can get some idea of the part played by inventories (i.e., stocks) in oil and gas markets. To begin, I am unfamiliar with any undergraduate or graduate textbooks except mine that bring stocks into the picture in a meaningful fashion. This is not to say that interested parties should search models of the type presented below for revolutionary concepts, but at the same time — if they read the popular economic and financial press — they should realize that the constant reference to stocks of oil, gas and other commodities has an important significance, and deserves to be approached in a scientific manner.

The materials on oil in my previous textbooks (2000, 2007) provide a better discussion of the pricing of oil and gas than the discussion below, but the diagram that I use in those books looks so much like those used in elementary electrical engineering classes that it frightens many potential readers, and so here I will employ Figure 5.

The first thing to be understood is that a disequilibrium in the stock market might be followed by what appears to be a strong reaction in the flow market, while a disequilibrium in the flow market may or may not be followed by what appears to be a strong reaction in the stock market. Where the latter is concerned, it depends on the availability of stocks and expectations about the future price of the commodity.

Note that the initial stock equilibrium in Figure 5 was at (I', p'), and this was disturbed by an increase in demand for stocks (i.e., inventories). To obtain this increase, price had to rise, perhaps to p'', but in any event it

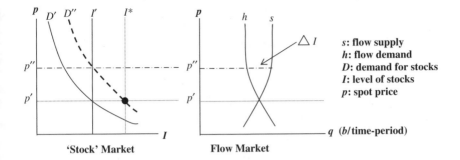

Figure 5.   Supply–Demand Relationships in Stock and Flow Models

stayed above p' — though not perhaps at p'' — until the new equilibrium at (I^*, p') was reached (via a fall in p and a rise in I). Note that while p is the same at the new equilibrium, there has been an increase in stocks (i.e., inventories).

Even in elementary mathematical economics textbooks, you have been exposed to information which suggests the formulation of an equation in which the change of price with respect to time is a function of the difference between actual inventory (AI) and desired inventory (DI), which means that the relevant model is a stock-flow model of the type I have frequently used to discuss the market for crude oil, and not the flow model that you mastered in Economics 101. The implicit form of the relevant equation for Figure 5 might be $dp/dt = f(DI - AI)$, and what the discussion here signifies is that if $DI > AI$, then price increases; if $DI < AI$, then price decreases; and if $DI = AI$, then $dp/dt = 0$ and price is constant. *Readers should make considerable efforts to comprehend this particular discussion!*

This analysis applies to the entire market. For instance, it was once said that a "comfortable" level of natural gas inventories for the U.S. during the 2009–2010 winter was 3.4 Tcf ($=DI$ in Figure 5), where T signifies trillion. To attain this it was suggested that injections $(s - h)$ should have been about 63 billion ft^3 per week ($=63$ Gft^3/w) for a period of 6 weeks. Since the 5-year average for that period was 57 Gft^3/w, the implication was that *ceteris paribus* the gas price would increase. However, with the U.S. economy in retreat, there was also a downward pressure on the price, in addition to the downward pressure placed by the rising production of shale gas.

There has recently been a great deal of talk about shale gas, and this has led to the conclusion that the price of natural gas is going to remain at a comparatively low level. Some of us are not as excited about shale gas as others, but there is no denying that the total natural gas production in the U.S. has increased to record levels, with the yearly average for 2011 reaching close to 68 billion cubic feet a day, as compared to about 66 billion cubic feet the previous year (2010). Gas at the benchmark Henry Hub (near New Orleans, Louisiana) is expected to average $3.50/Mcf, although recently it has been close to $3/Mcf ($\approx$$3/MBtu). A problem though is that stockpiles of gas may continue to increase, and this may press down the price to an extent that some producers will start to shut down their operations.

What about some arbitrage that alters this situation, for instance, shipping natural gas from oversupplied U.S. to undersupplied Asia (assuming that macroeconomic growth rates in that part of the world do not collapse)? The idea of a genuine international market for liquefied natural gas (LNG) has been discussed for years, and with the price differentials that have occasionally existed between various regions of the world, the prospect of increased gas-on-gas arbitrage seems to have increased the rate at which vessels for transporting LNG are being constructed.

Just as Henry Hub is the main trading hub for natural gas in the U.S., the National Balancing Point (NBP) plays this role in the UK. Many LNG cargos are imported into Europe through Spain, while it needs to be noted that the most friendly coastline for LNG installations in the U.S. is to be found on the Gulf of Mexico. Interestingly enough, Qatar Petroleum — one of the largest firms in Qatar, which is the largest exporting country of LNG in the world, with a production capacity of 77 million tonnes per year — has given some consideration to reconfiguring (or repurposing) its U.S. import terminal for LNG (in Texas) to export LNG, and therefore take advantage of what is sometimes called a 'gas supply glut' due to the increasing production of shale gas. As a former chairman of Qatar Petroleum pointed out, it does not make economic sense to ship gas to the U.S. when the rest of the world requires a great deal of gas, for which it is willing and able to pay comparatively high prices.

8. Storage, Hubs and Market Centers

At the present time I believe that just about everybody understands the great change that has come over the oil market since the price of oil seemed to be on its way to a price similar to that of a McDonald's hamburger in the last year of the 20th century. My argument where peak oil is concerned is often that if we had a peak in a huge and oil-rich land area like North America, then there is a distinct possibility that we will face the same situation globally. The truth is that I am amazed at the number of observers who have adopted my way of thinking over the past few years. The reason is not, however, that they have been studying my books and articles night and day, but an oil price that — after a collapse in the early autumn of 2008 — has once again cracked the $100/b barrier, and there is a constant threat that it will go higher if the global macroeconomy regains some of the strength it possessed before the oil market meltdown in 2008.

As shown in Figure 6, the natural gas production-consumption process begins with the lifting of gas from a 'field' or 'deposit', and proceeds to a large diameter transmission or 'merchant' pipeline, with usually some gas siphoned off to 'run' the compressors, and some gas diverted from so-called 'end-users' or 'final destination' (i.e., households and small businesses) into storage, further processing, and sales to very large consumers such as manufacturing industries and electric generators.

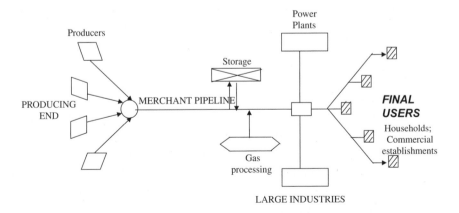

Figure 6.   Natural Gas Production and Consumption Process

Eventually the gas goes into a distribution system where the pipelines are smaller, and via these pipes to homes and smaller commercial establishments, which can be designated 'final consumers'. Do not forget though that instead of going to plants and homes, it could be going to very complicated processing facilities to be transformed into liquefied natural gas (LNG). LNG takes up only 1/600th of the volume of gas in the gaseous state, which makes it more economical to ship over long distances. I have discussed this at great length in my lectures on natural gas during the past few years, where I note that the transport of natural gas in specially designed cryogenic (LNG) carriers can make a great deal of economic sense given some of the prices paid for LNG in Asia.

Storage is another of those topics which submits to an interesting theoretical treatment. On this occasion the exposition will be non-technical, although readers who want to impress others are advised to pay close attention to the terminology. Strangely enough, storage is almost completely ignored in microeconomics textbooks, despite the importance of its presence or absence: *when it is absent, prices of (e.g.) gas often tend to be extremely volatile*, and since domestic supplies of gas from nearby sources are falling, this is one reason why more storage facilities are being constructed in, e.g., the UK.

At the present time, where storage is concerned, Spain is the star. A massive new storage facility called 'Castor' is nearing completion, and is capable of storing at least 50 days' worth of the present Spanish consumption of gas. Since Spanish gas consumption is increasing as fast or faster than any in the world, a large storage capacity is natural, but it should be remembered that gas is often put into storage during the off-peak season, and then removed to satisfy peak demand. For this reason, it may be so that to a certain extent, in some countries, storage facilities can be regarded as a substitute for peak-load transmission capacity.

Gas in storage is turning out to be a carefully watched statistic, particularly in the run-up to winters. Low storage levels mean that any shortages of gas that may appear during winter months could impact severely on gas prices, as well as on the availability (and price) of other fuels, such as heating oil. This is because certain other fuels are substitutes for natural gas in various uses. The strategy here is to buy gas when it is cheap and store it. A short, easily read and valuable article on this subject is the one Lee

Van Atta (2007) published in one of the best energy forums, *EnergyPulse* (www.energypulse.net). He mentions that the majority of present storage development in the U.S. has to do with salt caverns, while most of the rest is in depleted reservoirs.

Just as transport involves moving a commodity through space, storage performs a similar function with respect to time — "similar", but not identical, because time runs in only one direction. By putting goods into inventory, we move from the present to the future at finite cost; but returning the exact same goods to the present — and also recreating the background existing when the decision to store was made — is conceptually a much more difficult operation, and for the most part impossible. This suggests that we have a *consistency* problem: at time t we make a plan for $t+1, t+2, \ldots, t+x, \ldots, t+N$, where N is the terminal date, but it might happen that at, e.g., $t+x$, we perceive that the decision taken at t was suboptimal. A new plan can then be put into practice, but conceivably we would have been happier if we had gotten things right in the first place, or formulated a strategy that would have taken into consideration the possibility of making — and subsequently having to correct — expensive mistakes. This strategy might have featured storing more or less of the commodity, and relying more heavily on such things as futures markets. Obtaining increased flexibility could involve a heavy cost, which increases the attraction of 'derivatives' like futures and options.

An important and accessible article on storage is that of Benoit Esnault (2003), although it contains one implication that I have some difficulty accepting. This is that deregulation is a logical precursor to a decrease in prices and improvement in service. Such was the theory when electric deregulation was adopted; but if it is true that the ultimate object of deregulation is lower prices, then I take enormous pleasure in noting that (*ceteris paribus*) electric deregulation has failed, is failing, or will fail just about everywhere. What we also have here — at least in some countries or localities — is a nice example of the consistency problem mentioned above. By that I mean the absence of a strategy for rapidly reversing a venture that turns out to be sub-optimal (e.g., deregulation), and thereby mitigating the bad news that might unexpectedly appear.

A concept that is unique for storage is the *convenience* yield. This is explained in some detail in my previous textbooks, but roughly it is the yield (i.e., gain) associated with greater flexibility that might devolve on the owners of inventories. For example, the availability of inventories permits sales to be increased without incurring the expenses that are often unavoidable when or if it is necessary to resort to spot purchases in order to fulfil contract stipulations, or, for that matter, purchase of futures or options contracts at prices that are regarded as unfavorable. The theory here is straightforward: an additional unit put into inventory can provide a sizable marginal convenience yield if inventories are small, whereas with very large inventories, the *marginal* convenience yield (associated with adding another unit) might be zero (although the convenience yield of inventories would still be positive and could be very large). In the simplest of cases, inventory accumulation would continue until the *cost* of a marginal unit outweighed its marginal convenience yield, with both cost and yield measured in some convenient monetary unit. Another way of viewing this is to say that having access to storage encourages the transfer of consumption from periods in which its value is low to those periods when it is higher (e.g., *peak periods*).

In examining this issue, it can be argued that gas storage can not only moderate upward price movement, but also functions as an excellent *hedge* against price and volume uncertainty. With natural gas — as with electricity — one of the key issues is *peak demand*. If a storage option is available, the exposition above indicates that gas is stored during off-peak periods, and if peak demand (or a 'glitch' of some sort in transmission or distribution) jeopardizes the ability to deliver desired quantities of some good to end users, then gas is removed from storage. (Electricity cannot be stored, and so this procedure cannot be employed, but peak demand is satisfied by holding some equipment idle during off-peak hours.) An expression that might appear here is *peak shaving*, which sometimes brings a frown to the faces of energy economics students, but it means no more than releasing gas from storage into a pipeline during periods of maximum demand (i.e., peak periods). Possessing this option might make large investments in additional producing or transmission capacity unnecessary.

Quality can also be brought into the storage picture. Depleted reservoirs are often used, but withdrawal from those structures is relatively slow. Salt caverns are better and allow rapid injections and withdrawal, which, as Van Atta (2007) points out, makes them attractive for traders who want to "capture value from price volatility". What this means is that when they have an opportunity to make some serious money, they do not want to be hindered by an inability to obtain the commodity that they are holding in storage, and with luck can be sold at premium prices.

Hubs are physical transfer points that are sometimes called 'pipeline interchanges'. They make it possible to redirect gas from one pipeline into another. However, at the present time, I prefer not to accept a recent report which claimed that spot prices at Henry Hub, which is one of the largest and best-known gas market hubs in the world (and is close to the Lake Charles (Louisiana) LNG terminal), have assumed the role of international reference prices. This kind of claim is sometimes tied to the belief that a large expansion in the trade of liquefied natural gas (LNG) will eventually lead to an international market that is capable of replacing regional markets of one type or another. In the very long run, this hypothetical international gas market would comprise — via uniform net prices — both pipeline gas and LNG.

This chapter is not the place to speculate on a scheme of this nature; but if the demand for gas in the U.S. reaches the levels predicted by the U.S. Department of Energy, and if shale gas does not live up to expectations, it could mean that the movement of LNG toward the U.S. could increase to a point where there will be upward pressure on gas prices in every market. Moreover, this is only the beginning. According to one prediction, China and India are expected to double their use of coal by 2030, and their combined oil imports are expected to surge from 5 or 6 mb/d in 2006 to (an 'estimated') 19 mb/d in 2030 (which is probably wrong). It is sometimes conjectured that to offset the environmental deterioration this is liable to bring about, they will almost certainly be in the market for enormous amounts of natural gas. This may or may not be true, because in my opinion the (*ceteris paribus*) oil production for 2030 that has been predicted by the IEA and the USDOE cannot possibly be realized.

In theory it might be desirable to combine hubs with market centers, where either of these might provide facilities that permit the buying and selling of services such as storage, brokering, insurance and wheeling — where *wheeling* means the provision of pure transportation services between external transactors. For pedagogical reasons, hubs are often portrayed as displaying a radial system of spokes (i.e., pipelines), and conceivably these spokes could be joined by adding short links.

Market centers are supposed to be able to operate independently of facilities for producing, transporting or storing the physical product, but even so, it might be optimal if they provide a locale where shippers, traders, etc., can buy and sell transportation, gas, etc. To a certain extent, the layout of these establishments could take on the structure of trading facilities in the financial markets. If there are imbalances anywhere, then in an 'ideal' market center there will be a mechanism where they can be located in a very short time and rectified, which might include providing access to tradable pipeline space and storage capacity.

In the U.S., for example, market centers have direct access to almost 50% of *working gas* storage capacity and, in general, enjoy a special relationship with many of the high-profile storage establishments. (Working gas is the amount of gas in a storage facility in excess of the 'cushion' or 'base' gas that is needed to maintain facility pressure and deliverability rates.) Regardless of the actual configuration, it is hard to avoid the conclusion that market centers will tend to form at, or in the vicinity of, hubs, and that the number of arbitrage paths that can be utilized for obtaining uniform prices in a system are expanded if there is an increase in hubs, market centers and storage facilities. An interesting issue now is what will happen if shale gas (and oil) are sufficiently plentiful to provide the benefits that many Americans desire to decrease the reliance of the U.S. on foreign energy sources.

9. Prelude to a Blunder

In the last decade or two, great changes have taken place on the natural gas front. The growth in the demand for gas exceeds that of all energy media except some renewables, and unlike the situation 15 years ago, gas is now highly recommended as an input for power generation. (In the UK, more than 70% of power is — or was — generated by gas or coal-fired power stations.) A main reason for this is the advent of *combined cycle* gas-burning equipment with a very high efficiency. What happens here is that in addition to the gas turbine, there is a secondary turbine producing steam from the waste gases/heat of the main gas turbine. The kinetic energy in this steam is transformed into mechanical energy that turns a generator. This generator produces additional electricity for a given input of gas.

However, as often happens, there are very many misconceptions in circulation about natural gas, the most pernicious of which — at least in Europe — have to do with the *restructuring* (i.e., deregulation/liberalization) of gas markets. Some questions need to be asked as to why and how these misconceptions came into existence, and it appears that the answer has to do with the short time horizons of many industrial buyers of gas, as well as the short time horizons and carelessness of small business and households and voters. In some countries, gas producers have expressed themselves in such a way as to suggest that there is virtually an infinite amount of natural gas reserves available for exploitation, although in many regions demand can only be satisfied by very large imports from distant sources. For instance, in much of North America, exploration/production yielded exceptionally disappointing results for a long time, and expectations about, e.g., the Gulf of Mexico and imports by pipeline from Canada often have an air of unreality about them. Similarly, the UK was once a major gas power, but may now import about 40% of its requirements. Many countries like, e.g., the UK and Turkey are *locked into* gas due to past investments which emphasized acquiring gas-based generators and infrastructure.

With certain exceptions, many gas buyers are almost totally unaware of how supply and demand might develop in even the comparatively near future, and instead continue to make plans for an 'Age of Gas' (as it is called) in which they believe that they will have access to all the gas that they will need, at prices that will continue to resemble those of the recent past. This might be a good place to note that in Brazil, starry-eyed deregulators once counted on gas-based electric power being cheaper than hydroelectricity. As they now admit, this incredibly *gauche* supposition was completely wrong. At the same time, it may be true that shale gas will save the U.S. from having to rely on very expensive imports, and if this is true, it would be nice if decision makers regard natural gas as a kind of transitional fuel to a sustainable and healthy (distant) energy future.

According to the International Energy Agency (IEA), the Age of Gas is right around the corner, and by 2035 the share of natural gas in the global energy picture will rise from 21% at present to 25%. This may or may not be true, because when lecturing at the engineering university in Bangkok,

I convinced both myself and most of my students that the IEA estimate of crude oil consumption for 2030 was absurd. Global gas demand is expected to rise by 2.5–2.7%/y, with the big consuming area being Asia, where it has been suggested that demand will increase by an average of 3.5%/y between 2001 and 2025. Another estimate has the average global gas production increasing by 2.75%/y until at least 2025, and gas passing coal as the second most important energy medium.

Before shale gas entered into consideration, world gas prices were expected to eventually display an unambiguous upward trend. In picturing U.S. prices remaining flat until 2005, the IEA was clearly mistaken in earlier forecasts, but they might be correct in noting that U.S. and Canadian gas supplies will tighten if shale gas does not live up to the publicity it is now receiving. A wellhead price of $2.5/MBtu (in 1997 prices) for purely conventional U.S. gas in 2020 did not seem particularly realistic to some observers when it was first predicted, and the upward trend in gas prices of a few years ago would have been capable of cancelling out the favorable economics of gas-based power generation that resulted from advances in combined-cycle technology, if by chance that price trend had continued.

This brings us to the impact of liberalization/restructuring. The IEA has mostly got it completely wrong on liberalization/restructuring in the electricity sector, and as a result I see no reason to expect an improvement in their ability to analyze the economics of world gas. The thing they should understand here is that economics is subsidiary to the history of gas and electric restructuring. Even the experts of the IEA should agree that it might be in order to insist that everything possible be done to prevent unsound ideas about deregulation/liberalization and a few other things from interfering with orthodox scientific reasoning and sound engineering practices. This effort should begin with trying to avoid, and if necessary expose, grotesque errors associated with plans to restructure both gas and electricity markets.

To this it can be added that where gas reform is concerned, the economics debate is not particularly encouraging, and in some cases is conducted by academic economists without an adequate feel for either the economics or the engineering aspects of the natural gas sector. They have not bothered to find out, for example, that an important component of the financial sector — in the form of several leading investment banks that are heavily involved with commodities — once scaled down its risk management commitments in some commodity markets.

Warburg Dillon Read — the investment banking arm of UBS — closed down its energy and electricity derivatives business as early as 1999, and in that same year Merrill Lynch announced its withdrawal from over-the-counter derivatives in natural gas. While this was going on, a consensus of commodity traders and analysts were still willing to wager that derivatives activity in gas and electricity would take off once market liberalization achieved a *critical mass*, and as it turned out, in electricity that condition was not too long in coming, although it did not turn out to be durable: it barely lasted long enough for the most important commodities exchange in the world (NYMEX) to declare its electricity futures contract hopeless, and also to cancel one of its natural gas contracts. (Let me note, however, that these contracts may already have been resuscitated. Wherever there are people who are sufficiently naïve to buy suspicious assets, those assets are certain to appear.)

Now for some particulars. Natural gas deregulation began in the U.S. about 20 years ago, and it is easy to remain sympathetic to the natural gas buyers and others in the U.S. who felt that the regulatory climate at the time of the 'gas bubble' in that country did not correctly address either efficiency or equity concerns. What eventually happened though was that economists, consultants, and various 'researchers' were provided with a forum in which they could unleash a barrage of unscientific ideas for correcting what they construed as market shortcomings, while at the same time promoting a radical transformation of the entire natural-gas sector — from 'wellhead' to 'burner tip'. In a number of lectures, beginning in Hong Kong in 2001, I attempted to convey some of the total lack of realism circulated by deregulation enthusiasts, and I believe that I succeeded. After all, condemning one of the worst economic ideas introduced in modern times does not require any great pedagogical skill.

How should we treat a collection of misjudgments of the magnitude and extent involved here? Perhaps it is not necessary to treat them at all, because unlike the electric deregulation travesty, gas deregulation was never able to get up full steam. One of the reasons for this was that in the U.S., and perhaps elsewhere, some important politicians and industry people as well as genuine experts from the academic world took issue with gas deregulation proposals. For instance, they pointed out that the natural gas market in the U.S. is *not* informationally efficient. This means that gas prices at widely separate localities do *not* follow each other in a manner which makes it possible to conclude that — when transportation costs are taken into consideration — these places are in *one* market, and thus the kind of arbitrage can take place which allows consumers faced with high prices to gain by buying in markets with lower prices. And not just in the U.S. A former CEO of British Gas contended that the "half-baked fracturing" of the gas markets in order to bring about competition is essentially counterproductive, and a similar argument is apparent in the work of Professor Philip Wright (2005).

Someone else with an important observation on this topic is Professor David Teece (1990) of the University of California. According to him, market restructuring in the U.S. has already "jeopardized long-term supply security and created certain inefficiencies." He also notes that "While more flexible, a series of end-to-end, short-term contracts are not a substitute for vertical integration, since the incentives of the parties are different and contract terms can be renegotiated at the time of contract renewable. There is no guarantee that contracting parties will be dealing with each other over the long term, and specialized irreversible investments can be efficiently and competitively utilized."

Readers should avoid worrying about guarantees where this topic is concerned, given the bizarre intentions of the Energy Directorate of the EU. For instance, assuming a 'path' (in, e.g., the form of a pipeline) between two markets, and the cost of shipping gas is ρ per unit, then prices (p) in these markets should lie within a distance ρ of each other, or $d(p_1 \cdot p_2) \leq \rho$. If there are no paths however, then, at one time — though hopefully not today — energy experts associated with the EU Energy Directorate expected billions of dollars to be invested in creating them, although the thinking here reduces to ideology and not economics or engineering. This is why a scholar in Milan (Italy) once used the expression "Stalinist" to describe deregulation. Although the various misunderstandings about derivatives markets (e.g., futures and options) have often been fascinating to teachers of economics and finance, they are paltry in comparison to uncertainties created by the transition from what some observers call "planning" to what they interpret as the freedom of spot markets.

By way of a partial summary of this topic, large and complex gas systems operating in a climate of uncertainty are most efficiently run on an integrated basis that emphasizes long-term contracting. *This kind of arrangement promotes optimally dimensioned installations, and although this may not be mentioned in your economics textbook, if pipeline-compressor-processing systems which fully exploit increasing returns to scale in order to obtain minimum costs are to be readily financed and expediently constructed, then the kind of uncertainties associated with short- to medium-term arrangements should be kept to a minimum. Failing to do so could cause a reduction in physical investment, and in the long run, could lead to higher rather than lower prices.* It was the proposed shift from bilateral transactions to spot markets that contributed to what is sometimes called *deregulatory uncertainty*, and a possible shortage in local (generator) capacity in the California and Alberta (Canada) electricity markets. This, together with the move to deregulated *oligopolies*, was a principal determinant of the ruinous electricity price rises faced by many households and firms in California, and perhaps elsewhere.

In Europe, the EU Commission initially mandated gas market restructuring by 2005. While they were likely sincere when they concocted this absurdity, it is unlikely that they sincerely believe any longer that restructuring can or will be taken much further than liberalization, by which they mean that anyone, anywhere, should be able to buy anything that they can afford, and if this "anything" is not for sale, then the rules should be changed so that it can eventually be put on the block. The rest of the restructuring/deregulation package — bringing into existence what they originally announced would be the kind of 'gas-to-gas' competition that is supposed to provide consumers with huge savings — will have to wait, and probably indefinitely.

One of the reasons for this is almost certainly a morale problem among deregulation proponents due to the widespread and intensive failures of electricity deregulation, but another is the negative attitudes displayed by a number of high-profile industrialists and important economists. An example of the latter is Mr. Ron Hopper, who was with the U.S. government's Federal Energy Regulatory Commission (FERC) for 11 years, and as a private consultant was an advisor to the EU Energy Commissioner and also to the 'regulator' OFGAS (in the UK). Hopper calls himself a strong believer in deregulation, but even so he said that "It is difficult for me to see the potential for pipeline-to-pipeline competition" (Hopper, 1994).

Although this author lacks Hopper's insight into the everyday mechanics of this subject, it is not just "difficult" for me to see the beauty in restructuring, but completely and totally impossible. I also have a problem comprehending why local distribution companies and consumers in the U.S. have been unable to understand that they might be forced to pay billions of dollars in transition costs in order to go from regulation to reregulation, and once they did, refrained from making their annoyance known to every politician in the country. Note: *not* in going from regulation to deregulation, but buying into a different brand of regulation.

To a certain extent, these payments were exactly what happened. Consumers and distribution companies (i.e., utilities) *were* burdened with higher costs, *and* found themselves assuming additional increments of the price risk that accompanied the various changes that were initiated. One of the reasons why things did — and were intended to — turn out this way was because, according to the deregulators and their academic booster club, consumers and distributors were certain to be big winners once changes were made, although this windfall might appear later rather than sooner. (This is also the kind of curious reasoning that the European Union movers-and-shakers specialize in.)

As for the matter of *reliability*, this was simply overlooked or ignored, although as the leading business publication *Forbes* (January 22, 2001) intimated, deregulation has "whittled away" the guarantee that many gas users in California had of a secure gas supply, since pipeline companies no longer have an incentive to resort to as much expensive underground gas storage as before, nor to employ long-term contracts (with producers) to the same extent. Let me summarize the discussion above by saying that the talk about gas-on-gas or pipeline-to-pipeline competition in the face of monopolistic/oligopolistic practices by suppliers in many regions is, at best, eccentric.

The key word when dealing with this topic is HONESTY. There is a great deal of money available for persons and organizations with a talent for circulating lies and misunderstandings, and these persons and organizations now have a large audience because of the internet. I am tempted to discuss a few of the more dishonest specimens in the academic world, but instead

I will cite one of the most honest in the business world, by whom on this occasion I mean Paolo Scaroni, who is the Chief Executive Officer of ENI, the Italian energy company. He puts it this way: "... the liberalization — and therefore fragmentation — of national gas markets no longer works in favor of the consumer. In fact it does the very opposite, putting the consumer at the mercy of the producer. Happily, in Europe, not all countries have implemented European Union rules designed to liberalize the markets. And in some countries — for instance, France and Germany — there has been a general willingness to protect national companies strong enough to negotiate with all-powerful producers."

Hopefully, the discussion above should be more than sufficient to convince alert readers and others that the corporations that have provided European consumers with plentiful supplies of low-cost natural gas for the last 3 or 4 decades should be allowed to carry on their business in the traditional manner. According to Tungland (1995), eccentric attempts to manipulate the laws of mainstream economics might prevent the mobilization of sufficient capital to realize economies of scale and to shoulder the cost of projects with very long lead times. This was his response to Professor Peter Odell, who had somehow come to believe that such things as deregulation and fragmentation could compensate for rising production costs and, apparently, a decline in the physical availability of gas and oil.

10. Some Aspects of Merit Order

A brief look will now be taken at the subject of merit order, and by way of moving into this topic, a comment seems appropriate concerning the use of gas to generate electricity in a deregulated setting. Electricity deregulation has failed just about everywhere, because in countries where there is no excess generating capacity, prices tend to rise as demand rises, and since managers prefer more money than less, they do not bother to make the investments in additional capacity that deregulation enthusiasts said they would make, and which consumers thought would be the case.

But it is more complicated than that, because there are situations in which investments are made, but even so, prices still increase. A good example here is where investments are made in gas-based capacity instead of, e.g., nuclear or coal. The advantage with gas is that in the short run it is more profitable, since capital costs are lower for gas-based generating equipment; however, gas prices have a tendency to be very volatile, and in addition have been on an upward trend over most of the last decade. As a result, consumers often find themselves presented with much higher gas bills. Moreover, even if the daydreams of the deregulation booster club are realized and generators (logically) construct smaller facilities, and there are many of them — thus giving the appearance of a competitive market — it does not make a great deal of difference to consumers, because all generators might be exposed to high and volatile gas prices, which they would not hesitate to pass to their customers.

The issue being discussed here is the so-called *merit order*, and what it boils down to is that given the structure of energy prices, using gas for base-load generation — i.e., large loads that are always on the line — rather than just for peak periods could violate the so-called merit order, and result in consumers paying prices for electricity that are higher than if the base load had been generated with, e.g., nuclear, coal or hydro.

Assuming that nuclear and gas display the highest fixed and variable costs, respectively, we can note the following cost rankings:

Fixed Cost	Variable Cost
Nuclear (F_1)	Gas (v_3)
Coal (F_2)	Coal (v_2)
Gas (F_3)	Nuclear (v_1)

To reiterate, $F_1 > F_2 > F_3$, and $v_3 > v_2 > v_1$, and in plain language, nuclear and coal facilities are costly to build and equip and, as a result, it does not make economic sense to have them standing idle a large part of the time. This makes them prime candidates for carrying the *base load*. On the other hand, with comparatively inexpensive gas turbines that are easily switched on and off, but whose fuel costs have been comparatively high until recently, the ideal role is generating the peak load, which is the load that is on the line only a small percent of the time. Hydro is not discussed here; however, when available, it is often a very economical energy source for generating electricity. One of the reasons for this is that it is suitable for carrying both the base and the peak load: its variable cost, e.g., is usually lower than that of nuclear, and additional output can be made available almost as fast as with gas.

To illustrate the key issue here we can assume that we have only two energy technologies, nuclear — with its high fixed cost (F_n) and low variable cost (v_n) — and gas with its low fixed cost (F_g) and high variable cost (v_g). If we look at the total cost for these two, *and assume linearity*, it is clear that we start out with a lower *total cost* (C) for gas ($C_g = F_g + v_g t$) than for nuclear ($C_n = F_n + v_n t$). This is because as t approaches 0, $C_g < C_n$. However, since $dC_n/dt < dC_g/dt$ as t increases, the difference between the two costs decreases, and eventually they are equal. Further increases in t have $C_g > C_n$.

In Banks (2000, 2007), this issue has been treated graphically and mathematically, but not much algebra is needed to provide readers with a useful insight into the kind of generating equipment mix that we require. Something that should be understood is that the two cost equations above are for one unit of capacity — that is to say, 1 watt or 1 kilowatt (kW) or 1 megawatt (MW) — and when looking at dimensions for cost, we have dollars per unit of capacity (e.g., $/MW). The thing that makes the following discussion different from the one in Economics 101 is that time must enter into the analysis in an explicit manner: base loads are on-line for 24 hours a day, 365 days a year, or 8,760 hours per year. Peak loads would normally be on-line for a considerably shorter time, and what we should be concerned with is allocating available generating equipment in such a way as to minimize cost, which in turn means being cognizant of the peak periods and their extent. Merely observing marginal costs is insufficient.

Continuing, in a year of 365 days we have 8,760 hours, and let us assume that the time at which we have an equality of the gas and nuclear total costs (t^*) is less than 8,760 hours. Obtaining this value is a simple matter. It is when we have $F_g + v_g t^* = F_n + v_n t^*$. This can immediately be solved to give:

$$t^* = \frac{F_n - F_g}{v_g - v_n}, \quad \text{where } t^* \leq 8,760. \tag{5}$$

Now for a key recognition. *When $t < t^*$, and one unit of capacity is to be added, it should be gas; whereas if $t > t^*$, and one unit of capacity is to be added, it should be nuclear.* This can be easily proven. Let us examine the cost situations at $t^* - \theta$, where $\theta > 0$. Then we have $C_n = F_n + (t^* - \theta)v_n$

and $C_g = F_g + (t^* - \theta)v_g$. The next step is to compare these two linear cost equations:

$$C_n - C_g = (F_n - F_g) + (t^* - \theta)[v_n - v_g]$$
$$= (F_n - F_g) + (t^* - \theta)\left[\frac{F_n - F_g}{t^*}\right].$$

In the above expression the value of $v_n - v_g$ was obtained from Equation (5). This can be simplified to give:

$$C_n - C_g = (F_n - F_g)\left[1 + \frac{t^* - \theta}{t^*}\right]. \tag{6}$$

The right-hand side of Equation (6) is unambiguously positive, and thus $C_n > C_g$ when $t < t^*$. A similar manipulation will show that when $t > t^*$ we have $C_n < C_g$. As shown in Banks (2000, 2007), the reasoning in this example is valid even if we have more than two types of equipment.

11. Russian Gas, and Comments on LNG and Shale Gas

A year or so ago a scholar from Oxford University visited the Stockholm School of Economics, where he presented a markedly unscientific version of Russian gas intentions both in Russia and in regions west of the Russian border. So many dubious statements were launched by this Oxford person and local 'researchers' during this get-together, that some members of the audience found themselves painfully aware once more of the macroeconomic and political catastrophe that may someday arrive because of a sudden shortage of energy resources — a condition that at least partially has its basis in the grotesque failures to study and teach realistic versions of energy economics.

In the wake of the 'Georgia Incident' that began around the opening of the 2008 Olympics, it appears that there has been a large-scale pilgrimage back to cloud-cuckoo land — both to reveal and scrutinize Russian geopolitical intentions, and, to a lesser extent, to circulate some bizarre opinions about the availability or non-availability of Russian energy resources. Several years ago a short article appeared in *Newsweek* claiming that oil and gas passing through Georgia was supposed to "free Europe from Russia" but, according to its author, "NOT ANYMORE". How anyone could believe that a prevailing gas superpower, Russia, was capable of having its ambitions thwarted by a few pipelines from the interior of Central Asia, is something that deserves the attention of psychologists or psychiatrists, and not readers of a weekly news publication.

The American ambassador in Stockholm also went off the deep end during that period, serving up an overdose of Cold War rhetoric to his many friends in Sweden and elsewhere. Of course, it was a widespread conclusion that he did not know what he was talking about, although as long as he had the most impressive office in the American embassy, he could not be informed of this embarrassing fact. In any event, that *contretemps* soon blew over, and in due course he returned to the U.S. with his title and his memories.

There were many items in my natural gas book (Banks, 1987) that prevented it from becoming the favorite bedtime reading of various gas experts, but almost certainly one of them was the contention that there should be more cooperation between the producers and consumers of energy resources, including Russia and OPEC, because by around the middle of this century a shortage of natural gas could be on the horizon, and it is important to use what is left of these resources to smooth out the transition toward a new global energy economy — probably one emphasizing nuclear energy and renewables. A gentleman who apparently had some difficulty with this concept was former U.S. President Ronald Reagan, as well as his advisors, because instead of buying gas from the Soviets, those energy gurus thought that some effort should be made by European consumers to obtain the supplies they required from, e.g., Africa and Argentina, arguing that by doing so it would weaken the Soviet economy. Qatar, which became the largest exporter of LNG, was not on their radar, nor on that of the many energy experts and consultants seeing action in Washington at that time.

Obviously, for a number of reasons, the Chief Executive was unable to accept a sensible strategy, which was to contract for the largest possible quantities that could be obtained from the Soviet Union, and to encourage that country to invest in (and fill) the largest possible pipelines. The basic issue was not merely safeguarding and expanding Western Europe's supplies of gas in the years to come, but increasing the general accessibility of all energy materials, including those purchased by the United States and its allies from any supplier.

When I pointed out the advantages of doing business with Russia in a talk at Cambridge University many years ago, and in addition suggested toning down Cold War rhetoric, a number of observers — including the founder of the influential publication *Geopolitics of Energy*, Melvin A. Conant — assured me and everyone else within earshot that although the ideological commitment of the Soviet Politburo was ostensibly to Marx and Lenin, it held an equally high regard for dollars and deutschmarks, which made Soviet gas industry executives prone to discharge their business obligations in a civilized manner. In the *Newsweek* article referred to above, as an aside it was stated that European gas buyers have excellent relations with Russia and do not fear greater dependence. In addition, there was a reference to the pipeline running through the Baltic Sea (called Nord Stream) which provides Germany with an increased supply of Russian gas.

I know enough about that pipeline to believe that the persons who study this project should learn to ignore the precious wisdom dispensed by Oxford University pundits and certain journalists. There was some delay with that conduit that seems to have been partially due to strange ideas in Sweden as to the ulterior purposes of the Russian hierarchy, when the most likely agenda of those good people turns on collecting as much money as possible, and sooner rather than later. What needs to be understood is the relationship between the amount of Russian gas coming into Western Europe and the price of, e.g., electricity in most of Europe — a price that would be boosted by the absurd electric deregulation agenda of the European Union, in addition to the decision to promote the sale of electricity on the electricity exchange Nordpool, which is a sophisticated version of what George Orwell once termed an "indoor welfare scheme".

According to Jeffrey Michel, an MIT engineering graduate living in Germany, and among the most important energy analysts in that country, the underwater pipeline in the Baltic Sea will mean that possible disputes between Russia and Baltic states will not lead to a reduction in gas contracted for by Germany and perhaps other countries, or even another Cold-War burlesque. A further observation has been made by the important petroleum consultant Herman Franssen, who notes that Russia is not only an energy powerhouse, but also possesses an enormous amount of unused and underused agricultural land. The efficient exploitation of this land with the possible help of experts from North America and Europe might be essential for feeding hundreds of millions or even billions of persons outside Russia at some point in the future.

Several years ago the kingpins of the European Union (EU) held a meeting at which the availability of Russian natural gas and oil was discussed at length, and the *Financial Times* (March 23, 2006) suggested that the sale of Russian gas to China and Japan might have a negative effect on the energy prospects of Europe, which may rely on Russia for approximately 40% of its gas. By extension, in the long run, this could have a negative effect on North America, because the global gas scene has started to take on some — but not all — of the features of a mainstream textbook market, due (among other things) to the ability of huge liquefied natural gas tankers to deliver 'spot' cargos. The new carriers, known as the QFlex and MFlex tankers, are designed to take 210,000 tonnes and 260,000 tonnes of LNG respectively, compared with the traditional-sized tanker loads of 135,000 tonnes and 145,000 tonnes. A theory is now circulating that LNG production will continue its growth, and by 2015 will account for up to 16% of global gas demand.

In referring to a "textbook market" I mean a market with more flexibility than a conventional LNG market. For instance, instead of a portfolio of long-term contracts in which gas carriers are locked into predetermined routes, British Gas (BG) now tries to structure its operations so that gas can be diverted to buyers that are willing to pay premium prices. At the same time, it should be appreciated that the majority of LNG business must involve and will continue to involve long-term (and relatively inflexible) contracts, because otherwise obtaining financing for the construction of LNG vessels would be more costly, and also arrangements where huge LNG tankers sail around the seven seas without fixed destinations, or unbreakable contracts, waiting and hoping for unspecified buyers to order a few tonnes of their cargo, sound weird. Even the late Aristotle Onassis might have had some difficulty contemplating this sort of approach to business.

For readers who prefer to deal in cubic feet, 1 tonne of LNG is roughly 48,000 cubic feet of natural gas, and the liquefaction process that changes gas to a liquid involves a volumetric transformation of 1/600. Today there are at least several hundred LNG tankers, and Simmons International has calculated that a new project costs on average 1 billion dollars for each million tonne of LNG.

According to the *Newsweek* article mentioned above, a recent Rice University Energy Program modelling exercise found that Russian efforts to deprive Germany of gas would likely be futile, as market deregulation would allow other suppliers to fill the gap. What this half-baked conclusion by the Rice University researchers missed is that there are no other suppliers in the short run, nor perhaps equally inexpensive suppliers in the very long run. On the other hand, in the short run, Russians could rush to completion any pipelines that they are constructing in the direction of China, and possibly beyond, in which case Germany and perhaps other European countries — and if the shale gas ventures do not match expectations, the U.S. — would also have to become involved in bidding for progressively larger and more expensive increments of gas.

The running mate of presidential candidate John McCain, Governor Sarah Palin, was apparently very receptive to larger investments in Alaskan (and perhaps Canadian) natural gas, which would eventually find its way to the U.S. Midwest. This scheme was discussed in some detail 20 years ago, and the cost was generally considered excessive at that time. At the present time the estimated cost appears to be even larger — larger than the 10 billion dollars that several Russian pipelines toward Asia will ostensibly cost. Given these circumstances the bluster originating from various critics of Russia should be restrained, because where energy resources are concerned, Russia's position with regard to alternative markets is so favorable that its government does not have to pay serious attention to voodoo economics *or* voodoo politics launched by neurotic Cold War warriors.

If there are large amounts of exploitable gas in Alaska, Governor Palin's pet project may eventually make economic sense, but at the same time it might make sense for the best geologists in the U.S. to help determine the amount of gas that may eventually become obtainable from shale deposits in or near the U.S., and perhaps elsewhere. Although it might surprise some observers, geological conditions are not the same in every region. In the run-up to the 2012 presidential election in the U.S., some of the best brains in that country expressed the need for a comprehensive energy policy, and a first step in a project of this nature is to get the hype out of present estimates of shale resources. Either that, or certain shale publicists should be taught how to lie more convincingly.

One of the editors of the *Financial Times*, Martin Wolf, has often been eager to discuss energy matters that he does not fully comprehend. He even suggested that Russian "elites" should be punished because their government overreacted in the Georgia "incident". The only kind of punishment I could imagine imposing on those ladies and gentlemen would be to prohibit them (for a season or two) from enjoying the marvellous skiing in places like Courchevel (France) and St. Moritz (Switzerland). Of course, the French and Swiss governments would have to agree to this foolishness, which is not very likely. If it were likely, however, I hope that they and their cash pay a visit to the wonderful Swedish ski resort of Åre.

Something else that we do not hear much about is a sizable participation in the near future of the Russians in the growing liquefied natural gas (LNG) market, although that option has been raised by some observers. There has also been some talk about Russian gas exports from the new Sakhalin LNG scheme gaining access to Asia-Pacific markets, which could include utilizing terminals that might open in India, and also taking advantage of the fact that Indonesia's gas fields are ageing and consequently are less attractive to potential customers. It is also interesting to note that the Russians have decided to develop the giant gas field Shtokman without foreign help, and possibly switch it from a suggested source of LNG for the U.S. to a pipeline venture whose gas is destined for Europe.

In theory there should be a place for Russian LNG just about everywhere, because while LNG accounts for only a comparatively small amount of the gas used by the U.S. at the present time, the United States Department of Energy (USDOE) once suggested that a large commitment would eventually be essential. Mark G. Papa, an important American energy executive, once said that "Right now, on the supply side, LNG is the only lever we have to pull" — although this may no longer be true because of the appearance of large quantities of shale gas. Something to watch is the difference in price between conventional natural gas in the U.S., and LNG delivered to Asian locations: late in 2009 I remember the former being quoted at about $8/MBtu and the latter at almost $20/MBtu. This price gap may be due to an increasingly higher linkage between LNG prices and oil prices, as a result of the increased demand for and shortage of gas in Asia.

It should never be forgotten though that many Americans do not want LNG plants in or near where they live. Algeria was the first country to ship

LNG, but some years later a severe accident reminded environmentalists that because of its density, LNG has a very large explosive potential, and so they informed the general public that LNG might prove to be an attractive target for terrorists. California is a state where the opposition to new LNG terminals is very strong, and as a result the next terminal serving consumers in that state will likely be in (Baja) Mexico, not too far from the large California gas market. It has been suggested, though, that the optimal position for new U.S. terminals is the Gulf of Mexico, where they could tap into existing pipeline networks.

At this stage of the discussion it seems proper to remind readers of the process involved in the obtaining by households, small commercial establishments, power generators and heavy industries of conventional gas that at some point is transformed into LNG. The first step is production of natural gas in the manner described previously. Next is liquefaction, where the gas is chilled and compressed in a manner so that $600\,ft^3$ of gas becomes one cubic foot of liquid. After that, this liquid is shipped in special vessels, and when it reaches the country in which it is to be used it is regasified. It can then be put into conventional gas pipelines and moved to buyers. In the last decade or two, costs have been greatly reduced by economies of scale and technological advances, especially in liquefaction. As with pipelines and compressors, the larger the units, the greater the cost efficiencies — up to a certain point.

LNG plants have been constructed in the fairly recent past in Nigeria, Australia, Qatar, and Trinidad, and eventually Iran should become a major supplier. In the last five years LNG has grown by almost 50 percent, but apparently demand is still greater than supply. Rumor has it that foreign firms do not want to invest in Iran because it may someday be visited by foreign military aircraft carrying bombs, but if bombs begin to fall on that country the price of oil could move off the Richter scale. Besides, with the oil price at its present level, and with the impressive technical ability possessed by many Iranian engineers, foreign investment is an option that is far from essential. Algeria was the first country to ship LNG (to the UK, in 1964), while Qatar has apparently already displaced Indonesia as the world's largest LNG exporter.

According to the *Financial Times* (October 20, 2006), Qatar exports LNG to the U.S. on short-term contracts. Why short-term? The answer of course is that they expect the price of gas to increase, and in addition they and their colleagues in the Gulf are in a position to make a reality of this expectation. The International Energy Agency (IEA) has predicted that before the present decade is over, LNG will account for up to 16 percent of the global demand for natural gas, and there has been some talk recently of the possibility of a gas producers' organization along the lines of OPEC. Someone who disagrees with this is David Victor of the University of California (San Diego), but I have decided to discount Mr. Victor's argument on this subject, and I say that there is a finite probability that we could see a producers' organization (i.e., cartel) for gas some day. That probability may now be well over 50 percent, especially if shale gas lives up to its advanced publicity and puts a sharp downward pressure on the price of natural gas.

The *Financial Times* has also stated that Qatar intends for its natural resources to benefit that country for the next 100 years. 100 is a nice round number, but if Qatar is serious and the other Gulf countries join them in this agenda, then it seems likely that low natural gas prices cannot be extended indefinitely in any part of the world. The government of Qatar once indicated that it hoped to account for 20 percent of the world's global gas market by 2010, but it does not appear to have reached this level. Moving ahead a few years, the IEA estimates that global LNG capacity will reach 600 billion cubic meters by 2015, when LNG is supposed to account for up to 16 percent of global gas demand. I am not sure that this estimate is correct.

An interesting aside here is that the best provisioned countries where natural gas is concerned, Russia and Iran, do not seem to be especially interested in extremely large LNG commitments. One of the reasons of course is that LNG is extremely capital-intensive, and producers have to run LNG facilities at or close to full capacity, or be paid higher prices than pipeline gas, in order for investors to obtain a suitable yield on capital. As a result, as larger amounts of LNG enter the picture, the *overall* (average) gas price should increase. This is a very satisfactory prospect for a country like Russia which is moving into position to supply larger amounts of (pipeline) gas to affluent clients.

It has been possible to express some surprise at two phenomena. The first is the continued 'flaring' of large amounts of natural gas, initially in the Middle East, but now in Nigeria and perhaps also in Russia. Qatar has announced that it will make the necessary investments to cease the flaring accompanying its LNG activities. According to some observers, the total amount of gas 'burned off' in 2007 had a market value of more than 40 billion dollars. Some of us are also mystified by the failure of Iran to realize its output and export potential, because it is possible to remember large and important conferences where important executives and researchers predicted that Iranian gas could be of great value to European gas users, and the sooner the better. Just as the energy in uranium and coal cannot be ignored, the same can be said for the huge gas reserves of Iran.

According to the IEA, Europe's gas imports will double by 2030, with Russia often mentioned as the key supplier. By that time, it might also be true that Iran will have realized the natural gas promise that many of us have expected for many decades. Of course, it could happen that Cold War-type innuendoes by armchair political scientists with access to the media may incite Russian producers to progressively increase the amount of gas sent to Asian buyers, and once enormously expensive investments have been carried out for this purpose, it is unlikely that Europe will ever regain its status as a favored customer of the Russians.

Quite often in this chapter, reference has been made to shale gas. I am not recommending that this topic be studied very carefully by readers without a background in geology, but they should at least be aware of what is going on and what might happen in the long run. In 2004 the important consulting firm CERA, with the widely quoted Daniel Yergin at or close to the helm, stated that "Today the natural gas industry faces the same conundrum that oil faced in the 1950s. The nation is hooked, domestic supply can no longer meet demand, and imports are inevitable." By "imports" they meant costly imports from distant sources. I held the same point of view at that time, and as a result it is not easy for me to believe the latest gas tidings circulating on the Internet and elsewhere, which is that the United States has a 100-year supply of natural gas, or more.

China is supposed to have almost 50% more shale gas than the U.S., according to a recent report by the U.S. Energy Information Administration, and Russia may also be reasonably well-endowed. The question that has to be asked here — and hopefully answered in the near future — is why is China still involved in a frenetic hunt for energy resources in virtually every corner of the globe? The EIA has not made any conclusive statements about the situation in Russia, but everything considered, that country must be pretty satisfied with the natural resource hand that it has been dealt. So satisfied, in fact, that when all of the optimism about shale gas is summed up, and the hype squeezed out, it might be possible to conclude that our energy worries will soon be "over and done with", to quote a phrase from a song that I often heard in Chicago when I was a boy. At least, over and done with for the next decade or two.

Nevertheless, to repeat what I have said throughout this chapter, and cannot stop thinking, it may be too early to jump for joy and order the champagne. A year or so ago John Curtis, director of the Potential Gas Agency, said that some of the leading newspapers in the U.S. had "the wrong end of the stick" where shale gas reserves are concerned, and he named the fact that there has been a lot of "promotion", which implies that much of this promotion is hype. Similarly, in the (UK) *Financial Times*, John Dizard stated that the faith-based assumptions about the plenitude of shale gas in the U.S. need to be examined, and especially the crank belief that an unlimited supply of this gas can be extracted at a low price and for an indefinite period. A researcher who has done some important work on this topic is Eamon Keane, and he very definitely has turned his thumbs down on some of the estimates about shale gas — estimates which imply that virtually an unlimited supply is there for the taking, if only investors would come to their senses and believe.

On the environmental front, a new study by Cornell University ecology professor Robert Howarth focuses on the methane content of shale gas, and claims that its greenhouse emissions are as large as those from oil or coal. Criticism of this study has apparently come from United Nations sources, which say that Howarth's work is questionable because he has chosen the wrong time horizon for discussing the sustainability of methane emissions in the atmosphere. Howarth has apparently taken approximately 20 years as the time horizon, while the UN people think that the appropriate time horizon is 100 years. To be explicit, the UN claim is that it will take 100 years for emissions from present shale gas production to disperse or vanish. On reading that, I remembered some of the 'scientific' work done during my 'tour' in the Palais des Nations in Geneva (Switzerland), and despite the accepted greenhouse gas content of methane, I hesitate to accept UN conclusions in this matter.

12. Conclusions: The Golden Age of Natural Gas

At the present time, oil is still the main topic in the energy debate, and until recently the importance of gas was sometimes overlooked. In the U.S., though, more than 20 percent of energy requirements are provided by gas, which also provides close to 20 percent of electric power generation. Many households obtain heat from gas, and a large amount of industrial primary energy comes from gas. This is one of the main reasons why people like the former governor of the U.S. Federal Reserve System, Alan Greenspan, and the former energy minister in that country, Spencer Abraham, pointed out that a large, efficient and affordable gas supply is almost as important for the U.S. as a similar supply of oil. Moreover, because of the high price of oil over the past few years, many business leaders have put energy first on the 'to do' list that they would like the government to accept.

Recently there were two highly visible discussions taking place about 'sets' of pipelines. One of these was the set proposed by Governor Palin of Alaska, and which as far as I know is a 'singleton' (or a set with only one member), while the other involved pipelines from the 'East' (i.e., Russia) running toward what was at one time called West Europe. I believe that this long chapter contains enough background for readers to understand the arguments of Ms. Palin in favor of the pipeline from Alaska, and, at the same time, those of her opponents in the ranks of 'Big Gas' and 'Big Oil'. In addition, as an aside, readers should be capable of interpreting the position of persons like the former American ambassador to Sweden, Mr. Michael M. Wood, who on at least one level is operating on the same wavelength as a crank ensemble of 'cold warriors' attempting to sell the notion that the Russian government is up to no good where its energy ambitions are concerned.

With all due respect, it is possible — though not certain — that Ms. Palin is correct in believing that a pipeline from Alaska is economically justifiable, and if shale gas does not unambiguously deliver what its propagandists say that it can deliver, it is possible to argue that its construction should commence in the near future. Moreover, if Mr. (former) Ambassador could alter his cold-warrior stance, he might find it expedient to convince the movers-and-shakers in his 'network' that it is a wise move to increase the availability of Russian gas to West Europe. More gas might result in lower electricity costs, which is something that everyone in Scandinavia should appreciate.

Some elements of the argument in favor of bringing more gas from Alaska to the 'lower 48' of the U.S. can be found in a paper I published on various internet sites, but serious readers with a taste for algebra should make the acquaintance of the work of Professor Hollis B. Chenery. It might also be useful to refer to some remarks made by Mr. Ed Kelly that were reproduced on the important site *Seeking Alpha*. Mr. Kelly is a vice president of the consulting organization Wood McKenzie, and he believes that the (Btu) gas price is due to move closer to the (Btu) price of oil.

The logic is simple. The forecast of the National Petroleum Council (NPC) of the United States is that gas consumption in that country could increase faster than ever. If there is a shortage of domestic gas, then in theory more LNG will become attractive, but pricewise, LNG is often in a very different category from domestic gas prices in the U.S. Although it seems unbelievable, at the present time 'spot' LNG cargos are being sold for \$15–\$17/MBtu in Asia, as compared to domestic U.S. gas selling for about \$4/MBtu. If buyers in the U.S. find it necessary to bid for these Asian cargos, domestic sellers of gas will have an incentive to raise their prices. (It has to be admitted however that shale gas may change this picture, although it will take more time to ascertain all the details of the so-called shale gas revolution.)

On this point, Carol Freedenthal claimed in the *Pipeline and Gas Journal* (September 3, 2008) that American markets have become the "dumping grounds" for excess LNG only when the high-priced markets of Asia and Europe are satisfied. Interestingly enough, at one time it was believed that the U.S. would be the most important destination of vessels carrying LNG, but as things turned out, a key supplier, Indonesia, lost some of its status in the Asian market due to ageing gas fields, and as a result, cargos that (*ceteris paribus*) might once have been destined for the U.S. are now moving toward the main Asian buyers from as far away as the West Indies.

In light of these factors, it is easy to believe that considerable thought should be given to making it easier for larger amounts of Alaska's known gas reserves (=35 Tcf according to Ms. Palin) to move toward the 'lower 48', because Asian demand is not going to subside, and it is uncertain if the construction of LNG vessels will keep pace with rising demands.

In addition, from 2007 to 2008, global electricity generation rose 4.8%, while the nuclear share of electricity generation dropped to 14% from a near steady rate of 16%–17% between 1986 and 2005. One of the reasons for this decline could well be the intention to use more gas, which unfortunately was or is the case in, e.g., Sweden. Nobody really knows.

A problem commonly associated with the 'Palin pipeline' is the time that might be required for its construction. Most observers think that it will take at least 10 years, and having once studied very carefully the modelling alternatives for the proposed Alaska Highway pipeline, I doubt whether it could be much less. 'Big Gas' very definitely does not envision the U.S. as a destination for its gas, but their judgment is of course a function of the large and expanding market in Asia.

As I have learned of late, there are plenty of self-appointed experts on the present subject who strongly object to bringing more Russian gas into Western Europe, either via a pipeline from Russia that is partially located on the bottom of the Baltic Sea (Nord Stream), or via any other conduit or carrier. To their way of thinking, that country has become too prominent on the European energy scene. An attitude of this sort sounds like a Reagan remnant, since former president Ronald Reagan (or his advisors) once came to the absurd conclusion that Europe should regard gas from Africa and Argentina as preferable to Russian resources.

At the same time, it could be argued that putting a limit on the amount of Russian gas purchased might make economic sense if more gas could be obtained from Libya, and eventually from Iran for the south of Europe. (According to the latest estimate of the U.S. Department of Energy, Libya has 54.4 Tcf of gas reserves.)

Still, it must be recognized that according to contemporary estimates, global demand for gas will double by 2025, and European demand will display the same increase by 2030. In these circumstances, even if these estimates are just partially correct, it is difficult to see the wisdom in antagonizing the Russians — especially given the option they possess for replacing sales to Europe with an increase in supplies to Asia. Exactly how easy this would turn out to be is uncertain though, because if a supplier becomes heavily involved with sunk costs (for, e.g., pipelines), then — at least in theory — it should make the supplier more reasonable in price negotiations. I should perhaps mention that several years ago I refereed an article in which the author suggested that the time was approaching when gas sellers should consider forming a cartel. An action such as this would cause a great deal of change, because to my way of thinking, judging from what OPEC has achieved, international cartels provide a blessing for their sponsors.

Among the objectives of this chapter has been my desire to set the record straight on several of the misunderstandings associated with energy economics, including natural gas. For example, I want readers of the energy literature to understand that when they see references to the work of Harold Hotelling on resources, they should think about tuning out, and if their teacher feels otherwise, then they should think about dropping out. (And, incidentally, Hotelling was a brilliant economic theorist whose work on resources could not possibly have been intended for the nonsensical uses to which it has been put.)

Another controversial topic that I spent an hour or two examining when I was lecturing on development economics in Dakar and Stockholm, and whilst doing research in Australia, had to do with the so-called 'Dutch Disease', which refers to the ill economic health that can plague an economy when it finds itself with a bonanza that has its origin in mineral wealth. Somehow it was originally thought by some confused researchers, before or after the fact, that Holland (i.e., the Netherlands) would be brought to the brink of disaster because of macroeconomic distortions accompanying the exploitation of gas in the super-giant Groningen gas structure. What actually happened was that revenues from gas may have created a very small group of losers; however, those revenues were instrumental in boosting the prosperity of a majority of residents of that country, and the country in general. A better example of course is Norway, where enormous revenues from oil and gas have not only secured the well-being of a large proportion of the present generation of Norwegians, but should do the same beautiful service for the next. The exploitation of very large mineral resources will almost always increase aggregate prosperity, assuming that at the same time it does not increase the kind of gross corruption, stupidity and greed that leads to unnecessary waste.

In my earlier period as an energy economist I was especially concerned about the possible scarcity and rise in the price of natural gas, given the possibility of a rapidly increasing demand for this commodity. At this point in time I would like to confess that on many occasions, and in many places,

I have delivered aggressive rants about the scarcity of oil, and the movement of oil to a price greater than $100/b in 2008 seemed to have confirmed my predictions, but this was innocuous in comparison to my current thoughts about the curse of electric and natural gas deregulation.

With electricity deregulation (i.e., *restructuring*) obviously imploding in many countries or regions, it might be pertinent to mention those aspects of the electricity story that are most relevant for gas. As already noted, first and foremost, we should make certain that EVERYBODY understands that *restructuring increases uncertainty, and (ceteris paribus) uncertainty decreases physical investment*. This is a straightforward neo-classical result, and from the point of view of common sense as well as mainstream economic theory, it is undeniable. At one time in Europe it appeared that deregulation, carelessly mixed with bureaucratic blundering, was a far greater danger with electricity than with gas, given that the EU energy bureaucracy wanted enough investment to take place to avoid a California-type situation. But now, with the evidence in, virtually every intelligent observer realizes that it would be best for everybody on the buy side of the market if electric deregulation was *passé*, regardless of the reason for its appearance.

Moreover, in Europe and perhaps elsewhere, restructuring means that a competitive or partially competitive gas purchasing structure could find itself confronted by powerful external suppliers operating in a monopolistic or oligopolistic mode, and Russia is not the only supplier that I have in mind. As widely known, gas buyers in Spain were occasionally unhappy with the ambitions of Algerian suppliers. They sensed an escalating gas price that would be highly detrimental to large firms as well as to households and small businesses, and needless to say, given the present condition of the Spanish economy, higher energy prices might be catastrophic. In the U.S., both former Energy Secretary Spencer Abraham and former Federal Reserve Chairman Alan Greenspan had almost as much to say about escalating natural gas prices as they did about the dangers of an escalating oil price. A mitigating factor may have appeared in the form of a large increase in the production of shale gas; however, as far as I am concerned, the jury is still out concerning the exact dimensions of this shale bonanza.

Put more directly, it might be a mistake to become overly infatuated with shale gas, although it may turn out that its potential is real. A region to pay particular attention to here is California, where a very large percentage of electricity-generating capacity is apparently fired by gas. Moreover, faith in the availability of gas appears to have been so extensive that a large percentage of the new gas-based power plants lack fuel-switching capacity, and it unfortunately seems that the older facilities with a fuel-switching option are, on average, less efficient. Moreover, according to Professor Eric Smith of Tulane University, environmental regulations (about air quality) have helped to eliminate economic arguments for dual capacity. Thus, the efficiency and versatility of the entire system might be less than it should be had the natural gas booster club been kept in its place, and nuclear energy received a better hearing.

It is stimulating to report that the majority of energy professionals are coming to their senses where the above topic is concerned, and as icing on the cake, considerably less tolerance is being shown to the ravings of flat-earth economists and their adherents where future supplies of gas and oil are concerned. What is happening is that these ladies and gen-tlemen have started paying closer attention to reality than to the kind of bizarre economic theory that became popular in the U.S. during the presidency of Ronald Reagan and the heyday of his economics role model Professor Milton Friedman, who at one time thought that the oil price would descend to $5/b. The problem here was that some economists were so enchanted with Friedman's macroeconomics that they accepted his energy economics.

I would also like to float the opinion that Mexico is not going to provide much help to U.S. gas consumers. Mexico may be destined to become a large importer of gas. Something worth emphasizing is that even if a substantial Canadian export capacity becomes available, very expensive pipelines would be necessary in order to provide large amounts of gas to the U.S. It has also been argued that increased drilling in traditional gas-producing regions in Canada is not increasing production by the expected amount, and in the Western Canada Sedimentary Basin (WCSB), production from gas wells — on average — has been declining at almost 6%/year for the last nine years. It may also be true that in Canada, as in the U.S., large producers are more likely to busy themselves with cost-reducing mergers rather than devoting scarce time and money to expensive investments in new capacity. The managers of these enterprises learned long ago that large gas fields take a long time to develop, which is something that is often overlooked by those journalists whose attention is usually concentrated on listed reserves, and at the same time believe that a *flow* from a *stock* of reserves can be obtained in no more time than it takes to manipulate supply curves in textbook presentations.

What some observers might not understand — deliberately or otherwise — is that the natural gas industry is inherently less flexible than, e.g., the electricity industry. Because the electricity sector is subject to Kirchoff's laws, many students of deregulation think that it is easier to control flows in the gas sector and thus bring about the amount of network price adjustments required to obtain (via arbitrage) the utopian results promised by the deregulators. As mentioned earlier, however, spot prices at widely separated points in large gas networks are generally not related to each other in such a way that it is possible to claim that they are in one market, and this is largely due to coordination problems that are almost unavoidable due to erratic shifts in the demand for gas. In addition, time lags are unavoidable in scheduling deliveries, which results in a sub-optimal use of storage and transmission capacity that is further distinguished by the frequent appearance of transactional bottlenecks. Deregulation is not likely to improve this situation. Even the electricity market is more accommodating when it comes to avoiding 'glitches' of this nature.

John Stuart Mill, in his *Principles of Political Economy* (written in 1848), remarked that "the laws and conditions of production partake of physical truths. There is nothing arbitrary about them." Except, to partially quote U.S. Congressman Peter DeFazio, when we are dealing with people "who are going to make millions and billions". These are persons willing to do and say anything that will convince the consumers and businesses on the buy side of gas and electricity markets that restructuring will enable them to make hundreds and thousands of dollars, and now the presence of shale gas will make their countries energy-independent. This happened in Sweden where deregulation smoothed the way for power firms to invest in existing foreign generating capacity rather than new domestic facilities. Patriotism loses its relevance when there are "millions and billions" on the table.

Regardless of how we approach oil–gas–electricity markets, we inevitably see conduct which suggests that we do not have the perfectly (or even *partially*) rational transactors mentioned in your favorite economics textbook. In New Zealand, as elsewhere, there are a number of theories about what went wrong, or is going wrong, with the domestic energy supply. The simple truth is that nothing has gone wrong. Globally, oil and (in some regions) gas have become scarcer, and the consequences are that in most countries, consumers and producers will just have to learn to transact their business against that unpleasant background instead of the make-believe world of infinite supplies of energy that they were promised by the flat-earth economists.

When I first worked in Australia, the giant Maui gas field in New Zealand was considered a priceless asset, and virtually nobody took the trouble to think of it as having a finite 'lifespan'. Now it is in sharp decline, with apparently only a few years of what are sometimes called 'recoverable' reserves left, by which it is evidently meant reserves that can continue to supply gas at the production levels of a decade ago.

I can close this long chapter by suggesting that an understanding of the political and economic circumstances of that New Zealand gas decline would provide a valuable intellectual experience for anybody who believes that even today we are running 'into' rather than out of oil and gas. These pseudo-scholars and amateur researchers are making a dismal contribution to the traumatic situation that could occur should global oil and/or gas production unexpectedly begin a sharp decline.

I also want to note that there are academics and assorted paid and unpaid propagandists who have decided to inform everyone in their 'network' that the high oil prices now being experienced are irrelevant from a macroeconomic and financial market point of view. Ostensibly, today's economies have become so sophisticated when it comes to saving energy that a global oil price averaging $100/b and perhaps rising would not pose any threat to macroeconomic stability. There is a similar belief about natural gas, in that shale gas is supposed to alleviate the shortages that we read so much about just a few years ago. The best opinion on this subject may have been rendered by David Martin (2011) in a comment on *Seeking Alpha*: "The notion that gas, and specifically shale gas, will be in such plentiful supply that it is a 'get out of jail card' for decades to come is fanciful."

I dislike passing out free advice to persons who do not appreciate it, but I think that it is best to wake up and recognize the benefits brought to this world by a large supply of comparatively inexpensive energy. In a conference of EU movers-and-shakers several years ago, it was proposed that the EU countries should formulate a joint strategy for dealing with their energy vulnerabilities. I can sympathize with this to a certain extent, although I fail to see how this suggestion ties in with the deregulation nonsense that was launched by the EU Energy Directorate. A few years ago the most important EU energy bureaucrat was a man who believed that 'peak oil' is only a theory, and even worse, announced that electric and gas deregulation is a goal worth pursuing. Accordingly, I think that we would all be better off if we ignore wisdom of this nature until the movers-and-shakers who flaunt it absorb a few simple lessons from economic history.

Key Concepts and Issues

arbitrage	peak shaving
flaring	shale gas
hubs	storage
LNG	stranded gas
market centers	wheeling
methane	working gas

Questions for Discussion

1. Discuss Professor David Teece's observations on restructuring!
2. What are some of the objections toward LNG and shale gas?
3. Why are economists always saying "on the other hand"?
4. What is the importance of the R/q ratio for the supply of gas? The present global R/q ratio for gas is about 60 years. What does this mean, and what doesn't it mean? What is natural depletion?
5. Iran is not exporting as large an amount of gas as you might expect, given their reserves. Can you discuss this situation, perhaps with the help of some other books or articles on energy or natural gas?
6. The average price of gas today is about $6.5/MBtu. What is the oil-equivalent price? Comment!
7. What are hubs and market centers? Are inventories important for the price of gas? What is the marginal convenience yield?

8. Discuss the economics of gas pipelines, making sure that you mention compressors and looping!
9. President Reagan had some interesting ideas about European natural gas. Can you clarify this statement?
10. Provide a 'rough' outline of the theoretical production economics for natural gas, making sure that you mention a production function for gas.
11. What did Governor Sarah Palin have to say about natural gas? What about the people on the other side of the table?
12. How would we use a typical production function in discussing the movement of natural gas through a typical pipeline?
13. LNG has become an important market for gas suppliers. What is the explanation here?
14. If you were giving the orders in Japan, what would you like to see for an energy market?
15. Suppose that you were giving the orders in Russia. What kind of energy market would you attempt to promote?

Bibliography

Angelier, Jean-Pierre (1994). *Le Gaz Naturel.* Paris: Economica.

Bahgat, Gawdat (2001). "The geopolitics of natural gas in Asia." *The OPEC Review*, 25(3): 273–290.

Banks, Ferdinand E. (2007). *The Political Economy of World Energy: An Introductory Textbook.* London, Singapore and New York: World Scientific.

_____ (2003). "An introduction to the economics of natural gas." *OPEC Review*, 27(1): 25–63.

_____ (2000). *Energy Economics: A Modern Introduction.* Boston, Dordrecht and London: Kluwer Academic Publications.

_____ (1987). *The Political Economy of Natural Gas.* London and Sydney: Croom Helm.

Chenery, Hollis B. (1952). "Overcapacity and the acceleration principle." *Econometrica*, 20(1): 1–28.

_____ (1949). "Engineering production functions." *Quarterly Journal of Economics*, 63(4): 507–531.

Chew, Ken (2003). "The world's gas resources." *Petroleum Economist.*

Commichau, Axel (1994). "Natural gas supply options for Europe — are distant supplies affordable?" *The Opec Bulletin* (May).

Corzine, Robert (1999). "Battle with gas gets underway." *The Financial Times* (September 23).

Cremer, Helmuth and Jean-Jacques Laffont (2002). "Competition in gas markets." *European Economic Review*, 46(4): 928–935.

Dahl, Carol A. (2004). *International Energy Markets*. Tulsa: PennWell Books.

Darley, Julian (2004). *High Noon for Natural Gas*. London: Chelsea Green.

DeVany, Arthur S. and W. David Walls (1995). *The Emerging New Order in Natural Gas*. Westport, Connecticut: Quorum Books.

Dispenza, Domenico (1995). "Europe's need for gas imports destined to grow." *Oil and Gas Journal*, 93(11): 45–49.

Esnault, Benoit (2003). "The need for regulation of gas storage: the case of France." *Energy Policy*, 31(2): 167–174.

Happel, J. and D. Jordan (1975). *Chemical Process Economics*. New York: M. Dekker.

Hawdon, David and Nicola Stevens (1999). "Regulatory reform in the UK gas market: the case of the storage auction." Surrey Energy Economics Centre.

Hodges, R.E. (1985). "P.C. program selects gas-line sizes." *Oil and Gas Journal* (April).

Hopper, R. (1994). "Open access in Europe." *The Financial Times Energy Economist* (pp. 147–151).

Hotelling, H. (1931). "The economics of exhaustible resources." *Journal of Political Economy*, 39(2): 137–175.

Karplus, R.S. (1985). "Competitiveness of Norwegian and Soviet gas supplies" (Stencil).

Martin, David (2011). "Comment on Kanellos, Michael 'Cost: the great unknown in nuclear'." *Seeking Alpha* (April 11).

Paulette, C.H. (1968). "A new approach to use of the revised panhandle formula." *The Pipeline Engineer* (March).

Söderbergh, Bengt (2010). *Production from Giant Gas Fields in Norway and Russia, and Subsequent Implications for European Energy Security*. PhD Thesis, Uppsala University.

Späth, Franz (1983). "Die preisbildung für Erdgas." *Zeitschrift für Energiewirtschaft*, 3(4): 99–101.

Teece, David J. (1990). "Structure and organization in the natural gas industry." *The Energy Journal*, 11(3): 1–35.

Tungland, K. (1995). "Comment on Odell." *Energy Studies Review*, 7(2).

Urban, Julie A. (2006). "New age natural-gas pricing." *Journal of Energy and Development*, 31(1): 111–124.

Van Atta, Lee (2007). "Natural gas storage takes off on volatility." *EnergyPulse* (November 8).

Wiggins, Dena E. (2006). "The shifting sands of US legislative and regulatory policy: Implications for natural gas supplies from foreign sources." *IAEE Newsletter*, 1st Quarter.

Wright, Philip (2005). "Liberalisation and the security of gas supply in the UK." *Energy Policy*, 33(17): 2272–2290.

Chapter 6

Economic Theory and Nuclear Energy

1. Introduction

The first three chapters of this book are designed to provide a short introductory course in energy economics, while this chapter is intended as a more
extensive survey of the economics of nuclear energy that, when necessary,
employs slightly more mathematics than in earlier chapters, and a small
amount of repetition. At the same time there is nothing genuinely 'high level'
about this chapter, because my main intention in this chapter *and* this book
is to increase knowledge about the economics of nuclear energy. In what
might be termed an ideal and logical world, this chapter — like the two
previous chapters — can serve as partial backgrounds for long lectures that
readers of this book might someday give friends, neighbors, students and
anybody else requiring an introduction to a few of the mysteries of energy
economics, mysteries that in some respects are more meaningful than those
of nuclear physics.

As is well-known, nuclear energy is not a popular medium with
everybody. Even in France, virtually the capital of 'the peaceful atom', there
are many persons who hope that someday another energy source will replace
all or a large part of the almost 80 percent of the electricity supply that
originates with nuclear power. Frankly, even if there were a major accident
in France, or a nearby country, that yearning seems unrealistic. In nuclear-
intensive France and Japan, where adequate energy is recognized as indispensable for maintaining economic competitiveness (and consequently the
standard of living), the present energy configurations have never been seriously questioned by a majority of the population. "No oil, no gas, no coal,
no choice" is the way the French put it, and although the energy prospects
of many other countries may appear to be rosier, these other countries could
also find themselves warbling the same melancholy tune some day. Certainly the people who give the orders in Japan know this, and they know
it just as well after the tragedy at Fukushima as before, because they have

made it clear that it was the tsunami that caused the trouble. What the voters know or do not know is quite another matter.

This does not mean, however, that it makes economic sense to consign conservation, renewables and/or alternative energy sources and strategies to the margins of the energy scene. The ugly fact of the matter is that the world will probably be in a very bad way if these things do not become widely available in a few decades, or perhaps even sooner, because they might have to accommodate a very large part of the energy load in all except a few lucky countries. But one way to make sure that they will *not* be available is for a majority of the voters in a given country, or even a decisive sub-set of the voters, or for that matter just the decision-makers, to accept the twisted hypothesis that it is economical to introduce renewables and alternatives on a very large scale before present and future costs, and other constraints, are fully understood. By future costs it is meant the manner in which a rising price of, e.g., fossil fuels, and governmental policies concerning carbon emissions, will influence the present (opportunity) cost of nuclear energy.

An example involving nuclear equipment might be useful here. The most widely used nuclear power generator today is the light-water reactor (LWR), so named as to distinguish it from the heavy-water or deuterium oxide technology emphasized in, e.g., Canada. (The LWR comes in 'pressurized water' and 'boiling water' versions. There is a short glossary at the end of this chapter that is intended to clarify some of this terminology.) These reactors, their predecessors, and various 'spin-offs' are generally labelled first- and second-generation models. The most modern equipment available today are third-generation models (Gen 3), such as the recently designated 'Evolutionary Pressurized Reactor', which was originally developed in France and called the European Pressurized Reactor (EPR). The intention with the EPR is to achieve an unconditional safety, in that 'meltdowns' cannot take place. (If control systems stop working, the reactor is supposed to shut down automatically, dissipate the heat produced by reactions in its core, and prevent both fuel and radioactive waste from escaping.)

One of these reactors — the largest in the world (at 1,600 megawatts = 1,600 MW) — might still be under construction in Finland, and is designated Olkiluoto 3. Given the costly delays that have been experienced, and the high efficiency of existing Finnish reactors, a question could be asked as to whether, at the present time, it might have made more economic sense to construct two more LWRs (instead of Olkiluoto 3) if the same capacity was desired, taking special care to incorporate the latest technological virtues into these reactors.

It is possible that a study of relative costs by Guentcheva and Vira (1984) of the Technical Research Center (Helsinki, Finland) was important in convincing Finnish decision-makers that there were substantial economic advantages of an *increasing returns to scale* nature to be gained by constructing a very large reactor. The equation they suggest as relevant in these matters, and intended to be applicable to Finnish conditions, is [(specific cost 1)/(specific cost 2) = (size 1)/(size 2)$^{m-1}$], where the values of m ranged from 0.38 to 0.75. As an example, suppose we take $S = (S_1/S_2)$ as a size ratio, and $C = (C_1/C_2)$ as a cost ratio, and specify that we have $S_1 > S_2$. Writing $C = S^{m-1}$, and then taking the logarithm of both sides and differentiating, we get $dC/C = (m-1)dS/S$. With $0.38 \leq m \leq 0.75$, we have $m - 1 < 0$, and so a given percentage increase in size will give a smaller increase in cost. In other words, for $0.38 \leq m \leq 0.75$, we have increasing returns to scale.

Just as important, though not emphasized in your Economics 101 book, there is little or no talk about such things as increasing returns to experience. The Swedish nuclear (and aircraft manufacturing) experience provides a useful example, as does the U.S. aircraft industry during the Second World War. After a while, persons involved in these activities stop making mistakes, find it easier to carry out their work, and even learn to suggest improvements.

It should perhaps be noted that in the above equation, size is measured by megawatts (MW) of capacity; however, its applicability to what is largely a new technological situation is uncertain, because in new situations a number of things can go wrong for 'non-scientific' reasons. Olkiluoto 3, for example, might have been plagued by 'cultural differences' between Finns and their French 'partners'. But it should be appreciated that had construction matched intentions, with 'ground break' taking place in 2005 and 'grid power' achieved in 2009, Olkiluoto 3 would have been labelled a very profitable investment. The opinion here is that even with the unexpected increase in expense due to an increased construction time, in the long run the Finns will be grateful for the presence of this equipment, and apparently two more 'Gen 3' reactors (or similar equipment) are under consideration.

It is sometimes claimed that a fourth-generation (Gen 4) reactor is no more than a decade away, with research focusing on two types of reactors:

the very-high-temperature reactor (VHTR), and the sodium-cooled fast reactor (SFR). Both provide guarantees against meltdown, while the VHTR might be much more efficient than current equipment. The SFR though is not only very energy efficient, but ostensibly can burn spent nuclear fuel rods from other reactors, as well as depleted uranium from the later stages of the nuclear fuel cycle. This terminology will become more familiar in the sequel. Once 'Gen 4' equipment becomes available, decision makers might think in terms of a much higher electrification of the economy. For instance, for private or public conveyances, many observers believe that 'plug-in electric vehicles' ('PHEV' or hybrids) have much to offer. The problem is that aggregate electricity requirements could turn out to be very large in the event of a sizable resort to plug-in vehicles, and unfortunately the equipment supplying electricity for these vehicles might only be used a fraction of a day or a week. We should be careful in judging the efficacy of, e.g., hybrids.

It has been estimated by the German firm Siemens that there will be about 400 new reactors in the world by 2030 (in addition to the approximately 440 reactors available today), although it is impossible (and unnecessary) at the present time to judge their size and exact technological details, or for that matter the correctness of that numerical estimate. Siemens once announced that it will join the Russian firm Rosatom for the purpose of developing and selling Rosatom's pressurized-water reactor in Russia, China, India, Bulgaria and elsewhere. It is likely that the construction of new reactors in Russia will be good news for Germany and several other countries, because (in theory at least) it might increase the availability of natural gas for Russia's gas customers in Europe.

The early theorists of Soviet communism seemed to believe that a Soviet-like political structure, plus electricity, would create a heaven on earth. Similarly, the implicit assumption in Sweden after the Social Democrats assumed power was that something called the 'Swedish welfare state' would emphasize social democracy plus electricity. A low-level statistical and algebraic analysis makes it clear that in terms of reliability and cost, the Swedish nuclear sector may have been the most efficient in the world before the curse of (electric) deregulation arrived. It is due to a more rational concern for the economic future, instigated by rapidly rising electricity prices, that the irrational nuclear 'downsizing' in Sweden has been at least temporarily halted.

A significant episode in the Swedish nuclear history was the upgrading of the 10 remaining reactors so that they could produce at least as much electric *energy* (in kilowatt-hours $=$ kWh) as the original 12 reactors, which amounted to more than 50 percent of the total generated electric *energy* in Sweden (with most of the rest accounted for by hydro). The logic here is straightforward, and cannot be altered by ignoring mainstream economic history: *a high electric intensity for firms, combined with a high rate of industrial investment and the technological skill created by a modern education system, will in theory lead to high productivity for large and small businesses. This in turn results in a steady increase in employment, real incomes, and the most important ingredients of social security (such as pensions and comprehensive health care).*

This is exactly what happened, and a relevant question for present-day Sweden is whether a once magnificent welfare structure that — for a number of years — was a role model for the residents and politicians of many countries, can be kept afloat if some of the most modern electricity-generating facilities in the world are scrapped or allowed to depreciate for short-term political considerations.

For instance, in order to recruit supporters with anti-nuclear tendencies, a recent Social Democratic prime minister informed actual and potential voters that nuclear power was "obsolete". The truth is that nuclear reactors might turn out to be the most important scientific inventions of the 20th century, and in the light of human needs and desires, few energy vistas have more to offer than an optimal combination of renewables, alternatives, and the reliable and comparatively inexpensive electricity that is possible with a properly designed and managed nuclear sector.

For some obscure reason, in 1978 all the major political parties in Sweden agreed that the growing controversy over the future of nuclear energy should be settled by a national referendum. The electorate was subsequently asked to choose between nuclear acceptance, or the more-or-less immediate closing of as many nuclear facilities as possible, or a gradual phase-out that was to be completed by 2010. Confronted by a barrage of anti-nuclear fictions, the latter option was selected. Although not comprehended by many Swedes even now, a key factor in that pseudo-democratic travesty was an assumption that Sweden's macroeconomic growth rate could be maintained even if the country's nuclear assets were liquidated. We see the same sort of travesty being promulgated today in Germany and a few other countries, and to add insult to injury, it is often inferred that tinkering with a country's valuable energy assets would not endanger the industrial framework or jeopardize the support of aging populations.

The prediction here is that eventually it will not be difficult to convince at least a small majority of voters in most countries that a more rational approach to nuclear energy is in their best interests. In the UK, on the other hand, some polls indicate that many voters want to see nuclear and coal-based installations phased out in favor of renewables, while the previous UK government was in favor of a nuclear revival (and a possible doubling of the nuclear output of electricity). The latter may have been because former UK Prime Minister Tony Blair and his team believed that without a larger resort to nuclear power, it would be impossible to achieve large reductions in carbon dioxide (CO_2) emissions. James Lovelock, a prominent Greenpeace supporter and environmentalist, has said that nuclear energy is "the safest and most environmentally friendly source of that vital product, electricity."

No country has made as great an effort to include renewables in the energy mainstream as Sweden, but even so, the result in terms of the energy now being generated with renewables is insignificant. Many Swedes now realize that while technically renewables can be substituted for nuclear energy, the benefit-cost ratio is probably unfavorable, and will likely remain that way. But politics are important here, and many voters unfortunately believe that their living standard can be maintained if there is a resort to inferior energy sources. *This is the biggest mistake of all*, even if in the long run it will almost certainly be judged a mistake and abandoned. UK decision makers such as Mr. Blair faced this problem too, because much of their talk about the environment was derived from their concern about energy: if more nuclear energy is not provided, there could be a greatly increased resort to coal.

2. Introducing Capacity Factors and Base Loads

Before going into some fairly technical considerations, we should consider the expression *Capacity Factor* (CF), which has to do with the amount of energy that is actually produced over a given period as compared to the amount that could be produced if the facility had operated at maximum (or rated) output 100 percent of the time. This can be written as: CF = Actual Energy Output over a given period *divided by* Rated or Maximum Energy Output. *When you hear about wind or solar energy, make sure that you ask about the Capacity Factor!* Consider a wind turbine with a *power* rating of 100 kilowatts. In a month of 30 days, its maximum *energy* output is $100 \times 30 \times 24 = 72{,}000$ kilowatt-hours. However, its *measured output* during that period might be much lower because the wind only blew half of the time. Assume that it was 36,000 kilowatt-hours. Then we would have CF $= 36{,}000/72{,}000 = 0.50 = 50\%$. As it happens, for wind, a capacity factor of 15%–35% appears average; and Jeffrey Michel confirms a stable 0.17 average for Germany before 2007. As for nuclear energy, the 'down-time' caused by unscheduled outages and scheduled maintenance has decreased, and average CFs are above 85%. Also, if capacity factors are calculated *net* of *scheduled* outages, then it is not unusual to find that they are about 95%. Now we come to a very important diagram in Figure 1.

Figure 1(a) is not about capacity factors, but about the configuration of demand, with the story told on the basis of the allocation of production capacity. To be specific, the demand for electricity (or electric capacity) typically varies during a day (24 hours) in a cyclic pattern such as that shown in Figure 1(a). The *load* (i.e., the demand for capacity) is on the vertical axis in kilowatts (kW) or megawatts (MW) or something of that nature (where 'kilo' stands for thousands and 'mega' stands for millions). Here you can think in terms of the size and number of light bulbs in your residence, many

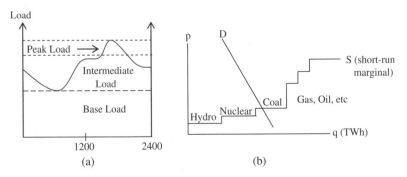

Figure 1.   (a) 24-Hour Load Pattern, and (b) Inapplicable Short-Term Supply-Demand Curves

of which are not on in the middle of the day, while all of them could be on in the evening. We might then say that your demand for electricity-based light 'peaks' in the evening.

On the horizontal axis in Figure 1(a) are hours, ranging from, e.g., midnight to midnight — a 24-hour period. (Ordinarily, as in Section 6 of this chapter, we use a one-year period, but for pedagogical reasons a one-day period is sometimes useful.) Thus, the 'box' that is designated "Base Load" is a portion of the *energy* that is expended during a 24-hour period, and this is measured in, e.g., kilowatt-hours (= kWh). The remainder of the energy for this 24-hour period (in kWh) is the remainder of the area under this curve. Continuing with the light bulb example, you do not pay the firm that supplies you with electricity for the size of your light bulbs, but you pay them for the amount of electric energy you actually require and that they provide, just as they pay the seller of, e.g., coal or gas or 'uranium' for the items they use to produce electricity (and in which the energy resides in the first place).

The base load here can be thought of as the load (in, e.g., kW or MW) that is always on the line, while the peak load is the maximum load on the line, and sometimes is only in place for a comparatively short period. (In Figure 1(a) the peak load is only on the line for a very short time, whereas in the ski area very close to my home the base load is theoretically 'on the line' from ten in the morning until eleven o'clock at night when there is adequate snow.) By definition, a base load power plant is one capable of providing a steady flow of power to a *grid* (i.e., a collection of power lines), and operates at all times except for unscheduled 'down-time' and scheduled maintenance.

Clearly, few readers would deny that the base load ought to be 'carried' by extremely reliable equipment. They should also add the expressions *base load* and *peak load* to their vocabularies as soon as possible, especially if they would like to participate in influential discussions on the present topic as an equal rather than an interloper. In addition, they should comprehend that we will not have much use for the highly stable demand curves featured in our courses in microeconomics, and in the algebra below the marginal cost sometimes appears to play a subsidiary role to the total cost. Figure 1(b) shows the kind of demand curve that is familiar to many readers, but where this topic is concerned, it unfortunately has a tendency to make appearances in important documents where it does not belong. Readers should make sure that they comprehend this matter.

Many readers probably understand that if generating equipment is extremely expensive — as is the case with a very large nuclear or coal power plant — it cannot be considered an optimal arrangement to allow it to stand idle during most of the day in order to be available during the period when there is a peak load to be serviced. Conventionally, nuclear, coal, hydro and — since the introduction of combined cycle equipment — gas have been important for the base load, while gas and to a certain extent hydro have been important for the peak load, since these can be quickly utilized when additional power is required. I have also heard it claimed that small *pebble-bed* nuclear reactors may be constructed in the future that are capable of supplying economic peak-load power. In fact, there is a claim that so-called *mini-reactors* employing this technology can be installed with no more effort or complexity than installing a large transformer. The beauty with the latter equipment is that — in theory at least — it can be hermetically sealed, and its power is available for 40 or 50 years. (*If* most of this information is true, I consider this a facet or an indication of the *flexibility* of nuclear energy.)

In the lectures I prepared for a short course in Spain, there was a mathematical exercise whose purpose is to expand the above discussion, but an important issue will be considered now on the level of simple algebra. Let us assume that we have two types of equipment, X and Y, where X has been designated as base load equipment with a high fixed cost, F_x, but a low variable cost, V_x. By way of contrast, equipment Y was purchased for peak load work, and so it has a low fixed cost, F_y, but in this case a high variable cost, V_y. The base load equipment might be nuclear or coal, while my favorite example for the resource carrying the peak load is natural gas. (In the U.S. just now, the variable cost — i.e., the cost of natural gas — is very low.)

Without being particularly concerned with Figure 1, let us call the total costs $C_x = F_x + tV_x$ and $C_y = F_y + tV_y$, where t is the time that this equipment is on the line. If we were thinking about a period of one day, then we would have $t \leq 24$ (hours), and if we were thinking about the more appropriate year we would have $t \leq 8,760$ (hours). Here we might write in a general sort of way, $t \leq T$ (hours). We might also have the case of a day in which the base load is 1,000 kW, which is on the line for 24 hours, and during 8 hours an additional 1,000 kW. The peak load is thus 2,000 kW, while the off-peak load is 1,000 kW. (Trivially, the base load plus the maximum value of the non-base load is the peak load, although obviously this can create some algebraic difficulties.) The peak load arrangement that we have in Figure 1(a) does not fit precisely into the discussion of electricity generation directly above, though it works for my local ski hill example where there is some flexibility in managing the (expected) peak on crowded weekends: *overloads* — i.e., long waits — can be tolerated to some extent on a ski hill, whereas overloads (excessive demand) in an electricity network can mean very bad trouble.

Now let us examine an expression for the *difference* in costs if we have base load equipment on the line, and peak load equipment either on the line or ready to go on the line when a peak load appears. The relevant equation can be written as $C_y - C_x = (F_y + tV_y) - (F_x + tV_x) = (F_y - F_x) + t(V_y - V_x)$. The first thing is to see what happens to $(C_y - C_x)$ when t increases (noting that at $t = 0, C_x > C_y$). A differentiation will immediately give the answer, which is $d(C_y - C_x)/dt = (V_y - V_x)$: as t increases, the (absolute) difference between $(C_y - C_x)$ decreases (and the cost of Y increases relative to X).

Think about that result as follows. For small values of t, on the basis of the assumptions above, $C_y < C_x$, and the difference between the two costs can be relatively large due to a large difference between their respective fixed costs; but as t increases, the cost of Y increases faster than the cost of X due to the difference between their respective variable costs. Thus, at very high values of t (for the period being discussed), C_y may be much higher than C_x. Now assume that at some value of the time, t^*, with $t^* < T$, the two total costs are equal. Then we have $F_y + t^*V_y = F_x + t^*V_x$. That gives us for t^*:

$$t^* = \frac{F_x - F_y}{V_y - V_x}. \tag{1}$$

We can now ask what is the cost situation when $t > t^*$, or, e.g., $t = t^* + \lambda$, where λ is a small positive number $(= t - t^*)$. Then we get:

$$C_y - C_x = (F_y - F_x) + (V_y - V_x)\left(\frac{F_x - F_y}{V_y - V_x} + \lambda\right). \tag{2}$$

This immediately reduces to $C_y - C_x = \lambda(V_y - V_x)$. In other words, if we require increments of electricity for longer than t^*, the cost of Y is larger than that of X. This makes sense! If X is nuclear and Y natural gas, we know that although X has a much larger capital cost than Y, its fuel — uranium — is usually much less expensive than natural gas for the same energy output. Thus, for operations over a long period — as in base load cases — X is preferable. (Remember also that in the very long run there is likely to be more fuel (uranium + thorium) for nuclear equipment than natural gas for gas-burning equipment, especially when breeders are used!) On the other hand, for peak loads, where the important thing is to start the equipment in a hurry and to take into consideration the limited time that the equipment will be used, Y is more suitable. After all, even if a reactor could be started as rapidly as a gas-based facility, there is no point in having an expensive nuclear plant standing idle a large part of the time.

Reasoning in this manner brings us to the important term *Merit Order*. Merit order is a way of ranking available sources of, e.g., electrical energy in terms of the *total* cost of their production. It means that ideally, energy sources are ranked according to cost and not subjective prejudices. In the above discussion, X can be shown to be more suitable for the base load than Y. Moreover, as I never tire of pointing out, Sweden once had very low-cost electricity, and one of the reasons for this was that with nuclear and hydro carrying the base load, and hydro — or electricity purchased from abroad — carrying the peak or intermediate loads, the correct merit order was being observed. What has happened of late is that the term "merit order" means little or nothing to decision makers in the Swedish and many other governments and their advisors, and so there are many suggestions and predictions for the future of the global energy structure that are completely illogical.

One final remark might be useful with respect to observations in the above and previous sections. Many years ago I was in the U.S. Army, and for a short time stationed near Yokohama (Japan), next to a small dam. I was therefore able to obtain some useful insight into the functioning of a hydroelectric installation. Even better, the last time I visited Japan I was told that large dams are often capable of 'returning' well over 50 times the energy invested in them — by which it is implied that if the money invested in constructing these dams were translated into energy units (which in theory is a simple algebraic operation) and compared to the energy generated by these structures over their expected lifetime, the ratio would be something around 50. Only nuclear energy approaches this impressive result.

3. Deeper Meanings

Among other things, this chapter is also intended as a comprehensive lecture on nuclear energy. One of the reasons for its construction in this form is that on several occasions I have been challenged to debates on the Internet about nuclear energy, but I always abstain. An Internet debate has no attraction for me. I prefer the real thing, because while my knowledge of nuclear physics is rudimentary at best, the knowledge of nuclear economics necessary to engage in a meaningful debate is something that is not on the radar of many physicists. Readers of this chapter and book should remember that.

As I previously made clear in Chapter 1, perhaps the most straightforward reasoning in favor of nuclear-based electricity is in the non-technical article by Rhodes and Beller (2000). They say that "Because diversity and redundancy are important for safety and security, renewable energy sources ought to retain a place in the energy economy of the century to come." The meaning here is clear, especially if you add that we probably will never possess what is known in intermediate economic theory as the *optimal* amount of nuclear power. Next, they unambiguously state that "nuclear power should be central. . . . Nuclear power is environmentally safe, practical and affordable. It is not the problem — it is one of the solutions."

Of course, not everyone welcomes this kind of reminder. The construction of the Swedish nuclear sector and its later development was one of the most impressive engineering phenomena of the 20th century; however, eventually a pseudo-scientific discussion began in which nuclear energy was characterized as just a "parenthesis" in world energy history, and after welcoming the opinions of various self-appointed experts, a recent prime minister called nuclear energy "obsolete". Just for the record, the Swedish nuclear sector — initially comprising 12 reactors, and probably supplying more than half of the Swedish electric energy (in kilowatt-hours) — was constructed in only 13 years (or perhaps 14 considering the usual preparatory activities). In the period before electric deregulation gained momentum, the cost of electricity generated in Swedish nuclear facilities was among the lowest in the world, and occasionally the lowest. In addition, to my great surprise, I discovered that the Swedish electricity price was also relatively low. This was especially favorable for the Swedish industrial sector, and it had a very favorable significance for Swedish employment and welfare expenditures.

In the most nuclear-intensive country in the world — which is France — the intention from the beginning was to create a nuclear sector that would provide some of the lowest priced electricity in the world, and to use that electricity to make it possible for the country to optimize its macroeconomic performance. Unlike the situation in Sweden, the French decision makers made plans to stay at the forefront of nuclear development, and in addition to provide both the industrial and household sectors with reliable and comparatively inexpensive electricity. Later, they expressed a desire to achieve and maintain a low level of carbon emissions.

Assuming that this is comprehensible, I would like to emphasize that nuclear cost issues need to be examined in greater detail for a meaningful discussion of electricity generation to take place. For France, the basic comparison initially was between nuclear and coal, and given the various costs associated with importing and using coal, it was easy to show that nuclear power was preferable. I have heard observers claim that this may not be true for the U.S., but I happen to think otherwise. A topic of this nature cannot be treated at great length in the present book, but the core of my argument turns on the supply of reactor fuel, the length of 'life' of a reactor, the lack of carbon emissions, the possibility of a radical improvement in reactor technology (such as a forthcoming *new* generation of nuclear equipment, perhaps including 'mini-nukes'), and the time required to construct reactors (which is important for the *investment cost*, which in turn is important for the *capital cost*) that will continue to decrease. I continue this discussion below, but let me note here that *I always insist that my students should be able to distinguish between investment cost and capital cost.* The failure to do so has led to some claims about the cost of nuclear electricity that are completely without any scientific value.

I have had the misfortune to see many estimates of the (investment) cost of a kilowatt of capacity of nuclear energy. Too many as far as I am concerned! They range from $1,500/kW by the director of an electricity-generating firm, to $9,000/kW by a gentleman who gratuitously stated that my work on nuclear economics was badly flawed. In my calculations I have often used about $2,500/kW as the cost of a reactor; however, I admit that I have not been able to obtain as much information about reactor costs as I want or need. By investment cost I mean the cost of the reactor *and* doing or having done the myriad of things necessary before it is ready to deliver grid power. Consider the nuclear intentions of the United Arab Emirates (UAE). For each nuclear facility, they give — in some way — the South Korean supplier 5 billion dollars, and in return they are supposed to obtain in five years a 1,000-megawatt reactor that is capable of producing electricity for a minimum of 60 years.

In a similar vein, the 'Gen 3' reactor being constructed in Finland at the present time has a capacity of 1,600 megawatts (=1,600 MW), which makes it the largest in the world from the point of view of capacity, and initially the intention was to construct it in 5 years. An early estimate of its investment cost was 5 billion dollars (which is also the apparent investment cost of each of the reactors to be constructed in the UAE). Now it appears that it will take 8 years to construct the new reactor in Finland, and it has been claimed that before grid power is attained, its cost may reach 8 billion dollars, and perhaps somewhat more. The question then becomes who pays the extra 3 billion dollars, and apparently this burden falls on the French firm Areva.

None of this reduces my enthusiasm for the nuclear option, because although much of this depressing news originates with engineers who have a far more thorough knowledge of industrial management and engineering than I ever possessed, I am satisfied with my ability to examine the issue on the basis of economics and history. First of all, we can consider the time from construction start to commercial operation of nuclear power plants in six important industrial countries. The figures that will be given below originate in the database of the International Atomic Energy Agency, and are quoted in an important article by Roques, Nuttall, Newbery, de Neufville, and Connors (2006). I have also questioned Fabien Roques — who wrote the chapter on nuclear energy in the latest International Energy Agency (IEA) survey — and he assures me that he finds them realistic. I quote the figures here employing the scheme [Country (Minimum Time, Maximum Time, Average Time)], where the times are of course construction times, and these are measured in years: China (4.5, 6.3, 5.1); France (4.9, 16.3, 7.1); Japan (3.3, 8.1, 4.7); Russia (2.1, 20.3, 6.8); UK (4.9, 23.5, 10.8); U.S. (3.4, 23.4, 9.2). In examining these, it should be clear that the average times are weighted in terms of capacity (i.e., power) or energy. Some of these numbers look odd to me, but I am prepared to accept them as approximately correct for the last few years. More importantly, the two CANDU reactors recently constructed at Qinshan (China) went from 'ground break' to grid power in 4 years. Remember that!

For a worldwide average since 1991, the figures given by Roques *et al.* are (4.0, 8.0, 5.2). With an average construction time of 5.2 years, it might therefore be possible to argue that taking 8 years or more to construct a nuclear facility is an aberration, and fairly soon the average plant should be constructed in about 5 years (which is the period named for the plants in the UAE that are going to be constructed by a South Korean firm). My viewpoint here is that when the nuclear renaissance gains momentum, the average plant will be constructed in about 4–5 years. I see no reason why the Chinese alone can apparently construct a nuclear plant in which the 'overnight cost of electric power' is about $1,500/kW, which suggests that the 'final' investment cost (per kilowatt) has nothing in common with some of the outrageous numbers that are in circulation in many countries, assuming a construction period under 5 years.

Let us use a simple analogy here. You go into a convenience store and buy a nuclear reactor for which the listed price is $1,500/kW, but to have the store set it up and put it in working order on the vacant lot near your house, you might have to pay altogether $4,000/kW in advance. This $4,000 is the (per kilowatt) investment cost, and, as in the case of the Finnish reactor, once that is paid, the responsibility is on the store to deliver grid power. It might take five years before grid power is achieved, but the guarantee given to you by the store manager is that once constructed, the reactor will function for 60 years if it receives the customary maintenance every year. (And if you bought this reactor one morning, and according to your agreement with the store it was ready to deliver grid power that afternoon at no extra cost, then the investment cost would be $1,500/kW.)

According to Donald E. Carr (1976), the Japanese were able to construct a nuclear plant within 4 years in the 1970s, which leads me to believe that they will be able to construct one in the future in 4 years or less when decision makers and voters comprehend what awaits their standard of living if they do not get the energy message. (I was also told some years ago in Vienna that the Japanese government was in no hurry to increase the size of the present nuclear inventory. What they wanted at that time was to utilize breeder reactors, which would enable a greater utilization of the energy in a reactor's fuel. Here I can say that the breeder is an entirely different proposition which requires a lot of thought.) It is possible to argue that if it takes 5 years or more to construct a nuclear plant, then coal or gas is a more economical resource for electricity generation than nuclear power; but since I expect nuclear plants to eventually take 4 years or less to construct, and nuclear fuel to be more plentiful, a sophisticated and comprehensive calculation would indicate that the nuclear option is less expensive.

Put another way, with total recycling and the widespread introduction of thorium, as well as the utilization of the latest reactors, reactor fuel may no longer function as a constraint over the next hundreds of years. As this is being written, the aftermath of the earthquake and tsunami in the vicinity of Sendai has lowered the enthusiasm for nuclear energy in Japan and many other parts of the world, but I regard this as a transitory phenomenon. A nuclear retreat has the kind of implications for standards of living that voters will not accept in the long run.

I can close this part of the exposition by suggesting that when the next (or 4th) generation reactors begin to appear, it probably will not make a great deal of difference if it does take 5 or 6 years to construct a nuclear facility. The incident in Japan, however, will not speed up the introduction of this equipment (and I was once informed by a gentleman who should know that 4th generation reactors are breeders).

As David Schlageter (2008) pointed out in *EnergyPulse*, "Renewable energy sources only supplement the electric grid with intermittent power that rarely matches the daily electrical demand." He continues by saying that "In order for an electric system to remain stable, it needs large generators running 24/7 to create voltage stability. Wind and solar generation are not always on-line when needed to meet energy demand, and therefore to help decrease system losses." In the promised land of wind energy, Denmark, voltage stability is attained by drawing on the energy resources of Sweden and Germany (and perhaps Norway). The Danes pay for the imported electricity, but not for the stability — which they would do in the great world of neo-classical economic theory. It can be suggested, though, that the Danes may be unable to afford more than the basics where electricity is concerned. According to NUS Consulting (of South Africa), the price of electricity in Denmark was the highest in the world in 2006 and the next highest in 2005. It can hardly be much lower today, since it is the highest in Europe.

Something else that you should find of interest, and which deserves close attention, is that Japan is one of the most nuclear-intensive countries in the world, but at the same time it has the longest life expectancy in the world. The life expectancy in non-nuclear Denmark (and non-nuclear Norway) is below that of nuclear-intensive Sweden and *very* nuclear-intensive Japan. Something must be drastically wrong in this world when voters and politicians ignore a reality of this nature! (The CIA 'fact book' has Monaco at the top of life expectancies, but tiny (and rich) Monaco is 'surrounded' by France.)

For geographical and industrial reasons, Sweden is one of the most energy-intensive countries in the world. Moreover, this high energy consumption should be considered a necessity rather than a luxury; it is the basis of a Swedish prosperity that once was the envy of the world. Before changing the subject, some information about the capacity factors of wind installations that was presented on *EnergyPulse* by Len Gould (2008) and Kenneth Kok (2008) can be cited. Unfortunately I cannot say whether these are extreme or typical cases, but they have one thing in common that all readers of this and other publications on energy economics should remember: the actual output from wind installations is often not just lower than the rated (or 'nameplate') output, but very much lower.

Len Gould (on *EnergyPulse*) cites an independent North American wind power company in which the actual capacity factor for 2007 was somewhere between 8.67% and 17.35%. This might be characterized as a revolution in energy technology in reverse. Even so, it was superior to a performance noted by Kok, in which a TVA facility on Buffalo Mountain (near Oliver Springs, Tennessee) registered a capacity factor considerably below the above figures. In these circumstances it should be easy to understand why it was possible to convince the voters and decision makers in Finland that in order to obtain the increase in electric energy that might be necessary for that country to maintain its standard of living, nuclear installations — with perhaps the highest (base load) capacity factors in the world — were preferable to wind turbines.

For those readers who have been exposed to secondary school algebra, the above reference to things like voltage stability is superfluous. Over the last decade, Sweden and Norway may have produced, on average, the lowest-cost electricity in the world. Norway, however, generates almost all its electricity with hydro, which is generally recognized as the lowest-cost power source, while Swedish electricity is produced in almost equal amounts by hydro and nuclear. On the basis of some elementary algebra presented later in this chapter, it can be argued that the unit cost of Swedish nuclear power is equal to the unit cost of Norwegian (and Swedish) hydro power. Of course, it needs to be recognized that numerical realities of this nature have little or no consequence in some of the higher councils of modern governments, since regardless of the realities of physics, economics and secondary school mathematics, they have decided that the illusion of inexpensive and plentiful energy from, e.g., wind and solar facilities will serve their personal agendas more satisfactorily than, e.g., mainstream science and economics.

What about nuclear waste, which is repeatedly portrayed as a malicious and unavoidable cost of nuclear-based electricity because, ostensibly, it will have to be locked up for hundreds of thousands of years? It is sometimes maintained that the *cost* of disposing of nuclear waste is balanced by the *benefit* of no carbon-dioxide (CO_2) emissions from reactors. For instance, the International Energy Agency (IEA) has calculated that in France — the country with the largest production of nuclear energy (as a percent of the total output of electric power) — the average person is responsible for 6.3 tonnes of carbon dioxide (per year), which, e.g., is one-third of the U.S. average.

The cost-benefit *trade-off* mentioned above is worth remembering; however, I prefer for students to know (and be able to explain) why France intends to treat 'waste' as a potential fuel. (A similar strategy has been proposed by a UK energy minister.) A law now exists in France stipulating that toxic waste is to be stored in such a way that it can be comparatively easily accessed and recycled if, at some point in the future, "new" technologies appear which will allow it to be classified as a preferable input in the nuclear fuel cycle.

The latter provision is, as the reader might guess, partially intended to appease or possibly bewilder nuclear sceptics, because technology is already available for recycling this *déchet*, and in the event that the price of newly mined and processed uranium escalates, it would almost certainly be utilized without further debate. French governments are seldom in the mood to play the fool for badly-educated foreign busybodies. A long exposition of how a nuclear fuel recycling program might take place is found in a conference paper by Kenneth Kok (2007); and in a paper published with Ricardo Lopez in 2008, Professor John Scire of the University of Nevada (Reno) proposes recycling as a substitute for storing the annual output of U.S. waste.

Moreover, as pointed out by James Hopf (2008), nuclear plants in the U.S. are charged a fee of 0.1 cents/kWh to pay for the nuclear waste program, and just as important it is unlikely that this cost would ever exceed 0.25 cents. This is a comparatively small amount of money, which leads Hopf to suggest that where the management of "waste streams" is concerned, nuclear advocates have a right to claim that it is a superior technology.

The same conclusion is reached by Rhodes and Beller (2000), and here it needs to be emphasized that nuclear technology is in its infancy. The plant being constructed in Finland at Olkiluoto is a so-called 'third generation' plant, where the emphasis seems to be on eliminating the likelihood of a 'meltdown', but the real prize should turn out to be a 'fourth generation' installation, which operates at radically higher temperatures that permit a more thorough exploitation of its fuel. It has been suggested that this kind of facility will be especially important if the supply of uranium is reduced, and as noted by many comments published in *EnergyPulse* and elsewhere, persons like Malcolm Rawlingson (2011) who have worked with or near uranium doubt whether there will be a shortage of this commodity in the foreseeable future, even if a nuclear revival eventually assumed the dimensions of a Manhattan Project. It can also be mentioned *en passant* that there is at least as much fissionable thorium available in the crust of the earth as uranium, and Norway may well be a source of some of the largest deposits. Consequently, the bad conscience that some Norwegians seem to display because of the wealth of their country will probably be shared by many future generations.

In the case of Sweden, the low cost of nuclear and hydro power, and fairly smart regulation, made it possible to provide electricity to the industrial sector at a comparatively low price — at least until the curse of electric deregulation was welcomed by politicians and their academic experts. This being the case, nothing is more offbeat than hearing about the "subsidies" paid to the nuclear sector. Cheap electricity meant the establishment of new enterprises, and just as important, the expansion of existing firms. The tax revenue that was directly and indirectly generated by these activities, and used for things like health care and education, more than compensated taxpayers (in the aggregate) for any 'subsidies' that might have been dispensed by the government.

An antithetical situation may prevail for wind and biofuels. In Germany the energy law guarantees operators of wind turbines and producers of solar energy an above-market price for power for as long as 20 years. This is an explicit subsidy, although it may be both economically and politically optimal due to the reduction in greenhouse gas emissions. More importantly, in theory at least, some inexpensive electricity for plug-in hybrids could be made available, since the batteries for these vehicles could be recharged at night.

An especially complex subsidy apparently accompanies the exploitation of biofuels. Research in the United States, and reported in the influential journal *Science*, claims that almost all biofuels used today result in more greenhouse gas emissions than conventional fuels if the pollution that is both directly and indirectly caused by producing these 'green' fuels is considered. In addition, there would be a substantial loss of 'consumer surplus' throughout the world due to a likely increase in food costs.

Some of the intricacies of this important issue have been examined on an elementary level by Clay Ogg (2008). Recent research from the University of Minnesota claims that ethanol from corn grain produces 25 percent more energy than *all* the energy that was invested in it, whereas biodiesel from soybeans returns 93 percent more. Moreover, compared with fossil fuels, ethanol produces 12 percent fewer greenhouse gas emissions, while biodiesel produces 41 percent fewer. The problem is that if all the corn and soybean production in the U.S. served as an input, it would only satisfy 12 percent of that country's gasoline demand, and 6 percent of diesel demand. Clearly, other biofuels are going to be necessary, and it has been suggested that prairie grass has a great deal to offer, including much larger environmental benefits.

Against this background it might be argued that a 'French' type acceptance of nuclear power makes a great deal of sense, and its details deserve more attention. As noted in the *Financial Times* (October 6, 2006), nuclear power has provided "an abundance of cheaply-produced electricity, made the country a leader in nuclear technology worldwide and reduced its vulnerability to the fluctuations of the turbulent oil and gas markets."

France can also supply some electricity to neighboring countries, which helps counterbalance the short-sighted foolishness being promoted by the European Union's Energy Directorate. (See also the survey by Murray Duffin (2004) and the comments on that survey.) Finally, energy from nuclear installations might prove to be an exceptionally valuable input in the production of biofuels.

4. A Minimal Outline of the Nuclear Fuel Cycle

"Satisfaction ... came in a chain reaction."
— *"Disco Inferno"* (*from* Saturday Night Fever)

In courses on energy economics and international finance, I have made a point of informing students that there are certain things I expect them to learn *perfectly*, and the same will apply to the items in this section the next time I teach energy economics. The reason is simple, and is based on the likely appearance of considerable new nuclear capacity (especially in Russia, the U.S., India, China and eventually Japan) that deserves to be studied and understood by persons who may find themselves in a position to influence the configuration of the energy structure in their country or their local community, or, for that matter, merely to comprehend and explain some aspects of nuclear energy to friends and neighbors. There will also be new nuclear capacity in localities where politicians and their foot soldiers have repeatedly taken what amounts to a sacred vow to never build or tolerate the construction of another reactor, because once voters fully grasp what the lack of abundant energy will mean for them *personally*, a more rational reassessment of the nuclear option will likely take place.

It is often said that the world's first (man-made) 'nuclear reactor' — which in reality was an experimental device of a very primitive sort whose function was to obtain the first man-made sustained nuclear reaction — was constructed by Enrico Fermi. This took place in the squash courts under the stands of the football stadium at the University of Chicago.

The first peacetime nuclear plant was used to power the U.S. submarine Nautilus — although some observers preferred the label 'wartime extension' to 'peacetime'. This was in January 1954. Six months later the Russians constructed a small nuclear-based installation whose purpose was to supply power to non-military users, and in October 1956 the first full-scale nuclear plant for civilian use was opened at Calder Hall in the UK. The first genuine 'civilian' power plant in the U.S. began operation in 1958 at Shippingport, Pennsylvania. Perhaps the most interesting event during that phase of the Cold War, however, was the launching of the submarine USS Sea Wolf in 1956, which evidently contained a liquid-sodium cooled reactor burning highly enriched uranium, and resembling the type of equipment that Ralph Nader once referred to as "maniacal".

We can now turn to the simple physics of nuclear energy. Energy produced from fossil fuel is the result of an uncomplicated chemical process; however, energy produced from nuclear fuel originates in the force binding the constituent parts of the fuel's atoms together, and its release features the alteration of the structure of the atom itself, and thus an enormous potential output of energy. This is one of the reasons why the Nobel laureate Professor Dennis Gabor called the nuclear reactor the most important scientific achievement of all time. There may be some question as to whether it deserves that lovely designation. However, since I was recently informed that one pound of nuclear fuel has the same energy content as 100,000 pounds — or more — of coal, then in many respects nuclear reactors are the most sophisticated.

Two terms probably already found in the vocabularies of readers of this exposition are *molecules* and *atoms*, but readers should be reminded that the latter is essential when examining the present topic. The expression "molecule" was coined by René Descartes in the 1620s, by which he meant an extremely minute particle. Molecules are made up of at least two kinds of atoms in a definite arrangement, held together by strong chemical bonds. For instance, the water molecule is composed of hydrogen and oxygen atoms, and designated H_2O. Atoms are generally thought of as 'indivisible', or the smallest particles characterizing a chemical element, but in reality sub-atomic particles have been identified.

Almost all of an atom's mass is found in its nucleus, which contains neutrons and (positively charged) protons, surrounded by swarms of (negatively charged) electrons, and the larger this nucleus, the easier it is to obtain the desired release of energy. Uranium is very important because one of the heaviest (and most complicated) atoms in nature is the *isotope* 235 of uranium, which is the only *naturally occurring* nuclear 'fuel' that will support a chain reaction. Its conventional designation is U-235, and it is important to know that different isotopes of an element occupy the same position in the periodic table, but they do not have the same weight. U-235 contains 235 'particles', with 92 protons and 143 neutrons. The other isotope of uranium is U-238, with 92 protons and 146 neutrons. (The difference in weight is attributable to the neutron difference.) There are also some 'trace elements' of U-234.

Fission is the breaking apart of a nucleus following the absorption of a neutron. If U-235 absorbs one additional neutron, it can become unstable and divide into two or more 'fragments' (sometimes called "atomic nuclei"), in addition to several neutrons. The mass of these fragments and neutrons is now somewhat less than that of the original nucleus and, most importantly, the reduction in mass corresponds to an increase in kinetic energy (i.e., motion), which is converted into heat as the fission products collide with surrounding atoms. Other U-235 atoms may absorb the neutrons released by a previous fission and may themselves undergo fission. A release of neutrons that leads to further fission constitutes a "chain reaction". What we have here is a mass-to-energy conversion of the kind associated with Albert Einstein's famous equation $E = mc^2$, where m is mass, c is the speed of light in a vacuum, and E is energy. This equation specifies that the amount of energy that can potentially be released by a very small mass is huge.

A complication, however, is that once a chain reaction develops, some sort of control is necessary to ensure that it continues at a steady level: sufficient but not an excess of neutrons must be obtained, and they must move at the right speed. On average this means that one neutron should lead to *only* one more fission. What is *not* desired is an *uncontrolled* exponential growth of fissions, which in the worst of cases could result in a *meltdown*, i.e., an overheating of a reactor core, or even an explosion. (It can also be noted that the neutron — discovered in 1932 by James Chadwick — is the key to nuclear fission, because as a result of being neutral, it is not influenced by the 'Coulomb' forces that are associated with atoms.)

Occasionally we hear the expression *critical mass* in the discussion of this process. (This is also used in socio-dynamics, where it means the existence of sufficient momentum in a system so that the process becomes self-sustaining in that it nourishes further growth. 'Bandwagon effect' was a popular expression when I studied economics, and it had to do with a kind of 'critical' stage in the economic development process.) It is important to know that U-235 is fissile, but not U-238; however, it is equally crucial to recognize that U-238 is *fertile*, which means that it can be the source of fissionable material not found in nature if it is bombarded with neutrons in a reactor. That material is plutonium (Pu, or Pu-239). Another fertile element is thorium, which was mentioned previously, and reportedly it is at least as abundant in nature as uranium. It can also be made fissile via neutron bombardment in a modified conventional reactor.

Natural uranium consists of 99.3% U-238 and only 0.7% U-235, and any variant of uranium ore or processed uranium with the same isotopic composition found in nature carries the delineation 'natural'. (There is a slight approximation here because there is a minute quantity (or 'trace element') of U-234 in natural uranium that is always ignored when discussing fission.) The relatively small amount of U-235 introduces a complication into obtaining a chain reaction, because enough enrichment must usually take place to raise the amount of U-235 in reactor fuel to at least 3%.

At the same time it should be understood that there are reactors — such as the Canadian CANDU reactor — where unenriched natural uranium is an input. What characterizes this equipment is a thorough removal of non-uranium impurities, which, together with the employment of suitably designed neutron reflectors and a *heavy-water moderator*, can provide a chain reaction. Although enrichment is a very costly activity, there does not seem to be any hard evidence that (economically) CANDU-type reactors are superior to light-water equipment, apart from CANDU managers not having to worry about an unexpected spike in the cost of enrichment, which is a discomfort that LWR managers might suffer if enrichment takes place externally.

The term "moderator" used above has to do with reducing the speed of neutrons, so that the main source of energy is the break-up of heavy fissionable atoms that are struck by relatively slow (rather than fast) neutrons. Often the moderator is water or gas, but in the CANDU it is heavy water, which means water containing deuterium atoms. By way of contrast, the breeder is often called the 'fast breeder' because neutrons are not slowed down, and in the breeding process, neutrons from the splitting of Pu-239 convert non-fissile U-238 to additional Pu-239. According to Barre and Bauquis (2008), a unit of fuel in a breeder can provide almost 100 times

more energy than it would in, e.g., a LWR. Here I want to highlight the book
by Barre and Bauquis, which is both non-technical and very informative.

Now for a brief resumé of the nuclear fuel cycle. The *front end* begins
with mining. This activity is not as straightforward as it sounds, because
the *ore* that is mined usually contains well under 1% uranium. In its pure
form, uranium is a silver-gray metallic chemical element that is approximately 70% more dense than lead (and weakly radioactive), and this metal
can be obtained by crushing and grinding the ore. However, since there is
relatively little demand for uranium metal, further processing in the form
of milling and leaching (with sulphuric acid) must take place. This results
in a substance called 'yellowcake', in which form uranium is usually sold.
(Remember that uranium metal and yellowcake are classified as natural
uranium.) In Professor Anthony D. Owen's seminal book on the economics
of uranium (1985), he indicates that there is much more uranium in the crust
of the earth than commonly believed.

The term "usually" was employed above because, apparently, in some
markets the relevant commodity might be sold under the designation U_3O_8
(or *triuranium octoxide*), where the isotopic composition is the same as in
nature, and which leads to it also being called natural uranium. The key thing
to understand here is that yellowcake tends to have a purity of 70%–90%
U_3O_8. The quoted price of yellowcake that is found in the trade literature
is thus (logically) a function of its purity (relative to pure $U_3O_8 = 100\%$).
Regardless of whether we prefer thinking about yellowcake or U_3O_8 (or
some other oxide of uranium), these are the basic materials for conversion
into *uranium hexafluoride* (UF_6), which is the form required by most commercial uranium enrichment facilities. In terms of isotopic configuration,
UF_6 is still natural uranium, but although a solid at room temperature, it can
be heated into a gas that is suitable for enrichment.

As has become well-known to television audiences, the purpose of
enrichment is to increase the percentage of fissionable U-235 in a bundle of
natural uranium from approximately 0.7% to about 3%, or perhaps slightly
higher. In other words, to raise the ratio of U-235 to U-238. This is a very
electricity-intensive and therefore costly activity. The higher the degree of
enrichment, the easier it is to maintain a chain reaction, and so the volume
of the reactor can be reduced. (There has been a great deal of talk recently
about certain countries taking enrichment to a point where they can obtain

weapons-grade uranium, which means enrichment to about 93% U-235.) Uranium enrichment services are sold in Separative Work Units (SWUs), which is a measure of the amount of 'effort' needed to obtain the desired intensity of U-235 in a 'package' (or 'batch') of nuclear fuel.

A useful term in this process is depleted uranium (DU), or *tails*, which is a by-product of enrichment. It is the uranium remaining after removal of the enriched fraction. The tails also contain some U-235, but much less than that found in natural uranium. The remaining proportion is described in terms of a *tails assay*, and according to Professor Owen (1985), is typically between 0.2% and 0.3% (as compared to 0.7% in natural uranium). The lower the tails assay, the more energy (in SWUs) required to produce a given amount of enriched uranium. It also turns out that producing a ton of (3%) enriched uranium requires about 6 tons of natural uranium in a suitable form, with the remaining 5 tons called depleted uranium (or tails).

Enrichment is obviously a very complicated process; however, its technology could be greatly advanced by moving from gaseous diffusion to the centrifuge system, and further improvements are almost certainly possible. Once UF_6 is obtained, the next stage is enrichment and the conversion into uranium dioxide (UO_2), which can be transformed into pellets. The pellets are loaded into specially designed tubes. In a light-water reactor, the rods are inserted into the reactor where fission takes place, and the ensuing heat raises steam in a boiler which turns a turbine-generator that produces electricity. From the boiler to the back end of the cycle, a nuclear power plant is the same as a plant operating on coal or gas, with approximately the same thermodynamic characteristics.

The thing to appreciate here is that a reactor is primarily a source of heat, and this heat turns water into steam that is converted into mechanical energy by a turbine, and into electricity via a generator. Much of this heat is lost, although ideally it could be a valuable input to households, industries and various commercial establishments.

When a reactor has been in use for a certain period, the percentage of U-235 in it has decreased, and because of this and the contamination of the fuel elements by fission products, the efficiency of the chain reaction is reduced, and eventually it cannot be sustained. *Spent fuel* is then removed from the reactor and 'fresh' fuel inserted. The spent fuel is mostly (but not entirely) U-238, and it cannot be used in 'slow' reactors, but if put in, e.g., breeder reactors and exposed to high-energy neutrons, it can be converted to fissionable isotopes of plutonium. (This expression "tailings" is also sometimes used with respect to spent fuel.)

A peculiarity with the cycle discussed above is that, theoretically, it is incomplete. The spent fuel that is taken from the reactor is usually stored. However, it could be reprocessed and in one form or another fed back into the reactor, thereby completing the cycle. If this is not done, what we have is a *once-through cycle*, where the spent fuel is put into temporary storage and kept there until consigned to permanent storage, preferably underground. *Put another way, however, this spent fuel is not 'waste' — which it is often called — but potential reactor fuel, because it contains an impressive amount of fissionable materials.* Were it not for political and other constraints, more of it would be turned into plutonium (which could be used in a breeder reactor) or a plutonium-laced mixture called MOX (mixed-oxide fuel) for reinsertion into modified conventional reactors. Perhaps the main bugaboo is that reprocessing involves the handling of a relatively large amount of plutonium, which is a substance whose presence in large quantities is not to be recommended at the present time.

It should be noted that two terms that often appear when the conversation turns to reactors are *thermal reactors* and *fast reactors*. Both require a fissionable fuel, which for a fast reactor can mean Pu-239 as well as U-235, and both require a coolant to counteract the heat that is created as a result of fission. Thermal reactors also require a moderator to slow down neutrons, as well as various components to ensure that the fission is controllable. The fast reactor also requires a mechanism for control, but it is very different from that employed in a thermal reactor. These are a few of the special items that deserve attention, especially if the fast reactor is also a breeder (i.e., creates more fuel than it uses), but they are best passed over in this short overview.

5. Some Basic Analytics

Readers of this chapter should now be in possession of sufficient terminology to convince fellow party animals that they have something useful to offer when the discussion turns from jazz or Frank Sinatra to nuclear energy; and for persons who feel at ease with what we are doing, the present section should enhance their knowledge of the subject. Particular attention should be paid to Equation (4) below, including the algebra needed to obtain this expression and the ensuing explanation.

Among the items that will be taken up below are the capital cost of, e.g., nuclear energy, and an analysis that emphasizes the so-called *burn-up* (which has to do with the efficiency of using uranium, since the larger the burn-up, the greater the efficiency). Finally, I want to discuss using a few equations a contention I have made in many lectures about the comparatively low cost of nuclear-based electricity in Sweden. To be exact, since Sweden and Norway may still produce the lowest-cost electricity in the world — although Norway employs almost exclusively hydro, and Sweden nuclear and hydro — some algebra shows that Swedish nuclear power has about the same low *cost* as Norwegian and Swedish hydro power. The price at which the generated electricity is sold is another matter.

Although the lowest-cost electricity does not mean that buyers of this good will always enjoy the lowest price, it is surprising (to me) to observe that over the past decade, the price of electricity in Sweden was among the lowest in the world. That price is rising now, largely because of the Swedish entrance into the European Union, which in turn seems to have led to an *extreme* and irrational form of electric deregulation — extreme because of the excessive liberties given to the executives of electric utilities. On the other hand, at the present time, Denmark may have the highest electricity price in the industrial world, which can be explained by its fairly large reliance on wind power.

Next we turn to a useful pedagogical first step for working our way toward a key economic concept. This involves a two-year situation in which $1,000 is borrowed and used to invest in an asset, for example, a mini-reactor that will be placed in the basement of your house, and which will be amortized (i.e., paid for) in two payments over (*an amortization period of*) two years. The rate of interest (r), i.e., the discount rate, in this example will be taken as 10% (or 0.10). (Amortization means repaying a debt, which in this example is tied to the purchase and cost of a reactor.)

It is here that we introduce the term *annuity*, which is the amount (A) paid at the end of every period (e.g., one year), and as will be calculated below, the annual amount A in this example is equal to $576. This means that in repaying the debt (= $1,000), we pay $576 at the end of the first year, and also $576 at the end of the second year. The debt 'today' is $1,000, and if paid at the end of two years, the lender would receive $F = PV(1 + r)^T = 1,000(1 + 0.1)^2 = \$1,210$ if 10% is the rate of interest. Let us put this as follows: $1,210 in 2 years has a *PV* of $1,000 if $r = 10\%$. Also, take note that $1,000 is *not* the capital cost; it is the *investment* cost. On the other hand, the $576 is the *levelized* capital cost. (Put another way, it is the *amortization cost* for a loan of $1,000, and an interest rate of 10%.)

We can take a closer look at this theme, continuing with the above numbers, and then moving on to some algebra that systematizes the discussion. There is a payment of $576 at the end of the first year, and this is equivalent to $576(1 + 0.1) = \$633$ at the end of the second year. If we add this to the annuity payment (*A*) of $576 at the end of the second year, it sums to approximately $1,210, or the same as the single final payment (*F*) above. It can thus be specified that, *ceteris paribus*, paying $1,000 now for the asset, or paying $1,210 (= *F*) at the end of two years, or paying $576 (= *A*) at the end of the first and second years, are (in theory) equivalent, given that 10% is the applicable rate of interest. Note the *ceteris paribus* criterion, because obviously in real life there are situations where this 'equivalence' would not be acceptable, particularly by a lender.

Something else that can be mentioned is that if the reactor had been paid for in cash removed from your wallet or purse at the time it was purchased, rather than borrowing, the concept of an annuity would still be valid. In this case the annuity payments would represent the *opportunity cost* of purchasing this asset instead of, e.g., lending the cash today to a person or a firm and earning interest (amounting to, e.g., $210 after two years).

Now for an algebraic generalization that has already appeared in this book, but which will be extended below. Perhaps the best way to begin is to note again that to pay a debt of *PV* (= present value) entered into at the beginning of the first period, is to pay $PV(1 + r)^T$ at the end of *T* periods, or equivalently annuities *A* at the end of each period, beginning with the *end* of the first period and ending (with *A*) at the end of the last period! Thus we get:

$$PV(1 + r)^T = A + A(1 + r) + A(1 + r)^2 + \cdots + A(1 + r)^{T-1}. \quad (3)$$

Table 1. Amortization and Interest Charges in a Two-Period Example

Year (T)	Beginning balance	A	Interest	Capital charge	End-of-year balance
1	1,000	576	100	476	524 (= 1,000−476)
2	524	576	52.4	524	0

Multiplying both sides of this expression by $(1 + r)$, we obtain:

$$(1 + r)[PV(1 + r)^T] = A(1 + r) + \cdots + A(1 + r)^T.$$

And subtracting the second of the above expressions from the first yields:

$$[(1 + r)^T]PV[1 - (1 + r)] = A - A(1 + r)^T.$$

From this we get an equation that we saw earlier in this book, and whose importance has been referred to on a number of occasions:

$$A = \left[\frac{r(1 + r)^T}{(1 + r)^T - 1} \right] PV. \qquad (4)$$

If we make $PV = 1,000$, $r = 0.10$ (i.e., 10%) and $T = 2$, then we can obtain $A = 576$ from this relationship. The cost A, which is the annual payment on a loan, or received from an annuity, is sometimes referred to as the 'levelized cost', or perhaps better the levelized (annual) capital cost, which takes into consideration both interest and capital charges. Now examine Table 1, which is self-explanatory (and as an exercise, explain in detail all the entries in the second row of Table 1).

Let us use the values in Table 1 to make a very important point: one which I was once asked about, and unfortunately gave the questioner a wrong answer. To begin, we can obtain the present values of the capital charges (476 and 524). These are clearly 476/1.1 and 524/1.21, assuming that the 'discount rate' (or rate of interest r here) is 10%. [Readers should make sure that they know what is happening, remembering that present value (PV) is defined as future value (FV) divided by $(1 + r)^T$.] The two present values are approximately 433 for both $T = 1$ and $T = 2$. Summed, they equal 866.

Next, we should obtain the present values of the interest charges. From Table 1 we see that these can be calculated as 100/1.1 and 52.4/1.21, and these sum to approximately 134. Readers should make sure that they understand these calculations. Finally, if the present values of capital charges and interest charges are added, we obtain 1,000, which is the investment cost. In other words, the $1,000 investment cost consists of discounted capital and interest charges.

This discussion can be concluded by summarizing and expanding the fairy tale presented before the derivation of Equation (4). The mini-reactor put in or near your house costs $1,000. This is the investment cost (I). The levelized capital charges (A) are $576 for both years, and from these we compute capital charges for the two periods, i.e., $476 and $524 (which are obvious from Table 1). The summation of the present values of these capital charges gives us approximately $866.

Although not often discussed properly, in this simple example the $866 is what is known as the *overnight charge*, which, according to Google, is the part of the capital cost that does not include interest (or $134). Thus, when you contact the seller of the reactor who explains that he will sell you a reactor and have it installed in two years (at which time it will be ready to deliver electric power), he is saying that the $1,000 he wants from you now (or $576 at the end of each year, or $1,210 at the end of two years) is for the reactor — whose purchase or construction and installing has a present value which is called the overnight charge, and in addition there is an interest charge that is an opportunity cost, in that it represents the income the reactor seller is giving up in order to do what is necessary to provide your reactor.

What about the profit of the reactor seller? That is in the capital charges, along with various other costs that may be involved in fulfilling the agreement with the buyer of the reactor. Of course there may not be a profit, because the reactor seller may have misjuged the cost of constructing or obtaining the reactor, as well as installing it. Something like this happened in Finland.

As was done earlier, Equation (4) can be easily obtained with some calculus, beginning with a basic (neo-classical) economic concept: the cost of an investment is the uniform return per period that an asset must earn, in order to achieve a net present value of zero. In other words, the asset price is the present value of future net yields (i.e., revenues minus costs). Notation in this derivation is changed somewhat in order to correspond to standard usage. Taking I as the asset price (i.e., the investment cost), P as the capital cost per period, and r as the market discount rate, we write for T periods:

$$I = \int_0^T Pe^{-rt}dt = \frac{P}{r}\left(1 - \frac{1}{e^{rT}}\right). \tag{5}$$

It takes a little manipulation to obtain $P = re^{rT} I/(e^{rT} - 1)$, and remembering that we can approximate e^{rT} by $(1 + r)^T$ for small values of r, we get Equation (4), though with a different notation. The discount rate here is the market interest rate, which assumes no risk/uncertainty on the part of lenders and borrowers. This is not the kind of recommendation to be taken seriously outside a seminar room.

Before continuing, a few more words dealing with the above might be appropriate. I neither have nor am interested in trade secrets having to do with actual firms in the nuclear business, but I know from reading the popular press that a South Korean firm has agreed to construct 4 reactors in the United Arab Emirates (UAE) for 5 billion (U.S.) dollars each, which to me means that they promised grid power in 5 years. (This was also the case in Finland, although that deal did not work out as planned.)

One way that this might function is that each reactor buyer gives the constructing firm 5 billion dollars, and that firm buys (or constructs) the reactor and does all the other work necessary to provide grid power in the agreed upon time. The 5 billion must not only pay for the reactor — in whatever shape it is delivered — but also cover the salaries of engineers, workers, managers, technicians, and all other inputs. The profit of the firm (or firms) constructing the reactors is also included in the 5 billion, and so if they construct a reactor in less than 5 years, then (*ceteris paribus*) their profit is greater, whereas if it takes longer, then instead of profits they might register losses. The latter is what happened in Finland, where a 5-year project became 8+ years, and the French firm Areva — which was in charge of the construction — had to accept all or most of the losses.

We can proceed by considering the difference between an ideal nuclear reactor (or, for that matter, an ideal furnace of any kind) and a real-world reactor. One of the differences is that in an ideal reactor or furnace, there is no heat loss. Many years ago I often shovelled coal into the furnace of a house in Chicago, and if that furnace had been an ideal furnace, all of the energy in the coal would have been transmitted — without loss — to the radiators and light bulbs in that house.

But heat loss is a thermodynamic fact of life, and so to have access to one million watts *electric* [= 1 MW(e)] we might need, e.g., three million watts *thermal* [= 3 MW (thermal)], where thermal relates to the fuel being used and the construction of the equipment in which it was used. In this discussion the fuel is uranium, and enough of it should always be available in the reactor to provide the desired energy output. Note that above we do not have megawatt-days, which is *energy*, but megawatts (= MW), which is *capacity*. Accordingly, to obtain 1,000 MWd(e) — or 1,000 megawatts for *one* day — we may need 3,000 MWd (thermal) — or 3,000 megawatts thermal for *one* day — assuming the capacity factor was 100%. However, it might happen that the reactor (or generating equipment) is 'down' an average of 15% of the time, so the thermal requirement for that day might only be $3,000 \times 0.85 = 2,550$ megawatts.

Heat loss is not the only bad news here, because the fuel that is inserted into a reactor is *not* pure U-235, but, as indicated earlier, it is a bundle of U-235 and U-238 that has been enriched from 0.7% of the former to at least 3%. As also noted, an item on the positive side is that some of the U-238 can be converted by neutron absorption into Pu-239 (plutonium) and fissioned, but this process is slower than the fissioning of U-235. It therefore turns out that to complete our calculation with the smallest amount of effort, we would like to have a technological parameter that will tell us about the efficiency of utilization of the (enriched) uranium fuel — i.e., fuel containing both U-235 and U-238 — and this is called the *burn-up*. *The burn-up is the total amount of heat energy created per unit of enriched uranium fuel (i.e., enriched uranium).* In economics this would be called an input-output coefficient, and a number that I have seen many times is 33,000 megawatt-days per tonne of enriched uranium (or 33,000 MWd/t). I have been assured many times in the past by nuclear engineers, however, that this number will increase, and this might already have taken place.

The use of burn-up enables us to skip the physics, and work with economics. Referring to the previous calculation, to include the value of the burn-up (= 33,000 MWd/t) and capacity factor of 0.85, we get for the annual input of enriched uranium required by a 1,000 MW(e) reactor for a *year*, with a thermal input of 3,000 MW for a year (= 365 days), the quotient $[(3,000 \times 0.85 \times 365)/33,000] = 28.2$ tonnes of enriched uranium. (Note:

if the capacity factor is larger than 0.85, more uranium is needed, whereas if the burn-up is larger, less uranium will be needed!)

Obviously, it might be useful to propose another input-output relationship between enriched and natural uranium. Using some numbers in the book by Professor Anthony Owen (1985), I calculate that 26 tonnes of enriched uranium corresponds to about 150 tonnes of natural uranium, and so the input-output coefficient is six. Thus, 28.2 tonnes of enriched uranium calls for 162 tonnes (t) of natural uranium.

Please note very carefully that we are dealing with approximations or averages above, and also in what follows. The relation between electric, or (e), and thermal could be less or more than 3 [i.e., 1 (e) = 3 (thermal)], while the capacity factor shows a tendency to vary, depending somewhat upon the skill of managers and technicians in a nuclear facility. The input-output coefficient in the above paragraph depends on the richness of ores and the efficiency of enrichment processes, while burn-up is increasing over time due to technological progress.

The last exercise in this section involves a statement about the relative cost of nuclear power — "relative" meaning that it will be compared to hydro power, which is the most inexpensive generator of electricity. According to easily available data, the (average production) cost of electricity in Sweden, which utilizes hydro (H) *and* nuclear (N), is about the same as the average cost of hydro-generated electricity in Norway. (The price to consumers may or may not be the same, which is irrelevant for the present discussion.) Calling Sweden country 1, and Norway country 2, the following algebra seems appropriate, where C/q (Cost/quantity) represents average cost:

$$\frac{C_{1N} + C_{1H}}{q_{1N} + q_{1H}} = \frac{C_{2H}}{q_{2H}}. \tag{6}$$

Taking q as the total output of electricity in Sweden, we rewrite the above as:

$$\frac{q_{1N}\left(\frac{C_{1N}}{q_{1N}}\right) + q_{1H}\left(\frac{C_{1H}}{q_{1H}}\right)}{q} = \frac{C_{2H}}{q_{2H}}. \tag{7}$$

Both sides of this expression can now be multiplied by q, and in addition, q_{1N} can be taken as θq and q_{1H} as $(1 - \theta)q$. Thus, we are able to write the above expression as:

$$\theta q[C_{1N}/q_{1N}] + (1 - \theta)q[C_{1H}/q_{1H}] = q[C_{2H}/q_{2H}].$$

The q's can be cancelled, and making the reasonable assumption that $C_{1H}/q_{1H} = C_{2H}/q_{2H}$, some simple manipulations will give us:

$$\frac{C_{1N}}{q_{1N}} = \frac{C_{2H}}{q_{2H}}. \tag{8}$$

What this says is that the average cost of nuclear-generated electricity in Sweden is equal to the average cost of hydro-generated electricity in Norway, which may be the lowest-cost electricity in the world. It appears also that students of this topic are unworried by possible yellowcake price increases. They suspect that if (or when) reactor fuel apparently becomes scarce, and/or the price of electricity begins to increase very rapidly and gives an indication that it will stay up, there will be more reprocessing of spent fuel, and — for better or worse — the breeder reactor will appear. Of course, some so-called experts believe that economical breeders will never appear, but I wish to join the ladies and gentlemen who say that this is completely wrong.

6. More Economics of Load Division

We can begin by specifying three kinds of equipment that can produce electricity. These are gas-based, with a low capital cost (F_1), but a high variable cost (v_1); coal, with a capital cost (F_2) that is higher than gas, but a variable cost (v_2) that is lower; and finally, nuclear, with the highest capital cost (F_3), but the lowest variable cost (v_3). This arrangement can be seen in the top diagram of Figure 2(a).

Once again the cost relationships are linear ($C = F + vt$), with the applicable part of these curves being the solid lines that form the aggregate cost curve. The capacity chosen is often 1 kW, but these screening (i.e., cost) curves can be used for any capacity. Figure 2(b) shows the load on the line in Norway in 1975. The maximum load was 11.25×10^6 kW, while during 2,000 hours of the year the load varied between 11.25×10^6 and 10×10^6 kW. Between 2,000 hours and one year (= 8,760 hours), the load was between 11.25×10^6 and 5.1×10^6 kW. Never during the year did it fall below 5.1×10^6 kW, and so it is clearly appropriate to designate 5.1×10^6 kW as the base load, although it could happen that the base load equipment supplied some of the intermediate load. What about the intermediate and peak loads? These are more or less subjective; however, I am sure that Norwegian engineers were capable of making the optimal choice.

The exposition will continue by extending the examples given in the previous section. The first step is to put the load diagram at the bottom of Figure 2(a) into perspective with the screening (i.e., cost) curves at the top of Figure 2(a). We can write as the values of the costs: $C_1 = F_1 + v_1 t$, $C_2 = F_2 + v_2 t$ and $C_3 = F_3 + v_3 t$. Something that can be noted again is the expression *merit order*, where the intention is to determine which type of equipment carries which load, and over what portion of the year (of 8,760 hours), on the basis of marginal costs.

What we are going to find out with some simple algebra is that in the arrangement in Figure 2, gas generates the peak load, coal generates a part of the intermediate load, and nuclear generates the remainder, which in this diagram means the base load plus a part of the intermediate load.

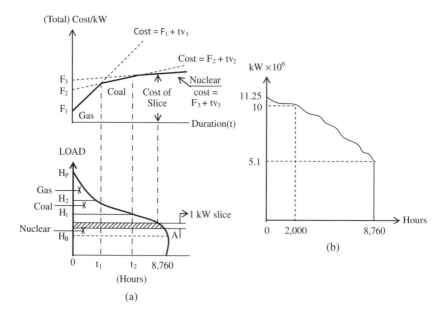

Figure 2.   Load and Cost Curves for Three Types of Equipment

Accordingly, during part of the time nuclear is idle, and so it might be designated a reserve, since after all there is considerable uncertainty associated with the *ex-ante* (i.e., before-the-fact or predicted) load. It might be argued that this is a bad example, since it is a waste to have expensive nuclear equipment standing idle; however, given the cost curves, the specification of cost-minimization leaves us with no choice. (It seems that a reserve of 15%–20% is usually deemed correct, although at the present time in the U.S. and perhaps other countries it has been suggested that there has been serious under-investment in power sources and especially transmission lines.)

Using the algebra and elementary logic from the previous section, it is a simple matter to find t_1 and t_2 in Figure 2(a). To get t_1, we start with $F_1 + v_1 t = F_2 + v_2 t$. This immediately gives us $F_2 - F_1 = t(v_1 - v_2)$, and so $t = t_1 = (F_2 - F_1)/(v_1 - v_2)$. Obtaining t_2 is just as easy, but it will be left for you to do as an exercise. Notice also that from the point of view of signs, the value of t_1 makes sense, because $F_2 > F_1$ and $v_1 > v_2$, and thus the quotient is positive.

Let us also note that in this example if we say that we have a peak, intermediate, and base load, only the base load $0 - H_B$ is unambiguous. The peak load definitely appears to be $H_2 - H_P$ and the intermediate load $H_B - H_2$, but aside from the algebra there is a certain ambiguity about these two loads which is not important for the readers of this book, nor for its author.

We can imagine a number of hypothetical situations which would have a place in this discussion. For instance, one might be that after we make our calculations, gas carries none of the load. Why? The only possible answer is that it costs too much relative to coal and nuclear. On the other hand, gas may carry the entire load. Why? The short answer is that it is less expensive than the other options. Consider the situation in California over the past decade. As a result of technology improving combined cycle gas-based equipment until it became very efficient, and the gas price remaining very low up until the last few years, gas-based equipment became sufficiently economical to generate a much larger share of the electric load in that state than previously contemplated. The same thing happened in the UK with the so-called 'dash for gas'. As bad luck would have it though, the gas price suddenly escalated, and while existing gas-based equipment could not be instantaneously discarded, when new generating equipment was considered, gas looked much less attractive. Accordingly, there is a difference between the optimal merit order and the existing merit order.

Readers who have come this far without feeling frustrated can feel very satisfied, because in truth I am always amazed by the shortage of knowledge about nuclear economics. It is interesting to note that in Australia, (non-nuclear) plants that are designated as intermediate load plants comprise about 40–45 percent of total installed capacity, as compared to 50 percent for base load plants, and 5–10 percent for peaking plants. This says something about the place of uncertainty in these matters and how it causes a deviation from the 'ideal' or optimal merit order of the type presented above where we get unambiguous answers as to the value of t_1 and t_2. Generally the peak load represents only a small fraction of the demand for electricity, and it can happen that only a portion of this capacity is in use, but since the available generating equipment must be able to satisfy the maximum demand that may appear in the system, capacity or *load factors* for peak load facilities are often quite small. As a result, considering the price of domestic and imported power, it may turn out that as much peak load power as possible should be purchased rather than generated in the system, and many utilities do everything possible to satisfy this requirement by purchasing from other cities, states, regions or countries. The expression "load factor" can be defined as the Average Load divided by the Peak Load. Ideally this is fairly large, but there can be some ugly surprises where electricity generation is concerned, and so in some regions considerable excess or reserve capacity must be available because there can be a simultaneous peak in neighboring regions.

In some classroom discussions, students want a formalization of the results presented above on 'load division'. Here I can point to an important article by Michael Einhorn (1983) whose mathematics I have altered somewhat to correspond to my classroom presentations of this topic. Taking note of Figure 2, we get:

$$TC = F_1(H_P - H_2) + F_2(H_2 - H_1) + F_3(H_1 - 0)$$

$$+ v_1 \int_{H_P}^{H_2} t(H)dH + v_2 \int_{H_2}^{H_1} t(H)dH + v_3 \int_{H_1}^{0} t(H)dH.$$

Integrations take place horizontally, and involve slices of the capacity H bounded by the vertical axis and the system's load curve. This is synonymous to Equation (9):

$$TC = \sum_{j=1}^{N} F_j(H_j - H_{j-1}) + \sum_{j=1}^{N} \left\{ v_j \int_{H_j}^{H_{j-1}} t(H) dH \right\}. \quad (9)$$

Next we differentiate TC with respect to, for example, H_j, and set the resulting expression equal to zero. This gives us:

$$\frac{\partial TC}{\partial H_j} = F_j - F_{j-1} + v_j t(H_j) - v_{j+1} t(H_j) = 0. \quad (10)$$

From this we obtain for a crossover time:

$$t(H_j) = \frac{F_{j+1} - F_j}{v_j - v_{j+1}}. \quad (11)$$

For instance:

$$t(H_2) = t_2 = \frac{F_3 - F_2}{v_2 - v_3}. \quad (12)$$

We have an overtone of (Albert) Einstein's equivalence theorem here in that two approaches (informal and formal) lead us to the same result, which suggests that the same fundamental law is operating, and the informal presentation was also about cost minimization. I will admit, though, that the informal presentation was the only one I was concerned with in the lectures I gave in Stockholm and Grenoble many years ago.

7. More Background for Curious Readers

One of the most brilliant appendages to the topic that we are discussing can be found in a non-technical article by the late Samuel Schurr (1984). Concentrating on the United States, Schurr demonstrated that the total energy use in what he termed the "business sector" more than doubled over the period 1920–1973, and, in relation to capital (= machinery + structures), increased by 50%. The observed slight fall in the energy intensity of *output* was then shown to be due to technological change (largely motivated by increasingly energy-intensive *inputs*) raising output by so much that, *percentage-wise*, output increased by more than the increase in energy consumption. (It was due to the failure to understand the details of this phenomenon that many concerned observers often found themselves elaborating at length on the bogus argument that output could be maintained or, for that matter, increased even if the input of energy declined.) Schurr also hypothesized that electrification meant a flexibility in industrial operations that would have been impossible with any other form of energy, and this was the cardinal reason for productivity growth. Equally as important, electricity would play an indispensable role in the employment of items such as computers, whose revolutionary promise was just being realized.

If this is the actual situation, then an intelligent economist living in complete isolation from the 'real world' — which, unfortunately, is a world in which security problems cannot be ignored — might suggest that the best strategy for fostering economic growth, as well as stabilizing or reducing atmospheric greenhouse gases, is a *massive* program of nuclear construction, and it should be commenced as soon as possible. This humble commentator is definitely in favor of more nuclear energy, but at the same time it seems appropriate to wait until security problems assume another (i.e., lower) dimension before anything remotely resembling a "massive" commitment is undertaken. In addition, it needs to be emphasized that an optimal energy 'package' for any region probably contains a very large component of renewables. Here some dynamic thinking is justified. Even if renewables are not as efficient as desired, investing in and experimenting with them might turn out to make economic sense in the long run.

A country that illustrates to some extent the last observation is, of course, China. In 2007, 3.4 gigawatts ($= 3.4 \, GW = 3.4$ billion watts) of wind-based capacity was ostensibly added to China's electric grid, while by 2020 the desire is to raise nuclear capacity from its present $10 \, GW$ to $40 \, GW$, which is the fastest proposed increase in the world. Similarly, the intention is to add $1,300 \, GW$ to China's total electric-generating capacity by 2020 (which can be compared with the total present U.S. capacity of $1,000 \, GW$). Unfortunately however, regardless of the amount of nuclear, wind and solar that the Chinese deploy, a large amount of coal will almost certainly be consumed. A question that should therefore be asked of the anti-nuclear (and anti-hydro) booster clubs is would they like the government of China to cancel its nuclear and hydro programs, because this would result in an even greater increase in the output of carbon dioxide generated in that country.

The problems brought about by coal have been touched on briefly in my previous textbooks and also in my book on coal (1985), but this is an issue that requires additional consideration, because despite increased concern over global warming, it is very unlikely that the growing world population will ignore the energy in that commodity. An important discussion of coal in Europe that applies to other regions can be found in a short paper by Jeffrey Michel (2008), and in a personal communication Michel informed me that Carbon Capture and Storage (CCS) would be extremely energy-consuming "and consume twice the water of conventional power generation, making its wide-scale implementation doubtful." Although it cannot be taken up here, the Swedish utility Vattenfall used the increased profits made possible by electric deregulation in Sweden and open borders in Europe, to raise electricity prices in Sweden, and also to expand their coal-producing operations in Germany. Despite assurances to the contrary, this latter commitment will more or less guarantee an increase in atmospheric CO_2 both in and exterior to Germany.

Observing the Swedish nuclear past provides a valuable but generally unappreciated insight into the mechanics and digressions of what Professor Ken-Ichi Matsui (1998) calls the "Seventh Energy Revolution", which he believes will be based on nuclear energy. In Sweden, natural gas was once suggested as one of the main replacements for nuclear. The main advantage of nuclear as compared to gas-based equipment was in the cost of the fuel, which meant that gas was at a disadvantage for carrying the base load, although the capital cost of a nuclear facility was much greater. When, however, combined cycle gas-burning equipment became widely available, and the price of gas fell, it was claimed by persons who should have known better that gas would *always* be much more economical than nuclear (for generating the base as well as peak loads). With the price of natural gas in the vicinity of $3.5 per million Btu (= $3.5/MBtu), it was easy to insist that gas was a better economic bet than nuclear, and a locale where this was argued at great length was California.

In Sweden, if an accurate (and comprehensive) calculation had been made, if notice had been taken of what was occurring in the rest of the world, and especially if there was less technophobia in and around the political establishment, the above claims made about gas would have been openly ridiculed, because it has long been obvious to many of us that the infinite supply of natural gas that a number of so-called energy experts were thinking of is actually finite — as everybody is either finding out now or will eventually find out — and since environmental considerations dealing with CO_2 are becoming more important, nuclear displays an indisputable social cost advantage over gas. Here I can mention that a few years ago, when the price of gas first touched \$7/MBtu (= \$7 per million British Thermal Units), many reputable sources claimed that it would reach \$10/MBtu by 2012. In point of truth, the price of gas could have reached that level that year had it not been for the macroeconomic meltdown and, some like to claim, the production of shale gas in the United States. But even with that price, there would still have been a large gap between the BTU price of oil and gas, and so the price of gas would almost certainly have continued to rise. Readers who want an up-to-date examination of the prices of energy resources should turn to the site *321 Energy*, which can be accessed via Google. They will also find on that site a large collection of non-technical articles published in the international press.

Another example might be useful here. The first time that I taught in Australia it was widely advertised that the Maui gas field in New Zealand was virtually inexhaustible. The reserve situation is quite different at the present time; however, that field is still spoken of as being extremely valuable. *Not*, it should be emphasized, for producing natural gas, but for storing — or 'sequestering' as they say — as much as possible of the CO_2 that will be generated in the coal-based generating facilities that might eventually be required to provide New Zealand with an increasing fraction of its electricity. It can also be mentioned that the New Zealand electricity deregulation, which at one time was praised as the most satisfactory in the world, was very likely based on beliefs about the availability of gas that were completely illusory. I suspect that this was an important factor in establishing a natural gas price for that country that was probably lower than what in economic theory is sometimes termed the 'scarcity price' (or the theoretically correct market price).

According to Torsten Gustafson, chief scientific advisor to the Social Democratic government of Tage Erlander, there was a positive attitude toward nuclear energy in Sweden until about 1970. After that time, two of the five major political parties in Sweden — the 'farmer party' and the communists — came to the grotesque conclusion that the 'friendly atom' was bad for Sweden and just about everyone else, although there were a number of opponents to nuclear energy in all political and social factions. There is no rational explanation for the strong aversion developed by any political grouping to nuclear energy, since, e.g., farmers and industrial workers are clearly important benefactors of inexpensive electricity.

This situation could probably be compared to something like the 'tulip *bubble*' in Holland in the 17th century, when intelligent people suddenly and inexplicably discerned enormously valuable qualities in the humble tulip, and paid fantastic prices for a commodity that eventually turned out to have no commercial and little intrinsic worth. The commodity in the present case is the belief in nuclear disengagement, which might provide a modest slice of the voting public with a tangible psychological satisfaction in the short run, but in the long run would deprive Swedish industries and households of an indispensable and comparatively low-priced input.

The final topic in this section is the almost unknown concept of the so-called 'backstop technology', which, as outlined by William Nordhaus (1973), is the technology used to exploit a resource or asset that will still be available when all or most of the conventional energy resources are history. Not much is heard of the backstop these days, and I doubt whether I mentioned it in my earlier energy economics textbooks; but there was a time when — in my lectures — I always brought it up in the context of discussing nuclear energy, although unlike Professor Nordhaus I treated it as a means for supplying the 'extra energy' needed to produce motor fuel or energy liquids from biological or other resources. For instance, although not widely known, Sweden (and probably other countries) possesses large quantities of low-grade uranium, and eventually science and technology will make it possible to utilize these resources in a politically and environmentally approved manner. As a result, they might make an impressive contribution to the Swedish economy.

For the purposes of this exposition, a backstop is a known technology for producing a very large amount of a given resource at a known price. (Actually, Nordhaus thought in terms of an infinite amount.) Suppose, for instance, that at a certain time only coal was used to produce electricity, but there is only a limited amount of coal in the crust of the earth. When this coal is exhausted, we will still desire electricity in the future, and so it is necessary to think about an alternative technology for producing electricity that is more than a 'stopgap'. One candidate might be uranium and thorium burned in breeder reactors, because given the likely amounts of these resources, they would be capable of producing electricity for many decades, or even centuries. Please be assured though that this statement is NOT an advertisement for the breeder.

Recently I had the misfortune to be informed by persons with advanced degrees in physics that my knowledge of nuclear matters is inadequate, because I often make a point of claiming that economics is just as important as physics where the issues taken up in this book are concerned. Actually, economics is more important, much more, because the hardware on the nuclear scene merely provides a background. Regardless of prevailing or predicted nuclear or non-nuclear technology at that time in the future when the rising cost of energy begins to accelerate, voters and their political representatives are going to find nuclear energy much more attractive than they had previously imagined. It could be argued though that the key issue here is neither physics nor economics, but psychology, because large numbers of persons — many of them highly educated — have been told things about energy that have absolutely nothing at all to do with the actual supply of and demand for energy resources in the real world, as well as the role that adequate energy plays in their quality of life.

8. Further Aspects of Nuclear Costs

One of the assumptions in my lectures is that even if natural gas were available at bargain-basement prices, it would be sub-optimal to unconditionally regard it as preferable to nuclear where new energy investments are concerned. The thing to be aware of here is that the global output of gas might peak in 30 or 40 years, whereas a new nuclear installation will be on-line for *at least* 60 years, and could still have access to *comparatively* inexpensive fuel. The peaking of gas during the lifetime of nuclear installations constructed, e.g., today will increase the opportunity cost of that resource in such a way as to make it extremely difficult to comprehend why it has often been considered as the optimal replacement for nuclear. Remember too that there is a fairly high environmental cost associated with natural gas, though not as high as with coal.

An important though straightforward discussion of the cost of energy resources can be found in Chapter 8 of Barre and Bauquis (2008). What I have done in various lectures is to recalculate the capital component of the energy cost using the techniques referred to above. Following that, I 'scaled' the values of the fuel and O&M components of the energy cost based on the work presented in that chapter. This provided me with a value for the energy cost of a kilowatt-hour of electricity close to that suggested by Jim Beyer (2008), who is one of the many important contributors to *EnergyPulse*. (Beyer's comment is surrounded by many other valuable remarks about the cost of nuclear energy, and that site contains extensive archives.)

Using some numbers from the chapter on 'economics' in the important book by Barre and Bauquis (2008), an amortization period of 40 years, a rate of interest of 10% (which is probably too high, but which is intended to compensate for some uncertainties), and a capacity factor of 0.85 (which is probably too low), I used Equation (4) in this chapter to obtain a levelized capital cost of $265 per kilowatt, which in turn gives for the capital component of the total cost, 3.5 cents per kilowatt-hour. Then, using the values for capital, fuel and O&M costs in Chapter 8 of Barre and Bauquis as a datum, I obtained 2.61 cents/kWh for the fuel component of the energy cost, and 1.40 cents/kWh for the O&M component of the energy cost. The total energy cost is the sum of these, or 7.5 cents/kWh, which comes sufficiently close to Jim Beyer's value of 8 cents/kWh for the U.S., and James Hopf's 6.8–8.2 cents/kWh (2008), to make it useful for this exposition. If I had to choose a value, I would take 8 or 8.2 cents/kWh.

The Economist (July 9, 2005) has presented some estimates from several sources for average electricity costs. For German utilities, the Union Bank of Switzerland (UBS) gives 1.5 cents/kWh for nuclear, 3.1–3.8 cents/kWh for gas, and 3.8–4.4 cents/kWh for coal. Similarly, they give 1.7 cents/kWh for nuclear in the U.S., 2 cents/kWh for coal, and 5.7 cents/kWh for gas. James Hopf (2008) has identified these as fuel costs, and in the same 'string' of comments he provides important information on O&M costs for nuclear energy. The International Energy Agency (IEA), employing a discount rate of 5%, argues that energy costs for nuclear are $21–$31/MWh, while gas ranges from $37–$60/MWh. In a summary of generating costs originating with the IEA and OECD, Tarjei Kristiansen of Statkraft Energi AS (Norway) claims that with a 5% discount rate, energy costs in $/MWh are 23–31 for nuclear, 25–50 for coal, and 37–60 for natural gas. The only comment that I have on these results is that — on average — they might be accurate for some countries and inaccurate for others, but in my opinion the best discussion of this issue is found in the comments on my 2008 article in *EnergyPulse*.

In considering relative or intertemporal costs it is necessary to have some idea of the resource base. According to estimates of the World Nuclear Association in 2000, the country with the largest uranium reserves is Australia, whose reserves at that time were 622,000 metric tons (= 622,000 tonnes = 622,000 t), and whose production was 7,720 t. In what follows I would denote this as (622,000; 7,720). This discrepancy seems very large, but not when I remember the negative attitudes toward the production of uranium by my mathematical economics students in Sydney and Melbourne. Professor Tony Owen, who is now director of the Adelaide branch of the London School of Energy and Resources, recently informed me that the world's largest mine is Olympic Dam, which is located just north of Adelaide (Australia), and that the full extent of that structure has yet to be determined.

The largest producer of uranium in 2000 was Canada, with a production of 12,520 t and reserves of 331,000 t = (331,000; 12,520). Other important countries were Kazakhstan (439,200; 2,018), Namibia (156,120; 2,239), Niger (69,960; 3,095), Russia (145,000; 2,000), United States (110,000; 1,000), Uzbekistan (66,210; 2,400), and others (306,940; 2,774). Total estimated reserves in 2000 were thus 2,246,430 tonnes, while total production was 35,767 tonnes. For technical details see Owen (1985), but the total input of uranium in, e.g., the production of electricity exceeded 35,767 tonnes because a great deal of the resource can also be obtained from the recycling of spent fuel and former military ordnance.

Sweden does not appear in the above because exploiting its low-grade reserves is uneconomical at the present time. Eventually this situation could change because of scientific and technological improvements in mining and processing. Something that might cause a quantum jump in the value of Swedish uranium, however, would be the breeder reactor becoming a commercial proposition, because in that case the output of energy (due to the exploitation of the plutonium that could be bred) might make it economical to use even low-grade uranium. As far as I am concerned, though, the Swedish government (and most other governments) is at present completely incapable of solving the security problems that might be posed by a greater presence of and/or reliance on plutonium. This may be the only point on which I happen to be in agreement with people like Ralph Nader and Amory Lovins. Unlike them, however, I believe that by rejecting the energy in uranium when it is used in 'conventional' reactors, the (psychological) conditions are being created for a panic-stricken rush into the breeder when the fundamental scarcity of oil and gas — and the present inadequacy of renewables — are made clear to the television audience.

An interesting factor here is that Sweden was at one time believed to be surrounded by comparatively unsafe reactors: a total of six were said to be found at Sosnowy Bor outside St. Petersburg (Russia), and at Ignalina in Lithuania. In the film *The Deer Hunter*, Christopher Walken sang a drunken version of the marvelous tune "I've Got My Eyes on You", and before its dismantling many nuclear experts in Sweden had their eyes on Ignalina as an installation (of the Chernobyl type) that could pose a danger to their country — but not the 'Greens'. Their eyes instead have been fixed on safe reactors in Sweden, as well as the new super-safe facility that is currently under construction in Finland and which will have a rated output of 1,600 megawatts, or as much as the two Swedish reactors that were closed at Barsebäck (near Malmö) combined.

Somebody else with a keen interest in reactors is Mr. Romano Prodi of the EU, who is one of the overseers of the ridiculous crusade to deregulate Europe's electricity and gas. Among the reactors in which he has taken a particular interest are those of Bulgaria, which the International Atomic Energy Agency (IAEA) considers to be on a par with the average in Western Europe. According to John Ritch, the U.S. ambassador to the IAEA, the European Commission has decided to "blackmail" Bulgaria in such a way as to make its entry into the EU contingent on its willingness to reduce its nuclear capacity.

Even a combination of John Maynard Keynes and Sigmund Freud would have a difficult time comprehending the reasoning here, although Mr. Ritch feels that this scheme originates with the "anti-nuclear environmentalists" that play an important role in the Prodi team. This may be true, but as I pointed out in a talk in Milan several years ago, it may also have to do with a belief by the Prodi brain trust that since half of Bulgaria's electricity came from nuclear reactors (as compared to 30% in Europe overall), electricity deregulation in that corner of Europe would be easier if Bulgaria's nuclear capability was reduced. Theoretically this may make sense, because in Sweden, competition — which was supposed to be the object of deregulation — *decreased* rather than increased after deregulation was introduced: e.g., large generators have been able to merge with smaller firms.

Ideally, this chapter would include a discussion of the optimal plant mix, and O&M costs would have been handled in a more sophisticated way than scaling the results of other authors. Where the first is concerned, I refer to my earlier textbooks; and as for the second, the scope and complexity of O&M costs are mostly unknown to myself. However, the bottom layer of comments on my 2008 article in *EnergyPulse* (www.energypulse.net) should illuminate a large part of this subject.

9. Nuclear Energy and the Kyoto Hobby-Horse

As I have found out, it would not be a good idea in Sweden (and probably elsewhere) to belittle the Kyoto Protocol if you are anxious to impress the 'broad masses' with your wisdom, or at least that portion of them with the typically "deep interest" in environmental matters that characterizes many of the ladies and gentlemen heavily involved in economic research in Sweden. The basic problem here is that this sub-set of the broad masses do not really understand the issue. They do not understand that, at bottom, the Kyoto Conference itself had little or nothing to do with reducing greenhouse gases. Michael Hanlon, the science editor of the *Daily Mail* (UK), puts it as follows:

> "According to the environmentalist gurus (*sic*), there is only One Solution to global warming, and its name is Kyoto. The Japanese city in which a rather shambolic agreement to curb carbon dioxide emissions was signed some years ago has acquired talismanic status among people who, one suspects, have little idea what 'Kyoto' is, would do or how it works."
>
> (Hanlon, 2005)

Among the "people" that Mr. Hanlon is describing were most of the 'delegates' to Kyoto, whose principal interest was to obtain tickets for the next climate warming jamboree. According to Professor Sven Kullander and several colleagues at the Swedish Academy of Science (2002), Kyoto was an important first step for reducing greenhouse gases, but *"helt otillräckligt för en reell förbättring"* (= completely insufficient for a real improvement). If readers can accept the latter portion of this judgment, then I accept the first part — although in reality I put the Kyoto meeting in the same category as the 'World Summit' in Capetown, where perhaps 60,000 heavy eaters and drinkers assembled to solve in their own gluttonous way the many and varied problems confronting contemporary societies.

I can also note that Professor Kullander's knowledge of nuclear energy has recently been brought into question by several members of the Swedish environmentalist elite who believe that new base load electric-generating capacity should be based on wind instead of nuclear energy. By way of examining this absurd proposal, readers should examine the comment made by Malcolm Rawlingson (2011) in *EnergyPulse*.

Swedes accept Kyoto for the same reason that they accept electricity deregulation and the European Union: they were told to accept it by celebrity politicians and journalists. The physicist Richard Feynman once said that in matters of the above nature, the logic of science is superior to that of the authorities — but a hypothesis of this nature has no place in the pretentious deliberations of assorted media luminaries, which assures that it is taboo for a large part of their audiences (at least when they are sober). Swedes are also great partisans of 'emissions trading', although an advisor to President Putin once called it a scheme to make money that is irrelevant for suppressing greenhouse gases. I have also published a number of similar comments on this topic.

10. Concluding Remarks

Let me sum up what I said in a recent article in the journal *Energy and Environment* (Banks, 2004). We do *not* know if global warming is the real deal, or just part of a cycle; but we do know that gas and oil are running out, although it may take a few decades before the depletion of reserves, or the increasing cost of maintaining or increasing the production of oil, registers with a majority of voters. In these circumstances the optimal behavior is to become friendlier with the 'friendly atom', and to do what former Prime Minister Blair of the UK and many environmentalists now suggest, which is to increase the use of nuclear energy. "Better nuclear than coal" is their mantra, because it will have to be something: the voters are not going to accept a sustained decline in their standard of living, and this prediction should be absorbed by everyone!

As suggested in this book, nuclear energy will probably be essential for supplying the 'extra energy' needed to, e.g., obtain the new (or, for that matter, the old) fuels that voters in the energy-importing countries have no intention of doing without, regardless of what they say. As Len Gould informed readers of the forum *EnergyPulse*, these voters intend to have enough fuel to continue their transportation activities — much of which is mandatory if they are to maintain the standard of living of themselves and their families — even if they must go to war to obtain this commodity.

There may be a sharp drop in the amount of electricity generated by nuclear power plants as a result of the Fukushima accident. However, global nuclear electricity generation in 2010 came to about 2,630 Terawatt-hours (= 2,630 Twh) according to statistics from the IEA, which means a 2.8% increase from the 2,558 Twh generated in 2009, which in turn was close to the peak year for nuclear output, which was 2006.

Germany is a country that, together with Switzerland, France, Japan, Sweden and perhaps some others, has expressed an intention to reduce or abandon its nuclear ambitions. After the widespread distribution of my short paper entitled "Some Friendly Economics for the Nuclear Energy Booster Club" (Banks, 2008), I received mails from several persons in Germany and elsewhere requesting that their names be removed from the list of persons directly receiving my papers. I was especially surprised by several of these 'Dear Johns', however …

"Wir Werden Wiedermal Marschieren" (= We Will March Again) was the title of a publication that gained considerable attention in Germany when I was in that country with the U.S. Army. It was widely read, and was about the retaking by the German Army of places like the Sudetenland (in Czechoslovakia) in the coming Third World War, which the author of that publication and many of his readers saw as inevitable and necessary.

Early in my 'tour', the armies of NATO countries participated in perhaps the largest peacetime military exercise held in West Germany up to that time, which was called 'Apple Harvest'. Toward the completion of that memorable exercise, the referees concluded that the Red Army had broken through the Fulda Gap and had almost reached Nuremberg, and it was judged that the only way they could be stopped was with nuclear weapons. For reasons that I did not understand and did not bother to ponder, I was told to make calculations for a simulated nuclear projectile that would be fired at the advancing Red Army. Had it been a real instead of a simulated projectile, the eastern suburbs of Nuremberg would have been removed from the face of the earth. After that 'simulated' outcome came to be known, German officers, journalists, book-club members, politicians and various decision-makers lost their appetite for marching. *The same kind of reversion will eventually happen when the German public comes to realize that abandoning nuclear energy could wreak havoc on their standard of living!* Among other things, this scenario could mean that many factories in Germany could become candidates for transfer to other parts of the world with an adequate and reliable supply of energy.

This is a reason why I want the nuclear capacity in Sweden maintained at its present level, or slightly higher. The key issue is not the choice of generating equipment, but my pension! It is also the key issue for many academics and energy professionals in this country and probably elsewhere, although they have been convinced by members of the anti-nuclear booster club and their favorite politicians that they would be doing themselves a disservice by understanding the easily understandable.

And last but not least, some brief comments referring to the tragic situation that recently unfolded in Japan. Across a total of 32 countries, there were approximately 14,000 cumulative reactor-years of commercial nuclear operation before the Fukushima incident. During that time there have been two major accidents: Chernobyl and Three Mile Island. Where the first is concerned, this was a major tragedy; fortunately, it is not likely to be repeated today in most installations because reactors of the kind installed at Chernobyl are no longer welcome in just about all countries. As for the Three Mile Island incident, this appears to have involved a *total* meltdown, or something close to one, as compared to the partial meltdowns that may have taken place in Japan recently at several reactors, but as far as I know there were no casualties. Readers can also look at the article by Micah J. Loudermilk (2011). He notes that the U.S. Navy has successfully managed 500 small reactors on ships and submarines. The thing to pay particular attention to is that after the Kobe earthquake, the Japanese began constructing buildings that are ostensibly survivable during an earthquake. Given that many Japanese reactors are located in or near to earthquake zones, one would have thought that the structures containing those reactors would have been rebuilt or strengthened, if that was possible. Of course, it is uncertain whether structures containing reactors can be protected from tsunamis of the kind that struck Japan, but it might have been sensible to 'relocate' facilities that are located in the danger zones after examining the tsunami that took place in Thailand a few years prior.

The unfortunate fate of some very important assets in the Japanese nuclear inventory will almost certainly lead to a debate in other countries, including Sweden. Earthquakes and tsunamis can possibly take place in any country where they have not occurred before, although the probabilities involved for most of those countries are very close to zero. As made clear by Rawlingson (2011), in almost every country that possesses the technology to construct and/or install nuclear facilities, this construction and installation will take place — though probably later rather than sooner, and regardless of previous nuclear experiences.

A Short Glossary

Base Load Plant: This is a plant that is normally intended to take a large part of the load, and usually produces electricity at a constant or near constant rate, and is in operation continuously (as compared to a 'peaking' plant).

Boiling-Water Reactor: A light-water reactor in which water is used as both coolant and moderator, and in which water is allowed to boil in the core. This water produces steam that drives the turbine, to which an electric generator is attached.

Breeder Reactor: This is a reactor that consumes fissionable fuel, but at the same time produces more fuel than it consumes (due to its operating on the non-fissile component of its fuel). This operating process is called 'breeding'.

Burn-up: A very important expression dealing with the amount of energy that can be produced per unit weight of fuel that is 'burned'.

CANDU Reactor: Essentially a Canadian reactor that uses heavy water or deuterium oxide rather than light water as the coolant and moderator.

Enrichment and Tails Assay: A measure of the amount of fissile uranium remaining in the waste stream from the uranium enrichment process.

Fissile Material: Material that can be caused to undergo atomic fission when bombarded by neutrons. The most important are uranium-235, plutonium-239 and thorium.

Fission: The process whereby an atomic nucleus captures a neutron, and then splits into several nuclei — usually two — of lighter elements, releasing a considerable amount of energy and two or more neutrons.

Light-Water Reactor (LWR): A reactor that uses water as the primary coolant and moderator, with enriched uranium as fuel. The two types of LWRs are the boiling-water reactor (BWR) and the pressurized-water reactor (PWR).

Natural Uranium: Uranium with the U-235 isotope present in a concentration of 0.711 percent by weight. The isotopic content here is exactly as it is found in nature. This can be compared with enriched uranium, which is uranium enriched in the isotope U-235 from 0.711 percent to at least 3 percent.

Nuclear Reactor: A device in which the nuclear fusion chain can be initiated, maintained and controlled so that the energy generated can be released at a specific rate.

Plutonium (Pu): A heavy, fissionable and radioactive metallic element, whose atomic number is 94. It occurs in nature in trace amounts, and in the present discussion is produced as a by-product of the fission reaction in a uranium-fueled reactor.

Uranium Concentrate: A powder produced from naturally occurring uranium minerals as a result of milling uranium ores or the processing of uranium-bearing solutions. Synonymous with 'yellowcake' or Uranium Oxide.

Bibliography

Albouy, Michel (1986). "Nouveau Instruments Financiers et Gestion du Couple Rentabilité Risque sur le Marché du Petrole." Stencil, Grenoble Université.

Banks, Ferdinand E. (2008). "Some friendly economics for the nuclear energy booster club." *EnergyPulse* (March 26).

———— (2007). *The Political Economy of World Energy: An Introductory Textbook.* Singapore, London and New York: World Scientific.

———— (2004). "A faith-based approach to global warming." *Energy and Environment*, 15(5): 837–852.

———— (2002). "Some aspects of nuclear energy and the Kyoto Protocol." *Geopolitics of Energy* (July–August).

———— (2000). *Energy Economics: A Modern Introduction.* Dordrecht: Kluwer Academic Publishers.

———— (1985). *The Political Economy of Coal.* Lexington and Toronto: D.C. Heath & Co. (Lexington Books).

Barre, Bertrand and Pierre-René Bauquis (2008). *Nuclear Power: Understanding the Future.* Paris: Editions Hirlé.

Beyer, Jim (2008). "Comment on Banks." *EnergyPulse* (March 26).

Braconnier, Fredrik (2005). "Utsläpps rätter trissar upp redan högt elpris." *Svenska Dagbladet* (August 20).

Carr, Donald E. (1976). *Energy and the Earth Machine.* London: Abacus.

Clemente, Jude (2009). "U.S. energy security and natural gas vehicles: a reality check." *EnergyPulse* (July 24).

Constanty, H. (1995). "Nucleaire: le grand trouble." *L'Expansion* (pp. 68–73).

Duffin, Murray (2004). "The energy challenge 2004 — nuclear." *EnergyPulse* (October 8).

Einhorn, Michael (1983). "Optimal system planning with fuel shortages and emissions constraints." *The Energy Journal*, 4(2): 73–90.

Goodstein, David (2004). *Out of Gas: The End of the Age of Oil.* New York: Norton.

Gould, Len (2008). "Comment on Banks." *EnergyPulse* (March 26).

Grunwald, Michael (2008). "The clean energy scam." *Time* (April 14).

Guentcheva, D. and J. Vira (1984). "Economics of nuclear versus coal." *Energy Policy*, 12(4): 439–451.

Hanlon, Michael (2005). "Why do greens hate machines?" *The Spectator* (August 6).

Hansen, Ulf (1998). "Technological options for power generation." *The Energy Journal*, 19(2): 63–87.

Harlinger, Hildegard (1975). *Neue Modelle für die Zukunft der Menshheit.* IFO Institute für Wirtschaftsforschung, Munich.

Held, C.M. (1983). "Evolution of size-scaling effects on electricity production during the 1970s." Paper presented at the Atomic Energy Agency, Vienna.

Hopf, James (2008). "Comment on Banks." *EnergyPulse* (March 26).

Kok, Kenneth (2008). "Comment on Banks." *EnergyPulse* (March 26).

_____ (2007). "Vision for the development of an international nuclear fuel recycling program." Paper presented at the 11th International Conference on Environmental Remediation and Radioactive Waste Management, Bruges, Belgium (September 2–6).

Kullander, Sven, Henning Rodhe, Mats Marms-Ringdahl, and Dick Hedberg (2002). "Okunnig att avveckla kärnkraften." *Dagens Nyheter* (April 7).

Loudermilk, Micah J. (2011). "Small nuclear reactors and U.S. energy security: concepts, capabilities and costs." *Journal of Energy Security* (May).

Martin, J.-M. (1992). *Economie et Politique de L'energie*. Paris: Armand Colin.

Matsui, Ken-Ichi (1998). "Global demand growth of power generation, input choices and supply security." *The Energy Journal*, 19(2): 93–107.

Michel, Jeffrey H. (2008). "German lignite usage: expensive and unsustainable." *Acid News* (www.acidrain.org).

Nordhaus, William D. (1973). "The allocation of energy resources." *Brookings Papers*, 4(3): 529–576.

Ogg, Clay (2008). "Environmental challenges associated with corn ethanol production." *Geopolitics of Energy* (January).

Owen, Anthony David (1985). *The Economics of Uranium*. New York: Praeger.

Rawlingson, Malcolm (2011). "Comment on Banks ('Germany and the nuclear future')." *EnergyPulse* (August 26).

Rhodes, Richard and Denis Beller (2000). "The need for nuclear power." *Foreign Affairs* (January–February).

Romm, Joseph J. and Amory B. Lovins (1992). "Fueling a competitive economy." *Foreign Affairs* (Winter).

Roques, Fabien, William J. Nuttall, David Newbery, Richard de Neufville, and Stephen Connors (2006). "Nuclear power: a hedge against uncertain gas and carbon prices." *The Energy Journal*, 27(4): 1–24.

Rose, Johanna (1998). "Nya Krafter." *Forskning & Framsteg* (September).

Schlageter, David (2008). "Comment on Alan Caruba ('Congress conjures up an energy deficit')." *EnergyPulse* (February 6).

Schurr, S.H. (1984). "Energy use, technological change, and productive efficiency: an economic-historical interpretation." *Annual Review of Energy*, 9: 409–425.

Scire, John and Ricardo Lopez (2008). "Yucca Mountain: Plan C." *The Nevada Appeal* (Carson City, Nevada).

Somsel, Joseph (2008). "Comment on Banks." *EnergyPulse* (April 8).

Stipp, David (2004). "Climate collapse." *Fortune* (February 9).

Tanguy, Pierre (1997). *Nucléaire: Pas de Panique*. Paris: Editions Nucléon.

Thunell, J. (1979). *Kol, Olja, Kärnkraft — en Jamförelse*. Stockholm: Ingenjörs-förlagen.

Yarrow, George (1988). "Nuclear power." *Economic Policy*, 6: 81–132.

Chapter 7

Economic Theory and the Great Coal Game

Several years ago I insisted that every student in my course on oil and gas economics at the Asian Institute of Technology (AIT) should master some important materials dealing with the availability of oil. By "master" I meant learn perfectly, assuming that they preferred a passing to a failing grade.

On the other hand, I have not dealt with *thermal* (or *steam*) coal in the classroom for many years, and as a result, I might be hesitant to insist that my students should always be ready to exhibit a profound knowledge of that topic to curious friends and neighbors. As many energy professionals know, thermal coal is a source of energy that has not received much notice in the 'learned' literature, and comparatively little on the blogosphere, despite its ubiquity in the provision of heat, light, transportation, and as a primary (intermediate) good producing the secondary (or final) good, electricity. In fact, the USDOE says that coal will supply 45% of U.S. electricity demand in 2035, and CO_2 emissions will increase by 5% (barring a cap on emissions). I think that the correct expression should be 'at least' 5%, as I should have clarified in my book on coal (1985).

I use the term "game" in the title of this chapter because I feel certain that — if it is available — more coal is going to be burned than commonly believed or desired by many; and despite assurances to the contrary by well-meaning (but slightly confused) decision makers, most of that coal will *not* be 'cleaned'. Thus, on one level or another, a *game* is in progress, which in the context of the present exposition means a medley of competitive situations in which strategic considerations (bluffing, disinformation, and exploiting the options provided by prevailing political routines and attitudes) have a key significance. Moreover, if the cost of cleaning larger quantities of coal is for one reason or another associated with a substantial increase in average energy costs and/or prices, then the tolerance for dirty coal is going to be much higher. Readers can work out for themselves what this is going to mean on the environmental front, since over the past eight years, coal has been the fastest growing source of 'fuel' in the world.

Perhaps the most recent example of the above is to be found in Australia, where coal supplies about 80% of the power for electricity generation. A coal power plant that was to be fully equipped with carbon capture and sequestration (CCS) facilities, ostensibly providing no emissions at all, and which was expected to cost 4.3 billion dollars, has been 'scrapped' after the spending of more than a hundred million dollars on preliminary procedures of one type or another. The reason given by the governor of Queensland was a lack of "viability", by which she meant that economically it was an unhappy mistake to have allowed that project to progress beyond the initial planning stage.

1. A Mandatory Background

For what it is worth, mandatory means obligatory. With all due respect, I have come to believe that energy economics is occasionally taught so poorly that the next time I present a course on this topic at any institution of higher or lower education in any country, all students will be held completely responsible for the information in this and the next section, regardless of what else they absorb about coal. Ideally, they must be able to express and expand somewhat on the observations just below at any hour of the day or night, or less dramatically, whenever they receive a question in class or on an examination.

(1) As pointed out by Joe Hung (2010), coal is the most rapidly growing fuel source in the world. Somebody else who makes this clear is Joseph L. Shaefer (2012). Coal is also broadly distributed, and Hung states that the energy in it exceeds that of all other fossil fuels combined. This can be shown by multiplying the (average) Btu (British Thermal Unit) content of a fossil fuel (e.g., coal) by proved reserves, as suggested in my earlier energy economics textbooks (2007, 2000), or in Chapter 3 of this book. Readers who want to extend this operation can examine the total Btu content of proved uranium and thorium reserves. If these reserves are used in the next generation of nuclear reactors (Gen 4),

then the energy they can supply will be greater than that in *all* proved *or* hypothetical fossil fuel reserves, as well as the *energy* that can likely be economically obtained from exploiting renewables during the first half of the present century.

(2) Where electricity generation is concerned, in the U.S., at the time this chapter was written, coal accounted for 48%, gas 20%, nuclear 20%, hydro 7%, and others — mostly renewables — for 5%. These numbers are highly suggestive where most industrial countries are concerned. However, China consumes almost half of the annual world coal output, and together with India — which also is a very large coal consumer — they are on a febrile hunt for more suppliers of that resource. Johanna Rose (2010) claims that China opens a new coal-based electricity-generating plant every week or two; however, this estimate is most likely erroneous if viewed on the basis of annual investments in coal capacity in China. Regardless, the Chinese, Indian, and U.S. consumption of coal will ensure that carbon dioxide (CO_2) emissions into the atmosphere overwhelm any technological countermeasures, or cap-and-trade foolishness, at least for the foreseeable future.

(3) Knock on any door these days and you might be informed of the beauty of reducing carbon dioxide (CO_2) emissions by turning coal into a gas and burning it in an ultra-efficient turbine, or burning it in pure oxygen without gasification, and then pumping the waste into the ground. The latter is termed 'carbon capture and storage' (or *sequestration*), or CCS. I will say more about this process below, but I first heard of it during my tour as visiting professor at Nanyang Technological University in Singapore, while over the past few years I have read or heard a great deal about it because of ambitions and efforts in that direction in Germany and Norway. Jeffrey Michel, an MIT graduate and one of the most important commentators on German energy matters, has referred to CCS as a "thermodynamic travesty"; however, *economic travesty* is a characterization that better appeals to my taste. In any event, CCS could double or triple the cost of a full-scale power plant — which admittedly is not of particular interest to, e.g., Norwegians, since in some respects Norway is the richest country in the world, and its governments (as well as some of its citizens) are always eager to flaunt their environmentalist credentials. However, the so-called

zero-emission plant that the Swedish utility Vattenfall constructed at Schwarze Pumpe (in eastern Germany) is only a pilot operation, which I think of as a publicity stunt rather than a serious attempt to solve a weighty problem. To get some idea of the expense that will be involved, CO_2 might have to be transported more than 300 kilometers, and then pumped into empty caverns 3 kilometers below the surface of the earth.

(4) Now for some repetition from Chapter 2 of this book. (The logic behind this departure is that the chapters of this book can, for the most part, be read separately.) Cap-and-trade is the method often favored (in theory at least) for combating CO_2 emissions, because it has a capitalist — instead of a *command and control* — flavor. Moreover, and more sinister, word has gone out that it has succeeded in Europe, where it is called the European Union's Emissions Trading Scheme (or EUETS). Actually it has failed for the most part in Europe, or is predicted to fail in that part of the world (and everywhere else) by almost everyone who does not have something to gain in financial or career terms if it is introduced. The essential thing that needs to be recognized and repeated whenever possible is that the economists who initially proposed this scheme have now declared it hopeless — that it was something for seminar rooms and academic conferences, but not the real world.

Then what is the problem? The problem is that a very large majority of the surviving recipients of the Nobel Prize once signed a document asking world leaders to act immediately to prevent a potential catastrophe due to global warming. Preventing disasters is not an easy thing to do, or even to think about doing. I believe that our leaders are anxious to adopt what might be called the *optimal* behavior in environmental matters, but they may be prevented from doing the right thing by the deluge of bad ideas that invariably surface when the environment is discussed. Perhaps the worst of these bad ideas is that an efficient abatement program can be constructed around emissions trading, because in the real world, such a market would be too difficult and/or expensive to organize on an *international* level, even though — at the alpine heights of pure economic theory — it once appeared capable of generating substantial welfare benefits. Unfortunately, it may be true that the emphasis will have to be placed on damage control, which is something that the engineering sciences should be able to tell us a great deal about, without having to put the issue into the hands of voters and politicians.

2. A Coal Primer

Coal is formed from the remains of trees that have been preserved for millions of years under special non-oxidizing conditions where, after falling, they either did not rot or rotted very slowly. (More generally, it is possible to speak of the *anaerobic* decay of all kinds of plant life.) Top-grade coal apparently requires a gestation period of millions of years, and scientists have calculated that a coal seam 1 meter thick might have been compacted originally from a 120-meter layer of plant remains.

A good example of what this is all about is the coal-rich state of Wyoming in the western United States (U.S.), which is the largest coal-producing state in the country. It has been estimated that the basis of coal seams in that region were formed tens or even scores of millions of years ago, and the dead vegetation was positioned in such a way that it did not rot or dry out. Perhaps 60 million years of this arrangement led to the thickest coal seams ever found — up to 60 meters thick in some places — with a low sulphur content. High-value resources such as this provided a strong incentive for further exploration. The U.S. has the world's largest coal reserves, and is the second largest hard coal producer after China, which is both the world's largest producer and consumer. The U.S., China, Russia and India possess 40% of the global population, but 60% of global coal reserves.

The U.S. coal industry is also on average the most productive (as measured in output/man-years), even though 'eastern coal' — from east of the Mississippi River — is largely from underground mines. In the 1950s, U.S. miners extracted less than a ton of coal per miner per hour. Today that figure is almost 7 tons/hour. The second largest coal-mining state is West Virginia, followed by Kentucky and then Pennsylvania.

Western coal production is generally an open-pit (or opencast) activity, where productivity is about 2.5 times as large as in underground installations. About 72,000 persons are employed in the U.S. coal industry, and at the present time the average weekly salaries for coal miners is $1,075. Similar progress is being made in Russia, and in the first 11 months of 2010, Russian coal output increased 6.3% to 287 million tons (including 61 mt of coking coal used in producing steel), and Russia exported 95 mt of coal, which was a 10% increase.

It is possible to distinguish a spectrum of coals, ranging from peat to anthracite. Peat — which is brown, porous, has a very high moisture content, and often contains visible plant remains — is the lowest class of coal, with an average energy content of 8.4 GJ/ton. (Here G signifies *giga*, which is a billion, and J signifies the basic energy unit *joule*. Thus, 1 GJ = 1,000,000,000 joules. The matter of energy units and equivalents was taken up in Chapter 3 of this book.) Next we come to lignite, which can be regarded as the transition link to hard coal (= bituminous + anthracite coal). Lignite also contains a great deal of water, and its average heat content is 14.7 GJ/ton. Bituminous coals, on the other hand, are characterized by a low moisture content, while the moisture content of anthracite coal is extremely low. Where energy values are concerned, we distinguish between sub-bituminous coal, with an average energy value of 25 GJ/ton, and bituminous coal, with an average energy value of 29.5 GJ/ton. Anthracite coal, which is jet-black and difficult to ignite, has an average energy value of 33.5 GJ/ton. Note here that thus far, the short ton (= ton = 2,000 pounds) is being used instead of the more common metric ton (or *tonne* or 't' = 2,205 pounds), and so 1 t = 1.1025 tons. Converting joules into Btu is explained below, and also in Chapter 3.

According to Brendow (2004), coal accounted for 37% of global electricity generation in 2000, and it will reach 45% in 2030. The power plants in which this coal will be used will, on average, be technologically superior to those in use today, but from an engineering point of view they will remain relatively simple affairs. Coal is burned in a boiler, and hot steam under high pressure is produced. This then goes to a steam turbine, whose mechanical work output takes the form of a rotational movement of generator shafts, which makes it possible to produce electricity. Students of thermodynamics know that energy losses cannot be avoided in this activity, but with a 'combined cycle' arrangement, some of the heat that might have been lost can be used to generate more electricity, which can sizably boost the overall efficiency of the installation. Brendow believes that by 2030, more than 70% of coal-based power generation will take place employing advanced coal combustion technologies. Obviously, for this prediction to hold, some gigantic financing problems will have to be solved.

In examining the energy literature on any level, we constantly encounter the word "primary". Primary energy is energy obtained from the direct heating of coal, gas, oil, etc., as well as electricity having a hydro or nuclear origin. Electricity obtained from the burning of substances such as coal is a secondary energy source. Something that should be appreciated is that the energy content of the coal used to, e.g., generate electricity is inevitably greater than the energy content of the electricity that is generated, because the coal-burning equipment does not possess an efficiency of 100 percent.

In some countries it is common to categorize coal as soft coal or hard coal. Soft coal consists of brown coals and lignite, whereas hard coal is bituminous coal and anthracite. In this system, peat is regarded as a fuel type in itself, and from a commercial point of view, is not particularly desirable any longer in many countries. Another system divides coal into two classifications: brown coal and black coal. Brown coal is geologically young and high in water content, while black coal is considerably lower in water content and contains much more carbon. Black coal ranges from sub-bituminous coals (often dull black and waxy in appearance) to anthracite, and is divided into two general categories: coking or metallurgical coal, and thermal or steaming coal. (Coking coal is largely ignored in this book, but is mainly used in the production of steel.) Brown coal is usually 'consumed' close to where it is mined, while steam coal exports are exclusively high energy-value coals.

The demand for coal (= hard coal + brown coal + lignite) grew by 62% over the 30 years prior to 2003, and the International Energy Agency (IEA) expects it to grow by another 53% from 2003 up to 2030. In addition, according to the IEA, in the most recent decade, coal accounted for about one-half of the increase in energy use worldwide. Moreover, the most expansive part of the world economy, Asia, accounts for two-thirds of the growth in global energy demand of all types, and of the countries in that region, China leads the world in coal production, importing, and consumption — despite all their fine talk about turning to renewables and alternatives. China may also be ready to launch a major effort to obtain large amounts of shale natural gas.

On the other hand, the U.S. has the largest coal reserves; Australia and Indonesia are the largest coal exporters; and the U.S. firm Peabody Energy is the world's largest private-sector coal company, and is a main reason why the U.S. is often described as the Saudi Arabia of world coal. These facts and figures make it very clear that coal is *not* on its way out, as many believe and/or hope. The world might be a better place if we learned how to use less coal, but in some respects, the electricity-generating sector is not a bad place for growth to take place: that sector probably has more experience in suppressing deleterious emissions than any other, and is better financed to make anti-emission investments.

It might also be useful to note that the average global power generation efficiency of coal is approximately 33%, while state-of-the-art efficiency is almost 45%. Considering that most of the power plants in existence now will be scrapped or upgraded by 2030, the aggregate efficiency in that sector should reach at least 40%. This will greatly favor coal as an alternative to nuclear energy, although by my calculations, in a carbon-conscious world, nuclear energy will definitely be a more economical source of electricity. On the other hand, on strictly private economical grounds, many buyers might be prepared to give up coal for gas as a result of the increased availability of gas (which is much larger today than a few years ago, due to the appearance of considerable *shale gas*).

With regard to the efficiencies mentioned above, these are so-called 'first-law efficiencies', after the First Law of Thermodynamics. Calling this efficiency E_1, we can write E_1 = (energy transfer achieved by system)/(energy input to the system). It would not be easy to challenge this definition on intuitive grounds; however, expanding a simple verbal assertion of the First Law to E_1 is too complicated to be done here. It can be mentioned though that the First Law is the well-known Conservation of Energy, which is usually stated as 'energy cannot be created or destroyed', and thus the total energy of the universe is constant.

It needs to be added that the icing on the thermodynamics cake is the Second Law of Thermodynamics, which happens to be a work of genius first proposed by the French artillery officer Sadi Carnot. One of the things it tells us is that there is an upper thermodynamic limit to efficiency, and technological progress involves no more or less than gradually raising the actual efficiency to that limit. It is difficult to say exactly what this limit will be for coal-based generating equipment, but Janssens and Cosack (2004), as well as others, indicate that 60% is the best that can be hoped for, although this will not be realized in the near future.

The most important exporting countries for hard coal are Australia, Indonesia and South Africa. The exports of these countries are close to 70% of seaborne hard coal. However, the U.S. is still regarded as the global *swing producer/exporter* of coal, occupying the same position with that energy resource as Saudi Arabia ostensibly does with oil. It also appears that exports of coal from the U.S. are increasing. Japan is the most important importing country, although most of its imports are coking coals (for making steel). It is forecast that Japan will account for 25% of total world coal imports in 2020. Coal consumption has declined in Europe because of environmental stipulations that favor gas, which at present is available in large amounts from the Norwegian North Sea, Russia, and North Africa — and perhaps eventually by pipeline from Central Asia and Iran via the former Soviet Union and/or Turkey.

The position of Japan in the coal import and nuclear picture is of considerable interest to me, since I spent almost two years in that country with the U.S. Army. Although not generally known or appreciated, the Japanese have (on average) the longest life expectancy in the world. This therefore suggests that nuclear and coal intensities may not be as important as often claimed in health matters.

Steam coal trade in the Pacific region surpassed the Atlantic market in the early 1990s, and by 2000 was 20% higher. Today a large number of firms/producers are active on the world market, which, together with domestic markets, gives the aggregate coal market the appearance of a competitive network — and according to some observers, considerably more than an appearance. At the same time though, reading the chapters on perfect competition in your favorite microeconomics or price theory textbook will not provide you with an ideal introduction to the kind of logic needed to understand the conditions under which this important resource is produced, bought, sold and priced.

For instance, it is impossible to conceive of *all* or most of those firms operating at the bottom of their long-run cost curves, as they would under ideal textbook conditions for other kinds of firms. Instead, equating supply to demand in the real-world coal market means the price rising to perhaps the bottom of the long-run cost curves of the highest-cost firms in the market, which in turn means that the 'intramarginal' (i.e., lower cost) enterprises will earn substantial economic rents (= profits greater than the amount needed to continue producing at the 'ideal' level). The key explanatory factor for this phenomenon is, of course, a difference in the quality of coal deposits controlled by individual firms, which is a condition that cannot be eliminated in the short run, nor perhaps the long run for many firms.

In the coal market, as everywhere else, there is a great deal of talk about replacing long-term contracts with spot transactions. This kind of aberrant thinking comes from the present urge toward what many observers think of as 'liberalization', and in some cases, makes absolutely no sense at all. The ostensible justification is that spot prices respond rapidly to the existing market situation, rising when the market is tight and falling when there are excess supplies, which is true. A problem here though is that enormously expensive investments are essential if markets for oil, gas and coal are to function in a manner that benefits households, small businesses and energy-intensive large businesses, and many of these investments will not be forthcoming if the managers of oil, gas or coal suppliers are constantly faced with highly volatile spot prices that provide mixed or misleading signals. With long-term contracts, this volatility can be much less malicious.

Let us put this a slightly different way. In finance theory, volatility is a common proxy for uncertainty. It can be easily demonstrated with some elementary algebra that, e.g., in the neo-classical models featured in conventional economics textbooks, a high volatility (and thus a high uncertainty) where price is concerned reduces physical investment. This is also common sense, and has to do with risk aversion. As it happens, although there is not a single world market for coal, nor a unique coal price, it is clear that coal has displayed a more stable price over recent decades than oil and gas, and as a result, we have not had to entertain the kind of complaints about inadequate investment that we constantly encounter with the other two.

The reference in Chapter 3 to energy units can be expanded on somewhat here. Several units are used to measure energy. Physicists seem to prefer joules, while engineers are often partial to British thermal units (Btu), or kilowatt-hours. (Generally, joules are preferable outside the U.S.) Another unit is calories (or kilocalories). The transformation between joules and Btu has been carefully measured: 1 Btu $= 1.055 \times 10^3$ joules. Since different coals have different calorific contents, a standard measure of energy content for coal can be extremely useful. This is the ton of coal equivalent ($=$ tce), which is defined as a *metric* ton ($= 1$ tonne $= 2,205$ pounds) of coal with a specific heating value of 12,600 Btu/pound. Consequently, more than one metric ton of coal might be necessary to produce the heating value of 1 tce. For example, 1 tce = 1.4 tonnes of sub-bituminous coal, using the heating value of 9,000 Btu/pound given earlier. (Note: 1 kilogram = 2.204 pounds.)

Consider also that in 1977, world coal production came to 3,400 million metric tons of raw coal, which was 2,500 million metric tons of coal equivalent (= 2,500 mtce), which in turn had the energy content (in Btu or Joules) of 33 million barrels of oil per day (= 33 mb/d). This last figure is obtained from the following equivalency between oil and coal: 1 tce converts to 4.8 barrels of oil, and 76 mtce/year is equivalent to 1 mb/d of oil. To a certain extent, 'tce' is an artificial unit, since its heating value is almost certainly higher than the heating value of an average tonne of coal extracted during any given year, but even so it is extremely useful.

It was mentioned earlier that the average global efficiency of coal using power generation equipment is 33%. Thus, a standard pound of coal equivalent functioning as an input in this equipment would have an energy output of only $12,600 \times 0.33 = 4,158$ Btu *electric* = 4,158 Btu(e). Readers should note the difference between 'equivalent' and 'electric'.

One more item can be mentioned here. Coal is sometimes referred to as a *backstop resource*, where the expression "backstop" (or even input into a backstop technology) was introduced by William Nordhaus in a brilliant article (1973), and involves the availability of a substitute to which no 'scarcity royalty' can be attached. For instance, at the present time, coal has been described as a backstop for motor fuel since it can be transformed into synthetic oil (as Marlon Brando assured us at considerable length in the film *The Formula*); however, hydrogen that is produced employing uranium or thorium in a breeder reactor probably comes closer to the strict definition, as perhaps does hydrogen obtained via electricity generated in wind installations. Moreover, coal might not be as plentiful as some people believe, and even if it is, it is not certain that using enormous quantities of 'uncleaned' coal is to be recommended.

As shown in Chapter 2, some very simple algebra will obtain the equation $T_e = \frac{1}{g} \ln \left(\frac{gX^*}{X_0} + 1 \right)$ for the approximate time to exhaustion of a coal deposit whose size is X^*. We can compare the difference between the static time to exhaustion (where $g = 0$) and the dynamic time to exhaustion (where, e.g., the value of g is taken as 2.5%/year). The static value $(= X^*/X_0)$ is approximately 260 years. Now, this can be very easily adjusted for growth by employing the above equation. We get $T_e = (1/0.025) \ln [(0.025 \times 260) +1] = 80.5$ years for the 'dynamic' value, which is a sizable difference — enough to make us wonder just how much coal our great-grandchildren will actually have at their disposal. Of course, by remembering that the total (or ultimate) amount of the resource will increase, and the rate of growth used to obtain this equation was taken as continuous rather than discrete, this 80.5 years can be extended somewhat, but very definitely not enough to come anywhere near the static value.

Normally, this would be the place for more algebra; however, in some respects the essential issues underlying actual production decisions for coal differ so fundamentally from those derived from standard economic theory, that playing with mathematical symbols often gets in the way of understanding real-world configurations. Instead we can begin by looking at Figure 1. The aim of that diagram is to make it clear how much coal there is in the world as compared to oil and gas, although it has been suggested earlier in this book, and also in other books, that there might not be as much as often assumed if the growth rate of coal consumption is correctly evaluated and entered into the discussion.

Moreover, while looking at Figure 1, readers can question the position of coal now that nuclear energy is in what might be called a 'partial retreat'. What we are told in some countries is that nuclear will be replaced by

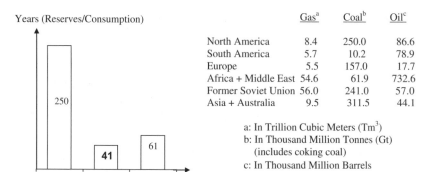

Years (Reserves/Consumption)	Gas[a]	Coal[b]	Oil[c]
North America	8.4	250.0	86.6
South America	5.7	10.2	78.9
Europe	5.5	157.0	17.7
Africa + Middle East	54.6	61.9	732.6
Former Soviet Union	56.0	241.0	57.0
Asia + Australia	9.5	311.5	44.1

a: In Trillion Cubic Meters (Tm^3)
b: In Thousand Million Tonnes (Gt)
   (includes coking coal)
c: In Thousand Million Barrels

Figure 1.   Global Reserve-Output Ratios for Coal, Oil, and Natural Gas

renewables, but to use the terminology of an American business executive I once encountered, "No one with his head screwed on correctly believes that." In the short run, more coal may be used, but the decision makers — at least those who can add and subtract — have no plans at all to give up on nuclear energy. This is because a majority of them, or their advisors, prefer facts to fantasy, at least where their professional activities are concerned.

According to the latest published scenario, coal will continue to play an important role in Germany after nuclear energy has been 'dumped', although over time that role is scheduled to be reduced. According to what in Germany is called '*Energiewende*' (= energy use), coal use will decrease from 31 percent today to 21 percent in 2022, to only 14 percent in 2032. This is some of the most grotesque nonsense foisted on the German people since the 'Thousand-Year Reich', and of course its purpose is to obtain for Angela Merkel the green votes needed to extend her vacation as head of the German government. As Joseph Shaefer (2012) points out, if Germany gets rid of its nuclear installations on environmental grounds, it will have no choice but to turn to the coal of Germany and other countries. The same is true of Japan, who might have to increase its coal consumption from 30 percent of the input for its electricity generation facilities to about 50 percent.

The numbers in Figure 1 are a decade old, but they indicate the relative situation where reserves of coal, oil and gas are concerned. In any event, it should be clear to all readers that while we may not have reliable estimates of uncertain or 'hypothetical' reserves, coal is a very plentiful fossil fuel, although at the end of this section, this belief is somewhat modified. In addition, coal has been relatively inexpensive, though coal prices now are higher than predicted in my coal book (Banks, 1985), and at present seem to range from \$70/tonne to \$120/t. Coal generates about 40 percent of the world's electricity, and this fraction may be sustainable for a while, even though the latest forecast is that global energy demand will increase by 50 to 60 percent over the next 25 or 30 years.

Five or six years ago it was claimed that natural gas would soon be close to coal in the electricity generation league, but on several occasions in the near past, a rapid increase in the price of gas brought its future into question by no less than the director of the U.S. central bank, Alan Greenspan. The price of gas is very low at the present time, but this may be due to a sluggish global macroeconomy. Perhaps the most dynamic selling point for gas these days is that it is 'cleaner' than oil and coal, and ostensibly large reserves of 'shale gas' can be exploited in the U.S. by a new technology that in theory can be applied elsewhere. As far as I am concerned, the long-term accessibility of economical (i.e., inexpensive) shale gas is still in question, and two of the most important oil and gas observers — Henry Groppe and Arthur Berman — have expressed reservations about shale gas, citing in particular the high natural depletion of that resource as production takes place.

The coal industry is going to great lengths to convince buyers and potential buyers that the cleaning of coal is a routine and relatively inexpensive operation. The technical aspects are interesting, and perhaps the most frequently cited are references to the gasification of coal in order to reduce its environmental impact. Moreover, the resulting 'syngas' works well in combined-cycle electric power plants, and essentially the same technology that transforms natural gas to liquids is capable of converting coal into high-quality liquid fuels.

What happens here is that pressure and high heat are used on coal in order to obtain combustible gases, which are then processed into 'clean' transportation fuels. There are essentially five stages in this Integrated Gasification Combined Cycle (IGCC) activity, and these can be aggregated as follows. During the first stage, coal reacts with oxygen and steam to form syngas, which is predominantly hydrogen. This syngas can be fed into a turbine whose revolving shaft generates electricity. To obtain the combined cycle effect, the residual gas can be burned to heat water, whose steam drives a separate turbine. It is also possible that someday coal will be a major factor on the petrochemical scene, since in the 19th century many inputs for the chemical industry were derived from *coal tar* (which is a highly viscous liquid that is a by-product of the carbonization of coal).

One way or another, when discussing coal, it is inevitable that carbon emissions will come into the picture. I recently heard some comments about CCS where 'emissions trading' was also mentioned, but regardless of the source, more scientific exposition is required. A non-technical introduction to the 'carbon trade' is presented in a short article by Victor and Cullenward (2007), with a focus on the European Union's Emissions Trading Scheme (EUETS), which is the world's largest carbon market, and presumably the best organized. It is often spoken of as a model for the carbon trading set-up that the U.S. government would like to see adopted in that country. It may indeed be a model, but not one that I would like to see installed in any country in the civilized world.

The basic mechanics of this model are simple, although the same cannot be said for some of its details. If, e.g., we have a firm (called X) that is releasing more CO_2 than the 'cap' (i.e., quota) assigned to it by the authorities (and for which it lacks the required number of emission permits), then the firm must either reduce its output of consumer and/or producer goods, or enter into some sort of market to purchase additional permits — for example, from a low CO_2 emitter (firm Y), which during the period in question has a surplus of permits.

If we examine the present-day situation in Europe, firms X and Y might be electricity generators, cement makers, paper and pulp companies, etc. Moreover, the intention is to systematically reduce the cap for various firms, which means that to avoid having to pay large sums for extra permits, firms must either reduce their output of investment and/or consumption goods, or invest in technologies that feature lower CO_2 emissions. *Ceteris paribus*, coal and coal-burning equipment would decrease in value. In the first phase of the European scheme (from 2005–2007), permits were issued free — or *grandfathered* — to firms, and generally these enterprises were supplied with all that they needed. In the next phase (2008–2012), firms were supposed to buy a fraction of their permits at an auction, and the same applies to the third phase (2013–2020).

The word "fraction" is obviously open to interpretation, and as a result, firms have often been provided with what seems to be an excess of permits for both economic and political reasons. This has apparently annoyed many environmentalists, who react by pleading for a 'tightening' of caps. What is often overlooked, though, is the amount of money that tightening caps could involve. Over the past few years, industry emissions of carbon in Europe averaged about 2 billion tonnes, and with average permit prices reputedly at \$17/tonne, even a relatively small amount of trading (i.e., buying and selling) could mean large gains or losses for some producers — but always large gains (i.e., profits) for the middlemen (i.e., traders) who conceivably handle most of these transactions. This might be a reason why some prominent climate scientists now believe that emissions trading is a gigantic mistake, and perhaps the most important of these scientists — Dr. James Hansen — went so far as to wish for a failure of the international climate jamboree in Copenhagen (which began on the 68th anniversary of the attack on Pearl Harbor). Fortunately, his wish was granted.

I once read that the price of emission permits has on occasion touched $40/t, and recognizing that this could have ugly financial consequences for some firms, arrangements exist for polluters to purchase a certain amount of 'carbon credits' from abroad, which take the form of certificates issued by local firms who become involved in emission reduction activities in developing countries. In dollars per tonne of emissions eliminated, this could cost much less than expenditures on market-based permits in Europe, or new equipment with a lower output of emissions.

As noted in O'Sullivan and Sheffrin (2008), a large enterprise in the United States elected to fulfil the environmental responsibilities mandated by the Climate Control Accord (of 1994) by paying for pollution abatement projects in China, India and other countries where abatement costs are lower than in its home state (Arizona). That firm also financed a reforestation project in Mexico, which ostensibly will help to suppress global warming because forests (and large bodies of water) absorb some of the CO_2 that is judged responsible for undesired climate change.

Ceteris paribus, abatement schemes of this nature are to some extent perfectly logical, since we have a continuous global environment. Thus, reducing pollution in, e.g., China or Mexico would unambiguously benefit the good citizens of the United States (and elsewhere). But regrettably, some of us recall that a possible outcome in an advanced "prisoner's dilemma" game is one in which restraint by one player could lead to excesses by others. Not "could", but probably does, because no matter what you have heard about the progress made in reducing undesirable emissions, the global output of CO_2 into the atmosphere increased by at least 2 percent during 2008 (and perhaps by 30 percent between 2000 and 2008).

According to Susanna Baltscheffsky (2009), a principal cause of this situation is economic growth in China and India, which is a secret that no longer has a great deal of value because it has been completely absorbed by all except the most drowsy members of the TV audience. What has not been absorbed is that regardless of what happens in the U.S. or similar locales, the above two countries are under no obligation to entirely or even partially reciprocate the good intentions of distant polluters by lowering (or at least not raising) their own output of deleterious emissions. It is also a reason why the cap-and-trade approach to emissions reduction that is said to be favored by the new U.S. government is unlikely to succeed: it cannot be enforced globally. Of course, both the Chinese and Indian governments have indicated that they will take steps to conform to the environmental stipulations agreed upon internationally, and I see no reason to believe that they are not serious; however, serious or not, this is unlikely to happen, since *the trade-off is lower economic growth, and for those countries, growth has been revealed as preferred to emissions suppression!*

Details of this nature are much less important for me than the often stated assumption that if the good citizens of the U.S. or Sweden pay higher prices for electricity — which they would almost certainly have to do regardless of how or where enterprises in their countries choose to abate a given amount of pollution — a (*ceteris paribus*) reduction in pollution in some exotic neighborhood on the other side of the world would subsequently alleviate pollution in the U.S. or Sweden by an amount close to that which would take place if pollution reduction expenditures had been made in or near New York City or Stockholm.

If this were *not* so, and foreign anti-pollution investments on the part of voters in New York City or Stockholm did not eventually provide palpable local benefits, it might become very difficult for the citizens of those two cities to maintain any enthusiasm they might possess for international pollution control. In addition, if anti-pollution expenditures were made in or near New York City or Stockholm, they might provide additional benefits in the form of wages, salaries, and possibly training. This is why it is often claimed, possibly correctly, that pollution control makes a certain amount of economic sense, as long as we do not focus on the initial expenditures. Or maybe I should say, it makes a certain amount of economic sense in classrooms, seminars, and various 'talkathons'.

Sweden deserves a special mention here, though not one that is particularly complimentary. Most of the electricity in this country is generated using nuclear energy and hydro, but even so, many Swedish politicians and environmentalists seem favorably disposed to the direct or indirect heavy taxation of a domestic output of emissions that happens to be one of the least intrusive in the industrial world. This official preference deserves special

Energy and Economic Theory

attention, because if the entire electricity-generating world imitated Swedish environmental practices, meetings of the Kyoto and Copenhagen variety could be cancelled or reduced to insipid talk-shops where delegates attend amateurish lectures, and busy themselves with obtaining invitations to the next climate warming 'happening'.

An examination of the energy economics literature makes it clear that this chapter is a necessary one. In the last 15 years, coal has tended to become a minor topic, with the possible exception of the attention that is paid to its environmental shortcomings. But even so, the consumption of that resource continues to grow at a rapid pace, and for good reason: there is an enormous amount of coal in the crust of the earth, and for many years it was comparatively inexpensive. In addition, large or fairly large deposits are found on 6 continents and in 50 countries, and this 'geographic balance' helps to solve a certain bothersome political issue.

A few more abstract observations might be appropriate at the present time. For instance, the really terrible thing about *excessive* global warming, if it is taking place or destined to take place, may turn out to be that it is NOT man-made, and thus absolutely nothing can be done about it. An extraordinarily bad climate event might be economically and socially devastating, regardless of the precautions that are taken to keep deleterious emissions out of the atmosphere. In fact, if an extreme range of unpleasantness seemed likely to pass across the globe, the lights in the Pentagon and similar establishments elsewhere would burn very late at night, as Bruce Willis spin-offs in Armani military creations try to figure out how to keep borders closed.

A sort of caveat may be appropriate here. The physical destruction inflicted on New Orleans by Katrina was clearly worse than that suffered by some German cities that had been subjected to repeated bombing during World War II. By the end of the Korean War (in 1953), regardless of the amount or type of damage that individual German communities had experienced, most of them functioned satisfactorily — at least in the opinion of the American military personnel who were lucky enough to be stationed near them. On the contrary, there is talk of it taking 20 years to rebuild New Orleans, assuming that the project is undertaken, and also assuming that the 'Green City' that former President Bill Clinton mentioned at the latest climate warming talk-shop will have the approximate dimensions and population of the 'old' New Orleans. The interesting thing in this case, however, is that if the probabilities had been correctly calculated from the beginning, and a modest amount of investment undertaken over the years, the kind of skillful engineering that has been practiced for decades in, e.g., Holland, could have turned Katrina into no more than a soggy happening.

One of the reasons for this chapter in a book that in some ways is already too long, is that I wanted to bring up-to-date several topics in my previous energy economics textbooks (2007, 2000) and in my book on coal (1985). Interestingly enough, I wrote another book on coal but never published it, because I mistakenly believed that hardly anyone would bother to read a work on that subject; however, once the production of oil has peaked, or shows signs of peaking, and should it happen that the future availability of natural gas is estimated to be less than that constantly claimed at the present time, a new coal book might find the readership it deserves — though hardly before. It might also be a good idea if readers examine a short, non-technical article by Murray Duffin (2004) in www.energypulse.net, and also the incisive comments on his work that are published at the end of his article.

3. Further Comments on the World Coal Scene

This section begins with a short review of the coal situation in various parts of the world — "short" because the rapid change that often takes place does not justify a more thorough perusal. More important for me, however, is the suggested 'commoditization' of the world coal market, which refers to the irrational desire by various buyers and/or sellers to greatly increase the use of 'spot' transactions whilst decreasing the employment of long-term arrangements. We have seen this sort of thing in other energy markets, and the results were not encouraging. Certainly, as Mr. Zach Allen (2005) pointed out to me, large mines may not be opened if producers/investors have to accept being at the mercy of spot prices.

Coal in North America is dominated by the large production and con-sumption by the United States. Based on statistics for the last decade, coal was the basis for slightly over 50% of U.S. electricity generation, and some of this electricity, together with the direct use of coal, heats and/or cools about 50% of U.S. homes. Moreover, the energy in U.S. coal reserves (measured in Btu or joules) is well in excess of the energy in Saudi Arabian oil. These are undoubtedly important (though apparently unspoken) reasons why the U.S. was not willing to sign the Kyoto Protocol — aside from the fact that the world would probably be better off without the Protocol, the conference in which it was produced, and similar conferences in the future. Coal can fairly easily be transformed into motor fuel, although — unlike natural gas — with present technology, this does not appear to be a very profitable activity. (And before the large-scale arrival of shale gas, it appeared to be unprofitable with conventional gas too, because according to information provided to me by Mr. Oliver L. Campbell, the price of gas in the UK once touched $15 per million Btu, which readers should turn into an equivalent oil price! Someone else who discussed this problem was the former head of the (U.S.) Federal Reserve System, Alan Greenspan.)

In South and Central America, only Colombia and Venezuela are major coal countries, and at the present time, only Colombia is making a large contribution to the world market. Most of the mining in those two countries is of the opencast variety, which suggests a high productivity, but this is not the case as yet. Much, however, is expected of these two countries.

In Europe (outside the former Soviet Union), Germany is the largest coal producer if lignite is counted, and lignite should be taken into consideration because it supplies the largest input for German power plants. Germany is similar to the UK in that it is a country where the 'quality' of coal produced may be increasing due to the closing of inferior deposits. I have also heard the remaining coal mines in the UK called the most productive in Europe, although not everyone may agree. Quantitatively, Poland comes after Germany, but according to the important coal expert Zach Allen, the Polish coal-mining sector has experienced severe problems with labor relations, resulting in comparatively expensive coal. Of course, if economic growth in Asia continues at its present pace, and oil and gas prices remain close to their present levels, then the financial prospects for all the large coal producers in every part of the world should be considerably improved.

As with oil and gas, Russia ranks close to the top of the coal production league. Although its productivity (in output/man-year) was well below international averages the last time I checked, the intention in that country is likely to produce and use as much coal as possible domestically, so that the highly profitable exports of oil and gas can be maintained or increased. The World Bank has taken a strong interest in Russia, ostensibly providing it with financial and technical assistance for so-called restructuring/privatizing purposes. I hope that the Russians are grateful for any financial help they receive; however, since I happen to find it strange that a country on Russia's technical level requires technical aid from an extravagant refuge for high-flown mediocrity, I prefer to conclude that the basic intention of the World Bank in this matter is to justify its budget in the eyes of its most persistent critic, which happens to be the U.S. government.

Two highly productive and large coal producers and exporters are Australia (which specializes in coking coal) and South Africa. Of late though, the progress of exporters like Indonesia and eventually perhaps the U.S. might raise problems for the enlargement of their market shares. Until recently it seemed that the U.S. had ceased to be the expansive force in the world export sector that it was during various periods of the last century, since there was a 20-year period of stagnation for the coal industry from 1980 to the turn of the century, and a decline in employment from about 1990 to 2003. It was occasionally claimed that the reason for this situation was that high wages and salaries decreased the international competitiveness of U.S. coal. But the recent growth in output and employment in the U.S. coal industry, as well as expectations about the future demand for coal in Asia, suggests that this situation may be changing.

Much more could probably be added to the above discussion; however, I think that everyone reading this chapter appreciates that, for good or evil, coal is extremely important in both the present and future energy scenarios. Globally, trade in steam coal is expected to increase at a fairly high rate between now and 2030, and because the price of coal is seen as stabilizing in comparison to oil and gas, increasing amounts of coal-fired generating capacity will likely be the rule in much of the world. Japan was briefly mentioned earlier as a major coal importer, especially of coking coal; however, steam coal generally has a poor image in Japan, and unless things have changed greatly in the last few years, I believe that the implicit desire of the Japanese energy establishment is to minimize the use of steam coal, while drastically increasing nuclear-based generating capacity — if (or when) that is politically and psychologically possible.

Unfortunately, it seems to me that it is highly unlikely that the huge amount of coal that is currently being used, and that will be used in future, can be 'processed/treated' in such a way as to substantially and efficiently reduce the amount of carbon dioxide (CO_2) that it produces. As you undoubtedly know, CO_2 is a key element in global warming, which in one sense happens to be good rather than bad, because without it the earth would be uninhabitable. But on the other hand, there is the possibility that too much of it is currently being produced, and perhaps this excess supply is due to man-made sources rather than the various quirks of nature. The opinion of this teacher of economics and finance is that regardless of the actual situation, it cannot be dismissed that the majority of the elite of climate scientists who claim that there are dangerously excessive CO_2 emissions know what they are talking about.

Furthermore, the excess production of CO_2 should be negotiated down by heads of states, and not jet-setters from the environmental bureaucracies. To me, the failure of the Kyoto exercise is precisely the inability of its participants to detect this option, and to recommend its immediate adoption. Of course, one reason they failed to do so is because half-baked talk-shops of the Kyoto variety are the lifeblood of many footloose busybodies whose speciality is pseudo-intellectual environmentalism, and the waffle at these congresses counts for much more to many of them than attempting to evaluate a topic whose details they are unable to understand. As for emissions trading, which is a highly advertised offshoot of Kyoto, it is best described as an elaborate scam. As an advisor to President Putin once remarked, it is more about making money rather than curbing emissions.

Something else that is about making money is the attempt to 'commoditize' the trading of coal. In the words of Robert Murray, president and CEO of Murray Energy — one of the largest independent, publicly owned coal producers in the U.S. — trying to make a true commodity out of coal is like "trying to fit a square peg in a round hole" (*Petroleum Economist*, October 2002). He continued by calling the proposed adjusting of coal trading "an unnecessary fad" and "a doomed concept". The economic issue here involves putting an intermediary between buyers and sellers in the form of a formal 'exchange' of one sort or another; however, for the time being, the idea is that 'over-the-counter (OTC)' establishments are to fulfil this function. Here I should make it clear that there is a very great difference between an OTC market and a genuine exchange — roughly the difference between the Fulton Fish Market and the New York Stock Exchange.

To me, the kind of language employed by Mr. Murray is perfect for describing electricity deregulation and the attempt to commoditize electricity, although that bogus escapade is rapidly losing popularity. The new-old argument being used in the case of coal is that both buyers and sellers would be better off if they accepted the beauty of OTC trading and short-term contracts because — as we teach our beginner students — genuine competition always provides better outcomes for all involved. This is undoubtedly true for many items, but I have grave doubts as to whether it applies to a market like coal, where tremendous amounts are involved under very special circumstances.

In an ideal situation, the OTC market would have many of the features of an auction market (like the stock exchanges), with full price transparency, and where the possibility exists for transactors to buy or sell almost any amount of the commodity at any time. It could then be argued that prices would correspond closely to the theoretically correct prices that would prevail in a textbook market. Moving beyond elementary theory, this would mean that the large inventories of coal held by, e.g., sellers could be reduced because these ladies and gentlemen would *always* be in a position to provide coal from their own mines or from the trading marketplace, and presumably any savings they achieved would be shared to some extent by consumers. Some consumers (i.e., distributors) also maintain large stockpiles, but these could also be reduced because they too could use the open market.

Here the reader should be aware that this kind of argument was employed in California when the electric deregulation fiasco was being sold to the television audience and their representatives in the California legislature. By putting in place an exchange or pseudo-auction market for large-scale trading between buyers and sellers, the theory was that it would be unnecessary for sellers to maintain a large reserve capacity, which in turn should eventually work to the benefit of everybody. The outcome of this less than brilliant gambit was the ruining of the state budget, an electricity price explosion in San Diego, and the then governor using the quaint expression "out-of-state criminals" to describe wholesalers (i.e., generators) who took the opportunity offered by deregulation to charge outrageous prices for filling the gap between local supply and demand.

Moreover, in a 'super-ideal' situation, some serious hedging (i.e., insuring against price risk) could take place, because the OTC contracts being used — or a spin-off of these contracts — could function in a manner similar to genuine futures contracts or something of that sort, which would allow buyers and sellers to 'lock in' present prices and thus avoid being faced with ruin in the event of having to fulfill any unfavorable commitments that they might have entered into. Naturally, all of this was 'hype', but as with the electricity markets in California and Scandinavia, it was treated with complete seriousness by some very intelligent and highly educated academics and businesspersons.

This brings us to a comment on the difference between real markets and ideal markets. In ideal markets there are large numbers of transactors on both the 'buy' and 'sell' sides, completely transparent prices, and a great deal of liquidity, which means that it is always possible to buy or sell any quantity without drastically altering these transparent prices. Furthermore, the prices that are formed are theoretically correct prices, which are sometimes called 'scarcity' prices, in that they accurately reflect the intentions and capabilities of buyers and sellers. In addition, in light of the bad news from, e.g., California, neither these prices nor the conditions under which they are formed encourage or facilitate 'gaming the market' by ambitious transactors.

Reality is very different from this, however. Although the physical coal market has many competitive aspects, various changes have taken place during the past few years, and in particular, some large consolidations (i.e., mergers) have undoubtedly reduced the degree of competition. Most important for this discussion, liquidity (and probably transparency) in the OTC market is too low to make it attractive for hedging large volumes. As with electricity, the best hedging item for buyers and sellers are long-term contracts. In addition, and this is crucial, Mr. Murray berates the (OTC) intermediaries for their lack of knowledge of the industry. The same is even more true of the electricity market, where the gap between 'quants' and traders in the exchanges, and the men and women involved with the physical market, is enormous.

Before ending this section, some remarks need to be added about the optimal deployment of coal-based power plants. The essential point here is that peaky, short duration loads should be carried by equipment with low fixed costs, since this equipment might be idle a large part of the time. Prior to the development of combined cycle gas-based equipment, the so-called 'merit order' called for natural gas to perform this function; but later it became conceivable that natural gas could compete with coal, nuclear and hydro for carrying the base load (i.e., the load that is always on the line). Accordingly, as long as the price of gas was low, it was perhaps the most versatile member of the merit order.

When I deal with the subject of an ideal electricity-generating system, I of course cite Sweden, where the base load is traditionally produced by nuclear and hydro. Hydro also carries most of the peak load, because it can be easily switched on and off, and its output raised or lowered. Naturally, it also produces a large part of the base load, and together, nuclear and hydro divide evenly most of the electricity output of the country amongst them. For what it is worth, Sweden has often had the lowest electricity-generating costs in the world, and is one of the lowest producers of CO_2 from its electricity sector.

Norway is the other winner in the low-cost league, and in that country almost all electricity production is hydro-based. Accordingly, for many of us who remember our secondary school mathematics, this means that since electricity costs in these two countries are almost equal, nuclear-based electricity is very inexpensive. I have unfortunately had to entertain arguments that nuclear energy is in reality very expensive for Sweden, and furthermore I have been assured that this will continue to be the case; however, to my way of thinking, if someone does not understand why this belief is incorrect, they would hardly be able to comprehend a simple argument to the contrary.

A good example here would be the so-called 'energy professor', Gordon McKerron. In an article in the *Observer* (November 4, 2005), he wants to know "who puts up the cash" for a new generation of nuclear power stations. The answer to that question is that it should be the persons who benefit from these facilities — whether they know it or not — and that means just about everybody. In the case of Sweden, one of the highest living standards in the world was created on the basis of the inexpensive power supplied by nuclear energy. However, this fact is largely unknown to the present Swedish government and the Swedish electorate, who have foolishly tied their economic future to the fortunes of the European Union.

Something that should be emphasized is that some of the logic being employed above is different from that provided in your microeconomics textbooks. Nuclear has a lower marginal cost than, e.g., gas, but if you construct a conventional supply curve and attempt to justify the use of nuclear to produce the peak load, you would be wrong: obviously, it does not make sense to construct a nuclear plant that might be idle for a considerable period. Hopefully readers of this book already understand this comment, because many non-readers do not have a clue.

Amazingly, I recently heard the expression "peak coal". What is the explanation here? The explanation is China and India! These countries need energy, and while I do not know about the ability of India to support its demand with cash and credit, China is on the hunt for coal as if it will be exhausted before the end of this decade. If there is such a thing as peak coal (in the near future), then the responsible consumer has to be China. The decision makers in that country know this, however, which is why the rate at which they introduce new nuclear capacity will probably break all existing records. China has also expressed the intention to satisfy 10% of its transportation needs by converting coal to liquids. This shines a very bright Chinese light on the Sasol Corporation of South Africa, which produces about 150,000 barrels a day of 'synthetic oil' from coal, and is a leader in coal-to-oil technology. Hopefully, before the Chinese put their large-scale coal show on the road, someone will inform them that environmentally, turning coal into liquids has been termed a "double whammy". This refers to the pollution produced during the coal-to-liquids activity, and on top of that, the pollution resulting when the liquids are burned.

4. Concluding Remarks: "Welcome to the Land of Coal"

"Coal is a dead man walking."
— Kevin Parker (Deutsche Bank executive)

Well, I don't know about that, Kevin. On a billboard greeting travellers at the airport in Ranchi (India), the capital of the Indian state of Jharkhand, "Welcome to the Land of Coal" is posted in block letters, and in case you are seriously curious, you will not have any difficulty finding someone to proclaim that coal is the way to go. Such will be the case for many years in the future. India needs more coal, and apparently needs it quick in order to maintain its momentum of economic development.

In the film *The Formula*, Marlon Brando informs a Swiss colleague that, "Today it's coal. In ten years it will be gold." As owner of a large part of the hard coal deposits in the U.S., and also of a superior process for producing synthetic oil from coal, the Brando character may well have known what he was talking about. Of course, his time frame was very likely wrong: it would not be ten years, but another decade or two before the billions of dollars started to roll in, although in terms of historical time it hardly makes a difference. Besides, in showing that they know something meaningful about the future importance and use of coal, the writers, directors, producers, and maybe even the actors involved with this film gave the impression that they were better informed about coal than many academic energy economists.

I am sure that my opinion of Hollywood is similar to that often expressed by Mr. Brando in his more articulate off-screen moments, but one thing the movie-men must be given credit for is that they understand the way that some political and industrial celebrities take care of real business: hypocrisy, public relations, bribes and taking advantage of the naiveté of drowsy voters. If you study game theory, an introductory course might emphasize players, payoffs and strategy, but more comprehensive game theory literature pays particular attention to information, and in particular *incomplete information*, which is a condition in which some players can have access to very important information which is not in the public domain. As David Lloyd George, prime minister of England during the First World War, said of the general public at one particularly traumatic point during that struggle, "Of course they don't know — how could they know." What he did not say was that this arrangement suited him perfectly.

As I often make a point of making clear when the subject is coal, the largest Swedish utility, Vattenfall, is building a pilot coal-burning installation in Germany in which CO_2 emissions into the atmosphere are supposedly close to zero. This facility of 30 megawatts — as compared to 1,000 megawatts for most new installations — may not be ready to be evaluated for perhaps ten years. If the news is good, a 250-megawatt demonstration plant will ostensibly be constructed. In other words, the financing of these new, quantitatively inconsequential installations will not interfere with the bonus program initiated some years ago by Vattenfall, and even more important, will not interfere with the flow of cash to the owners of Vattenfall — especially the Swedish Government — who need this money to pay their dues to perhaps the most grandiose parasitical organization in the industrial world, namely the European Union, as well as an absurd military commitment in Afghanistan. Exactly what contribution all of this will have to the reduction of 'greenhouse gases' remains to be seen, although, in all fairness, it may turn out to have a great deal if its admirers and taxpayers are prepared to wait a few decades to judge the results.

Now for the statement by Mr. Parker at the top of this section. It so happens that in the U.S. from 2009 to 2010, construction was not commenced on a single coal-fired power plant, and coal use (= 498 million tons of oil equivalent in 2009) showed an 8% decline since 2000. But in China and India, the "dead man" is in wonderful health, as discussed earlier. Thus, while many American officeholders, bureaucrats, and perhaps influential voters believe that coal's day is over and it is time to move on to something else, the truth is that there is no 'something else' where this issue is concerned. Globally, there is too much energy in coal for the remainder of that resource to be cast aside like clothes or shoes that have gone out of fashion.

A former Swedish prime minister called nuclear energy "obsolete", and the current U.S. president apparently has proclaimed similar thoughts about coal; but the objective macroeconomic situation in the world economy is going to change many minds in the U.S. and elsewhere, and perhaps sooner rather than later. Without adequate energy, the competitive strength of the U.S. economy could be drastically weakened, and since this is well-known to the decision makers and their advisors, coal is going to remain a major producer of electric power in the U.S. and elsewhere for the foreseeable future.

In economics, as compared to physics, there are many trivialities, and I am afraid that many of my students remain too occupied with them to get the information they might need about this topic. The crucial thing here is that: (1) there is going to be a huge increase in the use of coal, and (2) most of this coal may be an extremely large contributor to greenhouse gases. What happens as a result of this situation is left for students, teachers, and interested readers of the energy economics literature to read about and/or investigate, and to make their findings as widely known as possible.

Bibliography

Allen, Zach (2005). "Comments and observations on coal" (Stencil).
Baltscheffsky, Susanna (2009). "Koldioxidutsläppen når rekord höjder." *Svenska Dagbladet* (December 3).

Banks, Ferdinand E. (2010). "An unfriendly comment on another Green Fantasy: Roadmap 2050." *321 Energy* (April).

———(2008). "Nuclear and the new American president." *EnergyPulse* (December 31).

———(2007). *The Political Economy of World Energy: An Introductory Textbook.* London, New York and Singapore: World Scientific.

———(2006). "Logic and the oil future." *Energy Sources, Part B*, 1(1): 97–114.

———(2004). "Economic theory and a faith-based approach to global warming." *Energy and Environment*, 15(5): 837–852.

———(2000). *Energy Economics: A Modern Introduction.* Dordrecht and Boston: Kluwer Academic.

———(2000). "The Kyoto negotiations on climate change: an economic perspective." *Energy Sources*, 22(6): 481–496.

———(1985). *The Political Economy of Coal.* Boston: Lexington Books.

———(1974). "A note on some theoretical issues of resource depletion." *Journal of Economic Theory*, 9(2): 238–243.

Barbier, Christophe (2009). "Jean-Louis Borloo: Obama doit nous emboiter le pas." *L'express* (December 13).

Bell, Ruth Greenspan (2006). "The Kyoto Placebo." *Issues in Science and Technology*, 22(2).

Beyer, Jim (2007). "Comment on Banks." *EnergyPulse* (www.energypulse.net).

Brendow, Klaus (2004). "Global and regional coal demand perspectives to 2030 and beyond." In *Sustainable Global Energy Development: The Case of Coal.* London: World Energy Council.

Dales, H.H. (1968). *Pollution, Property and Prices.* Toronto: University of Toronto Press.

Duffin, Murray (2004). "The energy challenge — 2004." *EnergyPulse*.

Festraets, Marion (2009). "Jean Jouzel: 3 ou 4°C plus, ca change tout." *L'express* (December 13).

Goodstein, David (2004). *Out of Gas: The End of the Age of Oil.* New York and London: Norton.

Hansson, Bengt (1981). "Svalet kan elimineras." *Svenska Dagbladet* (July 24).

Harlinger, Hildegard (1975). "Neue modelle für die zukunft der menscheit." IFO Institut für Wirtschaftforschung, Munich.

Harvey, Fiona (2010). "European carbon trading survives key tests." *Financial Times* (April 8).

Hilsenrath, Jon (2009). "Cap-and-trade's unlikely critics: its creators." *Wall Street Journal* (August 13).

Hotelling, Harold (1931). "The economics of exhaustible resources." *Journal of Political Economy*, 39(2): 137–175.

Hung, Joe (2010). "Coal: the contrarian investment." *321 Energy* (April 10).

Janssens, Leopold and Christopher Cosack (2004). "Forging internationally consistent energy and coal policies." In *Sustainable Global Energy Development: The Case of Coal.* London: World Energy Council.

Johnson, Elizabeth (1972). *The Collected Writings of John Maynard Keynes, XIV.* London: Macmillan and Cambridge University Press.

Lave, Lester (1965). "Factors affecting cooperation in the prisoner's dilemma." *Behavioral Science*, 10(1): 26–38.

Montgomery, David (1972). "Markets in licenses and efficient pollution control programs." *Journal of Economic Theory*, 5(3): 395–418.

Nordhaus, William D. (1973). "The allocation of energy resources." *Brookings Papers on Economic Activity*, 4(3): 529–576.

O'Sullivan, Arthur and Steven H. Sheffrin (2008). *Economics: Principles and Tools.* New Jersey: Prentice Hall.

Petit, Charles (2005). "Power struggle." *Nature*, 438(7067): 410–412.

Roques, Fabien, William J. Nuttall, David Newbery, Richard de Neufville, and Stephen Connors (2006). "Nuclear power: a hedge against uncertain gas and carbon prices." *The Energy Journal*, 27(4): 1–24.

Rose, Johanna (2010). "Drömmen om rentkol." *Forskning & Framsteg* (March).

Schultz, Walter (1984). "Die langfristige kosten entwicklung für steinkohle am weltmarkt."

Shaefer, Joseph L. (2012). "A skål to old King Coal." *Seeking Alpha* (March 21).

Siebert, Horst (1979). "Erschopfbare ressourcen." *Wirtschaftsdienst*, 10: 1–7.

Thunell, Jan (1979). *Kol, Olja, Kärnkraft — En Jämförelse.* Stockholm: Ingenjörsförlagen.

Victor, David G. and Danny Cullenward (2007). "Making carbon markets work." *Scientific American*, 297(6): 44–51.

Victor, David and Varun Rai (2009). "Dirty coal is winning." *Newsweek* (January 12).

Yohe, Gary W. (1997). "First principles and the economic comparison of regulatory alternatives in global change." *OPEC Review*, 21(2): 75–83.

Zimmerman, Martin (1981). *The U.S. Coal Industry.* Cambridge: MIT Press.

Chapter 8

The Final Countdown

"Sun, sand and music
Flashing skis and winter scenes
But it's you I'm thinking of, in this Sweden of my dreams."

— *Beatrice Sylvan (and colleague)*

A Short Introduction

We are finally approaching the end of this book, and I sincerely hope that readers are now better equipped and motivated to comprehend the many energy issues that are destined to become more important in their lives in both the near and distant future. I especially hope that they read (or reread) — and believe — the statement at the top of the first chapter: "The production of energy is the moving force of world economic progress." The gentleman offering that assertion was no less than President Vladimir Putin of Russia, and I heard almost the same thing many years ago from Curt Nicolin, director of the Swedish firm ASEA, which played a key role in the construction of the first 12 Swedish nuclear reactors.

I also want to emphasize that the rhythm and/or content of this chapter is designed to imitate that of the superb online journal *Energy Politics*, as well as the forum *EnergyPulse*, and the important site *321 Energy*. Articles from these sites provide a valuable and uncomplicated way to expand and perfect one's knowledge of energy economics. (*Energy Politics* was started by Dr. Jenny Considine and Thom Dawson of Calgary (Alberta), Canada for the purpose of promoting the analysis of all aspects of energy during the 21st century. It features the work of leading energy economists.)

I also feel it necessary to confess that the logic and mechanics of global economic progress is a much more complicated topic than presented in this contribution, where the emphasis is on energy. In fact, it is more complicated than anything I studied and was occupied with in engineering, or taught in economics. I discovered that while writing my book, *Scarcity, Energy, and Economic Progress* (1977). I also see no reason to conceal that the icing on the energy and economic-progress cake has yet to appear in book form, nor is it likely to appear until the teaching and discussion of energy economics achieves the same status and scope as those abstract topics in academic economics that look impressive, but in terms of usefulness do not have much to offer!

In an informative article, Michael Parfit (2005) makes the following statement: "The trouble with energy 'freedom' is that it's addictive; when you get a little you want a lot." I failed to discover exactly what Parfit meant by "energy freedom", but when his article was published, he estimated global energy use to be 320 billion kilowatt-hours per day ($= 320$ Gkwh/d $= 320$ Bkwh/d), where G is 'giga', which is another way of writing billion. That was in 2005, but Professor Martin Hoffert and others contend that by the end of this century — if present trends hold — energy use could be three times as much as at the beginning of the century.

I think I am justified in stating that at the present time, *nobody* in the scientific (or any other) community has the slightest idea of how that much energy can be produced up to or at that point in time. And regardless of what they claim, no reputable scientist or economist can provide an estimate of the economic, social, political and environmental costs that would be involved! But even so, some provocative statistics have started to circulate.

According to some notes in a recent issue of *New Scientist* (November 17, 2012), between 1990 and 2010, global electricity production increased by about 450 terawatt-hours per year, which is the equivalent of adding the present electricity output of Brazil to the global supply every year. Furthermore, for what it is worth, the International Energy Agency (IEA) expects that this trend will continue until 2035. I suppose that many persons will be grateful for a forecast of global energy requirements by that organization, but on the basis of their prediction of future oil supply and demand that I unfortunately encountered when teaching at the Asian Institute of Technology (Bangkok), an estimate from that source covering about 25 years is something that I am reluctant to accept. It does seem reasonable, though, that at least one Brazil (of electricity production/demand) might be added to the global electricity scene each year for at least most of the present decade.

I am also unable to welcome another prophecy that is sometimes associated with the IEA, which is that Germany is prepared to abandon nuclear energy or coal in favor of solar power, regardless of the sincerity in which this absurd intention by Chancellor Merkel and her foot soldiers is framed. In fact, next year, as much as 4,000 megawatts of new coal-based generating capacity may come on stream in that country.

Here, I would like to recommend once more the book by David L. Goodstein, professor of thermodynamics at the California Institute of Technology, entitled *Out of Gas: The End of the Age of Oil*. It contains no mathematics, although it deals with most energy resources. *I regard that book as the energy equivalent of the celebrated Economics 101!* And please allow me to insist that you can gain a very useful insight into energy economics without delving too deeply into mathematics. In fact, learning energy economics without being tortured by an overload of irrelevant mathematics or econometrics is one of the aims of this book, and especially this long chapter, and I can hardly think of a single person — regardless of professional or intellectual preferences — who will not profit by absorbing as much of it as possible.

It is impossible to deny that renewables and alternatives will eventually have to bear a large part of the growing energy burden. Parfit (2005) expresses it correctly when he says that leading the way will be "a congress and not a king", which means that there will be room in this crusade for every energy mode that makes economic sense, as opposed to what Robert Bradley calls "the master resource".

Systematic and extensive reading by the rank-and-file as well as by officers and their advisors is the beginning of the solution to the energy quandary, and not the frivolous CNN 'mantra' about many (CNN-sponsored) solutions for a single problem. There is also a growing concern about the 'rebound effect', which I first heard about from the physicist Michael Dittmar, who is active at the CERN research establishment (in Geneva, Switzerland). If global energy use is greatly increased in order to support continued macroeconomic growth, what is this going to mean where the demand for clean fuels is concerned, or, for that matter, unclean fuels? And not just energy resources, but also non-fuel minerals, clean water and air, agricultural requirements, and all the other items that could be depleted as the global population increases?

And the population will continue to increase! When I published my first book, United Nations predictions focused on a sustainable global population of six or seven billion. The latest UN prediction has added a few billion to that figure, which leads some of us to wonder exactly what the situation will be at mid-century. I seem to remember informing my students that, according to the late Professor Joseph J. Spengler, we should not be surprised if the global population reaches 15 billion or more. What I forgot to add was that once we start talking about a potential global population of that magnitude, or somewhat less, it does not make any difference what the actual population turns out to be, because it will very likely be harmful for a great many residents in *all* the countries whose governments fail to formulate workable strategies for dealing with this issue.

Please excuse me for insisting that for most readers of this book, the present chapter — perfectly understood — is just as important as *any* article on energy economics that appears in *any* 'learned journal' ever published, even if that article is spiced with highly abstract mathematics and econometrics. For instance, you should be aware at all times that in 2013, OPEC produced about 40% of the global oil output, which means 40% of 88 million barrels a day (88 mb/d), or 35.2 mb/d. At the same time OPEC has approximately 80% of the global reserves of conventional oil (which is approximately 1,477 billion barrels = 1,477 Gb), with the largest share in the Middle East. Venezuela may have the largest OPEC *reserves* (or *confirmed oil resources*), but there are claims that shale oil will change the reserve picture, even if that is not certain, nor are there credible estimates of the size of this change. The above numbers are necessarily approximations, but as the great mathematician Bertrand Russell once said, "Although this may seem a paradox, all science is dominated by the idea of approximation."

My understanding of oil matters reached a peak in 2009, when the oil price fell from a high of $147 per barrel (= $147/b) to $32/b because of the global macroeconomic and financial market 'meltdown' that might have been primarily caused by the oil price escalation, which has often been suggested in my work, and more thoroughly in a brilliant paper by Professor James Hamilton (2009). It took OPEC only a meeting or two to fashion a strategy that caused the oil price to begin rising again, despite a continued weakness of the global economy that led to a fall in the demand for oil.

The core of that strategy involved no more than limiting the supply of OPEC oil.

The economics of nuclear energy can also be easily clarified without calling on higher mathematics or upper-echelon physics, and clarification is essential because an enormous amount of confusion exists about that energy source, and the same is true about electric deregulation. In point of truth, nuclear will be an invaluable ingredient of *any* energy future, while electric deregulation has turned out to be an unmitigated curse in many regions. At the same time I want to say that I have absolutely no problem with persons who prefer renewables to nuclear, although I dislike the belief that — at the present time — renewables can replace nuclear where cost and/or reliability are concerned, except under very special circumstances. These circumstances will probably never apply to *base load* power, which is the load that requires steady generation throughout a 24-hour period, and is discussed in detail earlier in this book.

There is also a great deal of illogical talk about natural gas replacing nuclear energy in this century. Natural gas is often described as a "game changer" because of the sudden appearance of large quantities of exploitable shale gas on the energy scene. In one publication, the contention was that for nuclear to be profitable, the price of electricity would have to be 12 (U.S.) cents per kilowatt-hour, while for natural gas a price of 6 to 9 cents would suffice. It was also claimed that constructing a large nuclear plant requires "the better part of 10 years", while the construction time for a gas-based facility of the same capacity was 2 years. I encountered this outlandish belief about nuclear at the Singapore Energy Week several years ago.

A large nuclear facility can reputedly be constructed and attached to a grid in China in less than five years (instead of the 10 years claimed by the Singapore person), and as a result I feel that I have the right to maintain that — on the basis of the kind of economics that I study and teach — the above cost figure for electricity from a nuclear reactor constructed in the not too distant future in every modern economy will very likely be no more than 6 or 7 cents per kilowatt-hour, assuming that the export of electricity is limited. Regardless of what it turns out to be, however, governments should sponsor the construction of a state-of-the-art reactor that produces power at the lowest cost in the world, or lower, and refuse licences to reactor manufacturers that cannot match that cost. The multi-billionaire Bill Gates is apparently financing such a reactor.

As we are constantly informed, the main cost factor of nuclear-based power has to do with the construction of reactors, which would not be changed by access to commercial breeder reactors that greatly reduce the cost of reactor fuel. Accordingly, the best way to handle high investment costs for reactors might be to build (in factories) the largest possible *transportable* components of reactors, which are then moved to the site of the nuclear facility by trains and/or boats, where the components are assembled into complete reactors. Something like this must be taking place in China and Russia today, just as some aspects of this procedure probably took place in Sweden many years ago, when 12 reactors were constructed in just under 14 years. Those reactors (together with hydro) gave Sweden some of the least expensive electricity in the world.

Some of the persons in forums to which I have contributed are not pleased by my tendency to constantly refer to the brilliant economic progress being made in China; however, I mean no harm. The same is true of the language found in some of my publications, where I am mainly interested in ridiculing the attempts of self-proclaimed energy experts to convince their audiences that nuclear reactors cost too much to be constructed in, e.g., the United States. As far as I am concerned, it might be a good idea if those excellent audiences study in detail the planning and construction of the American navy during the Second World War, which was one of the supreme engineering (and educational) feats of modern times.

It can also be mentioned that since the launching of the first nuclear-powered submarine — the USS Nautilus in January 1954 — there have been approximately 500 reactor-years accounted for on U.S. Navy ships without serious incident. I have a conclusion to offer here that will not be welcomed by some readers, and it goes like this: given the importance of energy, and the lack of importance of much of the sailing back and forth on the Seven Seas by the navies of various countries, it might be a good idea to devote some of the money set aside in the event of a Third World War to increasing the number and quality of reactors in countries like the U.S. and Sweden. Apparently the Russian and American governments have come to a conclusion of this sort, because they have agreed to cooperate on civilian nuclear matters. And why shouldn't they cooperate on nuclear (and other) matters? After all, they (and the UK) carried the majority of the load in the Second World War.

Note the title of this chapter. "The Final Countdown" is the title of a 'pop' tune sung by the Swedish rock group, Europe. The band also produced an album with that name. This business of a countdown — usually starting with ten and moving to zero — is something that seems to have been introduced by the military, perhaps during the First World War, although it was popularized by the launching of space shuttles. I start below with '12', finish the main business of the chapter at '1', and in the final section of this chapter (i.e., "Finishing Touches"), I provide some useful advice for all readers, especially my students.

12. In the Head of U.S. Energy Secretary Chu

As many readers of this book probably know, Dr. Steven Chu is the former Energy Secretary of the United States, a physicist, and a Nobel Laureate. *Discovery Magazine*, in a recent issue (2011), selected what it called the "100 top stories of 2010", one of which was authored by an editor of *Discovery*, and whose main purpose was to verify Dr. Chu's green credentials. As good luck would have it, that discussion was only two pages long, and did not contain any elementary physics or mathematics — two subjects that I failed twice in my first year at engineering school, and as a result of which I was duly expelled. On the other hand, some important observations of the Economics 101 variety were missing from Dr. Chu's answers to editor Corey Powell's questions, which unfortunately prevents me from recommending his 'piece' to all of my friends and neighbors.

Mr. Powell began this Q&A with a reference to the (U.S.) Gulf Coast oil spill, asking how an accident of this magnitude could happen. I will not bother to discuss Dr. Chu's answer, because both question and answer are irrelevant. Statistically, accidents of that type are unavoidable, and have always taken place. If we go back to the Second World War, we can look at, e.g., the unnecessary and appalling attack on Manila, the failure to clear the approaches to the port of Antwerp as soon as possible, and — perhaps the worst blunder of all — adopting the Sherman as the main American battle tank. Compared to those 'accidents', the Gulf Coast tragedy was small beer.

For long-term energy investments, Dr. Chu pictures the U.S. moving toward the electrification of personal vehicles. I can accept this proposition, only I do not have a clue as to the details, such as how rapidly a large-scale electrification could be completed if deemed necessary, nor the maximum number of vehicles that could be accommodated in the long run, nor what is regarded as an acceptable cost of this electrification. I therefore wonder if the Secretary and his successors, and their foot soldiers, could provide us with the kind of information that we can use in our teaching and publications, and to do so as soon as possible — assuming that they, unlike my good self, have examined this issue in great enough detail to tell us something not encumbered with public relations hype.

Dr. Chu's thoughts on nuclear energy bother me somewhat, because he states that large reactors will cost 7 to 8 billion (U.S.) dollars. I regard that estimate as completely and totally wrong, and suggest that he should have a talk with the former director of the French firm Areva, Anne Lauvergeon, and in particular ask about the information she has concerning the new Chinese reactors. In terms of cost, those Chinese reactors are probably capable of producing the most inexpensive electricity in the world. The problem here is the widespread failure to teach the kind of economics that would enable the advisors of politicians and bureaucrats to provide accurate calculations for their employers, and to identify and reject the vast amount of inaccurate calculations that are always in circulation.

If his French is not up to scratch, Dr. Chu can obtain and read some of the materials in this textbook, because evidence from the nuclear past and present leads me to insist that "large" nuclear facilities, whose construction is organized by competent managers, will soon cost a maximum of 5 billion dollars. In other words, the 'buyer' of a nuclear plant might borrow 5 billion dollars (perhaps in installments), give it to a firm like Areva or the South Korean firm that is constructing reactors in the United Arab Emirates, and about five years later they can begin selling electricity. This can be rephrased: the time span from 'ground break' to grid power will probably be less than five years when the nuclear renaissance moves into full swing, and perhaps much less.

"Future-gen" (in the form of *zero-emission* coal power plants) evidently plays a prominent role in Dr. Chu's vision of an optimal energy structure. It plays none whatsoever in mine, however, and I never use the expression "future-gen" nor listen to anyone discussing it. I also think that it is necessary to be realistic about the use of coal in the U.S. (and also other countries, particularly China). In the U.S., coal-fired power plants account for at least 45% of power production, which is slightly more than that of natural gas and nuclear combined. Moreover, estimates of U.S. coal resources (or *hypothetical* reserves) — as compared to *verified* reserves — are huge.

Carbon Capture and Sequestration (CCS) is found close to the top of Dr. Chu's wish list, and where that topic is concerned, the directors of the Swedish firm Vattenfall once made certain optimistic promises to the German government and newspaper readers in several countries concerning the wonderful intentions of their firm. Jeffrey Michel, an MIT graduate and important energy consultant living in Germany, calls CCS a "thermodynamic travesty", and recalling my own long and delightful study of thermodynamics and engineering economics causes me to say that Michel's judgment is much too mild.

As abstract as it sounds, the notion of carbon-free regions seems to be making the rounds. At a recent large energy meeting in Berlin, the emphasis was on the role that solar and wind will play in Germany's energy future. Where wind and solar are concerned, readers should turn to John Droz for some useful information. With all due respect, the announced German wind and solar aims are strictly off-the-wall, and place an unreasonable economic burden on German industry, while Japan's recently adopted solar ambitions give the impression of being beyond the scope of economic logic. My prediction is that both Germany and Japan will bet strongly on nuclear energy, and in the long run, will reveal a strong preference for breeder reactors. I can only hope that the security problems associated with these reactors are approached systematically and logically, because if the present panic continues to bloom, somebody could be in for a world of hurt.

Finally, Dr. Chu mentioned that "there is no law of physics which states that the whole society can't benefit", and unlike the contention of, e.g., Gordon Gekko (in the film *Wall Street*), he says that "there is no zero-sum game here". It was really very decent of the Secretary to inform us of his interest in the subject of game theory, because in a world of 9.5 billion souls — which is Dr. Chu's prophecy for 2050 — a complicated version (or extension) of the zero-sum paradigm is going to be the order of the day, and there is very little that he or all the Nobel Prize winners since Adam and Eve can do about that.

For his information, as well as yours and mine, for the first time in human history, people aged 65 and over will soon outnumber children under the age of five. This is sometimes pictured as posing enormous problems for the health and general welfare of those 'pensioners'. Unfortunately, even if the supply of youngsters were to greatly increase, it might not improve the situation, which is something for persons who are reading this book to ponder and discuss. Efficient labor markets, education, and governments that know how to distinguish fact from fantasy are the keys to obtaining the correct answers on how to deal with population and other dilemmas, just as pensioners who want more consideration should vote for politicians who are sincerely interested in providing this service.

11. On Libya and Oil

"Where there's war there's hope."

— *General George Patton*

Many years ago in Chicago, a man that I knew abandoned his bed in the middle of the night, rubbed the sleep from his eyes, put on his clothes, and left his apartment for the purpose of making a "citizen's arrest". Exactly why he regarded that kind of behavior as meaningful is something that I never found out, nor was I interested in finding out, nor was I interested in where he went to carry out this sacred mission, although I know where he did *not* go. He did not go to the neighborhood where the first sergeant of G Company in my infantry regiment was raised, and where Sergeant P. had learned to treat men like animals. That neighborhood was often called Dodge City. Nor did he accost the young lads in front of the Four-Thirty Club, a few blocks from where I lived, nor did he make his way to Drexel Square, because if he had and mentioned arresting somebody, that might have been the last that anyone saw of him.

Decades later a French president gets out of bed, rubs the sleep from his eyes, and announces to the world at large that the time has come to protect some civilians. He was not talking about the Ivory Coast, which once was a quasi-protectorate of France, and where the defeated Ivorian president had declared that he was not ready to leave office, and anyone in his country who thought otherwise was asking for trouble. Nor did the French president mean one of the 'suburbs' falling under his jurisdiction, e.g., Grenoble, and particularly La cité Mistral (commonly known as *"le dépotoir de Grenoble"*, or "the dump of Grenoble").

Something I can almost guarantee is that the foolish interference into the internal affairs of Libya took place because of the oil in that country. There might be some oil off the coast of the Côte d'Ivoire, but there is none in the Mistral, which is why the prime minister of the UK endorsed the goofy position of Mr. Sarkozy, and told his colleagues in the British government that when the people of a country want to get rid of a dictator, good men and women will come to their aid. What about getting rid of liars, fools and hypocrites who started or approved of another war for oil? I mean of course Iraq, where the pretence for that destructive exercise was not protecting civilians, but a lie about weapons of mass destruction.

The common denominator where oil is concerned is that governments who correctly judge the importance of that commodity are prepared to do anything to obtain it. As Len Gould said in a comment in the forum *EnergyPulse*, at some point in the future, voters might be fully prepared to go to war to obtain oil; however, their political masters have been prepared for this eventuality for a long time. One of the big mistakes in my life was to lose track of a map that came into my possession, which showed landing zones in the Gulf for marines and paratroopers in the event that the oil price went into orbit. According to Professor Douglas Reynolds of the University of Alaska, this option was discussed openly by Henry Kissinger, and mentioned in one of the weekly news magazines in the United States. At the same time I should perhaps mention that I eventually became convinced that the map in question was not 'the real deal', but was intended as a warning.

Someone who has a problem with my reasoning is Karel Beckman, formerly editor of the *European Energy Review*. He thought that Colonel Gaddafi had been tolerated long enough. From a scientific point of view, the *European Energy Review* is a half-baked product of bloggomania, which is splendid with me, because my books and articles have always been able to draw on many excellent articles and comments in several Internet forums. (Here I am primarily thinking of *EnergyPulse*, *321 Energy*, and *The Energy*

Tribune.) My present grievance with the good editor Beckman is not about being denied exposure in his mail-out. Instead, it pains me to know that the ignorance about energy matters that I have tried to help cure is as pervasive as ever.

Mr. Beckman was evidently concerned with whether the Swedes were prepared to join in this Libyan thing. Well Karel, I hope you remember that I told you they were. The gorgeous disco and bar life in the Mediterranean, the sun, sea, beaches, lovely human beings and wine were soon at the disposal of the gentlemen dispatched to that wonderful part of the world. Americans were also in action, which is to be expected.

In an article published in the *New York Times*, and reproduced in *The Observer* (2011), Erhard Stackl describes Colonel Gaddafi shaking hands and hugging such high and mighty paragons of the European political scene as Nicolas Sarkozy, Tony Blair and — with "special affection" — Prime Minister Silvio Berlusconi of Italy. This probably happened, although a video that was referred to was likely a silly pastiche of some sort, because in the background was the tune "Save Your Kisses for Me", performed by a former British winner of the Eurovision Song Contest. Quite naturally though, Herr Stackl lacked the sophistication to comprehend the quintessential logic of the Libyan 'fling', because since 1973, where oil is concerned, military responses *or* thoughts of military responses have virtually been a reflex action, and so regardless of how hot and bothered the above three gentlemen were when they were clinching with the colonel, he had to be spurned.

I have received many comments on my work dealing with Libya and oil, and I have come to believe that this issue can be greatly simplified. *Can anyone really believe that it would have been possible to form a coalition of the gullible to go to Libya had that country not possessed the largest reserves of crude oil in Africa (and also considerable natural gas)?* At the same time it needs to be recognized that when a government possesses an elite military asset called 'The Foreign Legion', taking part in small wars or police actions is something that is described as "easy-peasy" in Australia.

And by the way, before I forget, according to recent reports, political and economic chaos is expanding in Libya. Libya's el-Sharara oil field that once provided 350,000 b/d of high-quality sweet light crude now produces nothing, and estimates are that the daily loss of revenue is close to

150 million dollars a day for Libya's oil sector. Please excuse my pointing this out, because as far as I am concerned, Libya possesses everything required to become the richest country in Africa. Oil, gas, a moderate population and a long marvellous coastline not far from Europe spell prosperity in any language.

Perhaps before changing the subject I should mention that the latest announcement that I have seen by OPEC claims that oil market *fundamentals* (i.e., supply and demand) are "good", and apparently the chief oil executive in one of the key OPEC members has said that any oil supply deficiencies because of troubles in the Middle East will result in his country raising its supply. Hmm. The average oil price — WTI and Brent — is about $103, which suits OPEC fine, and if it was less than fine and the oil price showed signs of collapsing, the decisions taken in a short meeting at OPEC's headquarters in Vienna should soon make things 'as good as they can get' once more!

10. Emissions Trading and Climate Change: A Brief Negative Résumé

"The population of the planet has quadrupled over the last century, and the sheer mass of humanity is generating more CO_2 than the planet can process."

— John Petersen

Perhaps the statement above by Mr. Petersen can be modified to: *If there is bad climate news ahead, which may be true, then we can blame it on excessive population and not 'Big Oil' or The National Association of Manufacturers!* This does not mean, however, that I am ignoring a statement by Robert Frank (2007) in his important textbook, which is that "if a single agency had the power to enact globally binding environmental legislation, it would be a straightforward, albeit costly matter to reduce the build-up of greenhouse gases. But in our world of sovereign nations, this power does not exist."

I can add to that my belief that if a miracle had taken place, and the delegates at the first big climate conference in Kyoto had specified that climate issues should be exclusively dealt with by heads of governments and senior civil servants from the major greenhouse gas emitting countries, meeting in appropriate venues several times a year, we might already be in possession of the *optimal* environmental legislation, instead of the half-baked bunkum focusing on items like emissions trading that were eventually put into circulation. Moreover, the cost mentioned by Professor Frank might have been quite tolerable. (And please note, "optimal" means the best that we can do, given the constraints placed on us by things like the shortage of information and/or the political background in which this task is undertaken.)

It is quite likely that wind and solar make more sense than fossil fuels when confronted with a real or imagined deleterious climate situation, but what we do not know is whether wind and solar technologies can eventually be exploited to yield more than modest amounts of electricity, despite countless claims that they already make engineering, economic and political sense. According to the Canadian engineer Len Gould, solar thermal power already makes sense, but whether it makes sense for places like Scandinavia, Russia and Northeast America is quite another matter, although it might be able to supply a large amount of electricity to Spanish cities and perhaps even Las Vegas before this century is over. One thing though is certain: I wish that I had never heard the expression "emissions trading", because of the lies and misunderstandings spread by its propagandists.

My favorite approach to the interior mechanics of emissions trading almost always begins with a perusal of the exchanges for electricity pricing, which provide market-based schemes that attach a cost to the excessive generation of, e.g., greenhouse gas emissions. One of the institutions mentioned as being willing and able to provide the necessary trading facilities is the Nordic Electricity Exchange (Nordpool), whose endeavors I usually describe to my students as a 'scam' and deserving of the attention of serious fraud researchers. Let us consider a simple example.

Over a previous year when I was paying close attention to Nordpool, electricity prices in Sweden at least doubled. One of the explanations provided by that organization and its propagandists turns on the increased cost for emissions permits by electricity generators in the north of Europe, as well as the increased price of energy inputs for power stations. But since all except about 12 percent of Swedish electricity is generated using nuclear and hydro, emissions permits and higher oil and coal prices are, *ceteris paribus*, largely irrelevant for a majority of Swedish electricity suppliers.

Even so, Swedish taxpayers are faced with higher prices due to the presence of (generating) equipment in other countries associated with Nordpool, and because of the occasional appearance of a very high demand for electricity in those countries.

A spike in the demand for electricity outside of Sweden raises the price of all electricity due to the marginal cost pricing foolishly accepted by Swedish politicians and bureaucrats, most of whom receive (or received) advice from ladies and gentlemen in academia with about as much knowledge of this branch of economics as I have of brain surgery. Yes, I too praised marginal cost electricity pricing when I began teaching at the University of Stockholm, but I soon smartened up when I realized that this was just another scheme to make the wealthy wealthier, and to hold Swedes accountable for the shortcomings of persons and firms hundreds or thousands of miles away.

Now consider the situation in the future where emissions permits are supposed to play a major role in the fight against environmental deterioration. Regardless of good intentions by fossil fuel users and politicians, fossil fuel consumption will (*ceteris paribus*) still increase by a large amount because estimates are that in the next 25 years, the global output of electricity might increase by 50 percent. Even if Sweden were to reassemble the two nuclear reactors that were shut down and scrapped, increase the number of windmills in the country by the absurd amount desired by various know-nothing politicians and commentators, begin a Manhattan Project-type approach to developing and introducing new and more efficient technology, and if — while these measures are taking place — a big slice of the electricity-intensive Swedish industry moves to another country, an increase in the demand for fossil fuel-based electricity exterior to this country, and a possible rise in the price of emissions permits to various enterprises outside Sweden, would almost certainly boost electricity prices for Swedes as a result of having their electricity priced in an exchange of the Nordpool variety.

Given these circumstances, it should be made clear to all interested persons that, at best, emissions trading reduces to a highly efficient way to get rid of excessive carbon dioxide (CO_2) emissions in what the game theorist Ken Binmore calls a "toy market", by which he means a textbook market that is devoid of annoyances like risk (or uncertainty), monopoly, irrationality, spillovers (i.e., externalities), dishonesty and anything else that prevents a few simple equations from being put on a blackboard for the delight of drowsy teachers and students. Despite a statement in *Newsweek* that in order to suppress excessive CO_2, the United States needs a cap-and-trade system of the kind providing "magnificent results" in Europe, the sad truth is that the European arrangement is a cynical deception that mainly benefits the brokers and 'intermediaries' who expect to get rich by playing games with 'emissions credits' or 'carbon trading' or whatever bizarre and/or misleading scheme that has been pieced together in order to make this counterproductive activity appear socially beneficial.

Given my 'attitude' on this issue, it might be asked what kind of system do I feel is suitable to eliminate all or a part of the dangers to the environment? I am tempted to conduct enough research to give a satisfactory answer to that question, but I doubt whether I am more competent where this issue is concerned than President Obama's energy team, even though I persist in calling them an 'environmental team', since energy appears to be a subject beyond their comprehension. Moreover, if they should undertake to provide us with a 'world-class' solution, they can also produce a calculation which shows that when all relevant factors are taken into consideration, including macroeconomic factors, every country should be in possession of an easily understandable (though perhaps not comprehensive) energy plan that everyone in a responsible position that has anything to do with energy anywhere in the world can understand perfectly.

9. OPEC's Strategy: An Update

An article that was published in the *Infantry Journal* (U.S.) many years ago contained the following exotic question: "If you only had an hour in which to turn civilians into soldiers, of what would your instruction consist?"

Despite being congratulated for supplying what at that time was called the "perfect school solution" — whether it was correct or not — I was still expelled from the Infantry Leadership School at Fort Ord (California), though for reasons that I was not informed of then or later. However, these days I am interested in an equally important query, which has to do with OPEC's intentions where the production and export of oil are concerned. This time I am certain that I not only have the correct answer, but it will not take an hour for it to be presented and elaborated on to my students and colleagues, nor will I be rewarded with expulsion from 'the academy'. It begins with this: OPEC intends to export (and perhaps produce) as little conventional oil as possible. It ends with this: OPEC intends to export as little conventional oil and natural gas liquids as possible, *regardless of what they say or do!* It might also be wise to repeat this mantra a number of times in the event that your friends and neighbors want to discuss this topic for an hour or so.

The logic in play here is an extension of the work of three brilliant economists: Professor Gunnar Myrdal, who was one of my teachers at the University of Stockholm; Professor Hollis Chenery, who organized a small conference to which I contributed in Paris in the 1980s, and whose book (together with Paul Clark) I used when I taught at a UN Institute in Dakar (Senegal); and, of course, the superb article by Professor A.A. Kubursi (1984) entitled "Industrialisation in the Arab states of the Gulf: A Ruhr without water".

A short comment about these gentlemen is probably relevant. Gunnar Myrdal, Nobel laureate in Economics, and known and famous throughout the world, was unbeatable in any seminar room or conference, or formal or informal debate (as he demonstrated at the Nobel Banquet in the Stockholm City Hall in 1974). His belief was that the study of development economics should be based on a study of the economics and sociology of successful economies. The two countries he recommended for study were the United States and Sweden.

Hollis Chenery, a professor at Harvard, went almost unnoticed by the Nobel Academy, which was another example of its characteristic misunderstandings, although he was a leading mathematical economist in the study of economic development (and on the 'applied' level, was equal to the first winners of the Nobel Prize in Economics, Ragnar Frisch and Jan Tinbergen). More remarkable, however, is the neglect by development economists of Professor Kubursi (of McMaster University in Canada), although the big mistake here might be the failure of Professor Kubursi to adequately market his theories of economic development to the 'academy'. The only place that I have seen or heard his name called is in my books, articles and lectures, although it is not impossible that he discussed his work in OPEC councils, because somebody (or persons) in those councils listened to or read what he had to say, and we see the results today. By that I mean high oil prices, and an expansion or proposed expansion of refineries and petrochemicals in some OPEC countries.

Where the latter is concerned, it is liable to become one of the most important activities in Abu Dhabi, which is the richest of the United Arab Emirates. That state is also interested in a series of energy-intensive projects that include the world's largest single-site aluminium smelting plant, and given the electricity requirements for a facility of this nature, it is easy to understand the decision to utilize nuclear energy. As for paying for nuclear equipment, this hardly brings a frown to the faces of Abu Dhabi's decision makers, since the last time I inquired, I was told that perhaps 500 billion dollars (or more) belonging to Abu Dhabi were parked in various financial institutions around the world. The four reactors that are scheduled to be constructed and attached to the grids in the UAE are said to cost a total of 20 billion dollars, although they will probably cost more.

Of course, if for some reason the UAE are short of ready cash, they could always turn to their customers in the oil-importing countries for aid, by which I am not thinking of turning to them for charity, but influencing their OPEC associates to boost the price of oil being exported by a small amount. According to Thomas J. Barrack, Jr., a financial mover-and-shaker, for every $1 increase in the price of oil, there is a $9 billion per year windfall profit to oil producers in the Middle East. From the beginning until shortly after the shooting stopped during the NATO military venture in Libya, the aggregate price of oil increased by enough to pay for those four reactors in the UAE, and a great deal more. Viewed from another angle, that senseless military venture cost the U.S. many billions of dollars, in that a rise in the price of oil almost immediately resulted in a rise in the price of items like motor fuel. Remember that the U.S. consumes 20% of the world oil supply, or 18.5 mb/d.

The explicit observation by Professor Kubursi — and implicit in the work of Gunnar Myrdal and Hollis Chenery — was that instead of exporting oil in its crude form, if the development process is taken to an optimal conclusion, that oil should be used in OPEC-owned refineries, and a large amount of the refinery output should be used in the production of petrochemicals.

More importantly, simple mathematics leads us to conclude that investing in facilities to produce refinery products and petrochemicals in, e.g., the Middle East without having enough crude to utilize these facilities for very many years is a serious mistake! (The book edited by Professor Jean-Marie Chevalier (2009), and used in his institute at the University of Paris (Dauphine), deserves mention here.) Thus, even though the U.S. can continue to raise its output of shale oil, they will not be able to punish the OPEC countries in the Middle East, because those countries are going to take steps to add value to their oil by exporting more of it in a more profitable form (e.g., petrochemicals). By "profitable", I feel that *social profitability* should be mentioned and emphasized as much as private profitability.

Rather than confess that I could be completely wrong above, because it may be true that the OPEC leadership does not operate on the same logical and/or psychological 'plane' as myself, I would like to present the opinion of a distinguished oil economist who is definitely *not* a friend of OPEC, Professor Morris Adelman of MIT, and his colleague Martin Zimmerman (Adelman and Zimmerman, 1974): "In the production of petrochemicals, most LDCs (*less developed countries*) are at a severe and permanent disadvantage for lack of know-how, and the high opportunity cost of capital and feedstocks. Other countries, particularly OPEC members, who do not face these obstacles are expanding their petrochemical capacities. This too will drive prices down, lower the profitability of all plants built today, and force losses on many investors. Few can compete with those that get their feedstocks at a fraction of world prices, and are willing to earn low or negative rates of return."

Earning "low or negative rates of return" is not (and probably never was) the goal of new or potential OPEC refiners and petrochemical producers, since with the huge amounts of feedstocks at the disposal of these firms, acceptable margins should be available for a very long time for these firms, as well as impressive national incomes for the countries in which these firms are located. Moreover, even if desired outcomes are not immediately achieved, major oil-producing countries can look forward to returns that result from transforming inexpensive refinery products into high-priced petrochemicals. This was pointed out by the Nobel Prize (in chemistry) winner Sir Harry Kroto many years ago.

It cannot be overemphasized that since energy costs are a severe financial burden for chemical industries, the combination of inexpensive energy and the ability to choose the optimal technology that results from large and perhaps increasing investments in education will theoretically ensure that the center of gravity of the global petrochemical industry will move closer to the 'least-cost' Middle East. According to the time-honored theory of *comparative advantage*, that is where it belongs.

Regardless of the countries that may be involved, "center of gravity" does not mean complete domination. At any time the global petrochemical industry is a mixture of small and large, low and not-so-low cost, new and old, etc., and so the OPEC commitment might be principally concerned with helping to keep the price high enough so that some of the less favorably endowed plants can remain in operation in order to supply a certain portion of total demand. Even so, many firms in this line of work have become unpleasantly aware of the realities brought about by the cheaper methods of production at the disposal of countries that no longer want to be a hostage to unfavorable oil or gas prices, and who moreover have access to state-of-the-art technology. Exactly how traditional firms will react to this challenge is uncertain, particularly in the short run, although in the long run many of them may be left with no choice but to cut-and-run, to use one of President George W. Bush's favorite expressions.

After reviewing my thoughts on this subject, and once again reading the paper by Professor Kubursi, I would like to say that while you may not find the observations directly above relevant to your interests, or attractive, it is impossible to deny the bottom line: OPEC strategy is going to turn on producing and exporting as little crude oil as possible, and their business and political acquaintances abroad will simply have to get used to and adjust to this arrangement.

Here I can add something that John Maddox, a theoretical physicist who was once an editor of *Nature* (which, along with *Science*, was the foremost scientific journal in the world), pointed out at a lecture in London around 1975. The first oil price shock was a part of the "general realignment" of the relationship between the industrialized countries and OPEC. That sounds right to me, and I think it fair to add that OPEC is in an even better position today, because unlike their customers, they are doing everything possible to master the details of the economic progress game.

This textbook contains a long chapter on oil, but I have not bothered to take up the correct strategy for OPEC customers. As I indicated at the beginning of this chapter, we do not require game theory and this matter of zero-sum games if we are tempted to ask political masters why they could not protect us from escalating motor fuel and electricity prices. Professor Richard Bellman (1957) had the answer: "An optimal policy has the property that, whatever the initial state and initial decision are, the remaining decisions must constitute an optimal policy with regard to the state resulting from the first decision." In other words, forget about the ignorance and irrational behavior that got us into this deep trouble, and concentrate on finding the optimal future program.

8. Energy, Macroeconomics, and Economic Growth

Many years ago, one of my international finance students at Uppsala University (Sweden) — who is a gifted musician — wrote a song called "Macroeconomics is a Way of Having Fun". It was dedicated to yours truly, and occasionally played on the radio in the Stockholm-Uppsala region of Sweden, as well as at some of the brilliant parties organized by my students. I can begin by saying a few words about what the "fun" in the title of that delightful melody means by comparing macroeconomics with physics, since many physicists believe that economists often suffer from 'physics envy'.

If you read an excellent introductory textbook in macroeconomics, and learn what you read *perfectly*, you can tell academics, business persons, journalists, break dancers and rappers that you are a brilliant and highly educated macroeconomist, and, with a little luck, enjoy the 'fun' in this deception. Try the same routine with an introductory physics book, and you will likely be called a fool (or worse). I have taught macroeconomics on many levels in many countries, and eventually learned everything I wanted or needed to know about that subject. But of late I make a point of avoiding advanced macroeconomic presentations, and especially the elegant models that fill the so-called scholarly journals, because many (if not most) of them are useless. Even more annoying, most of those journals which display an interest in energy might best be described as an insult to students and teachers, though possibly unpremeditated.

To put this situation into relief, I suggest going back almost 20 years and perusing one of the best papers on oil that I had encountered anywhere prior to that date. It was by Richard Teitelbaum (1995) in *Fortune*, and had to do with four multi-millionaires who had decided to ignore the sanctimonious opinions of 'economic science' and go 'long' in (i.e., buy) oil properties. These were Philip Anschutz, Marvin Davis, Carl Icahn and Richard Rainwater. Of this cash-strong foursome, Mr. Anschutz declined to be interviewed; however, somewhat later I was informed that he was the kind of investor who preferred to let his money do his talking, since it was an open secret that he had participated in several big-ticket shopping missions in the oil and gas markets.

The other gentlemen were very much upfront as to why investors should take a bullish attitude toward oil. In fact, the late Mr. Davis supplied one of my favorite quotations on this subject: "You don't have to be a cockeyed genius to see this coming" (and to this I can add that you don't have to be any kind of genius to see what is going to happen on the nuclear energy horizon). The future billionaire Mr. Rainwater was also explicit about his estimate of the oil future: "The increased global demand paints a picture for me that doesn't have any other outcome. The price of oil is going to come up." I was amazed at the difficulty I had in presenting that message in a dozen lectures, many of them at international conferences frequented by high-level academics and experts from the media and oil firms.

The interesting thing about those ladies and gentlemen is that all of them were aware of the demand for oil that would be forthcoming in China and India, and surprisingly long before this matter was considered a worthy topic for mainstream energy economics publications, conferences and those drowsy academic seminars where high-flying amateurs do their song-and-dances while clutching lightweight 'scientific' contributions as if they were the gold to which an English economist once compared them. This might be the place to mention that everyone is now talking about China and its requirements, but few observers realize that India is also going to have a huge appetite for oil. Any decline in the Chinese demand will be more than compensated for by the increase in Indian consumption. Going a step further, the only explanation that I have for the present aggregate price of oil — WTI + Brent — is the increasing demand from Asia (and, of course, the perspicuity of OPEC).

The last time I stood in front of a macroeconomics class was at the University of Technology in Sydney, Australia, and since then I have adopted a very special approach to the topic. For me, macroeconomics involves *everything* that has to do with the aggregate standard of living, and thus includes items like energy, population, and education, although the *emphasis* is still on the kind of elementary conventional macroeconomics presented in the better textbooks. Excluded are lectures of the type presented in Sweden by some of the recent Nobel laureates in economics, and similarly 'junk science' discourses on energy economics that are often delivered to graduate students at various universities, and sometimes to readers of the best known and more renowned business press.

The Oil Price and Macroeconomics

In the silence of my lonely room, and sometimes in crowded seminars, I tell myself that I know a few important things about energy economics, and among other things I feel that this gives me the right to describe Professor James Hamilton as the leading academic oil economist in the United States.

Hamilton has carefully examined the relationship between increases in the oil price and the negative effect they have on the U.S. macroeconomy, beginning at the end of the Second World War (WWII), until the early years of the last decade of the 20th century. His results are similar to those of Professor Andrew Oswald of Warwick University and myself, but much more thorough, and covering a longer period. The main thing that my future energy economics students will kindly be asked to remember is Hamilton's claim that "all but one of the recessions in the United States since WWII were preceded — typically by about 9 months — by a dramatic increase in the price of oil."

From the formation of OPEC in 1961, until the beginning of the 21st century, it was the intention of that organization to manage not only the oil in their countries, but eventually, the global oil price. In order to do this efficiently, complete (or nearly complete) unanimity among the directors of that cartel was required, and as far as I can tell they did not obtain that like-mindedness until the price of oil fell below $10 a barrel (= $10/b), and the amateur energy experts in the oil-importing world began talking foolishness about it reaching $5/b. That was when even the 'independent thinkers' in the OPEC executive suite in Vienna saw the light, and fell into line with OPEC's main men.

Econometrics is a topic that I taught for a few years in Stockholm and Uppsala, but eventually abandoned; however, some simple calculations that

I made in 2004 indicated that the oil price had started to accelerate upwards. A few years later, while I was giving a long talk on oil at the Ecole Normale Superieure (Paris), the oil price was on its way into orbit and eventually it reached $147/b, which provided OPEC countries with the income they had been dreaming of since the formation of that organization. Fortunately, a high degree of intelligence and rationality prevailed in the OPEC executive suite, and so there was no attempt to over-exploit a good thing. Unfortunately however, according to myself and Professor Hamilton, the macroeconomic damage had been done. As much as I hate to say it, the machinations of speculators and the clumsiness of bank directors and politicians had very little to do with the bad economic news that began in 2008, which is often described as the onset of the most serious economic downturn since the Great Depression (which began in 1929).

Future students of mine will have to understand the above perfectly if they prefer a passing to a failing grade. They will also have to understand the power of OPEC. The recession triggered by the oil price escalation cut the ground out from under the global macroeconomy, and as a result, the demand for oil fell in such a way that the oil price bottomed out at about $32/b. OPEC then simply reduced production by a small amount, and the oil price quickly climbed to $72/b. This would have been unthinkable a decade earlier. Shortly after — with the global macroeconomic apparatus still in disarray — the oil price kept moving up, until finally the aggregate oil price exceeded $100/b, although the demand for oil was almost stagnant (or only increasing slowly).

It is also useful to cite what happened when the war in Libya began — a war, incidentally, that was about oil and not about protecting civilians, as the NATO president had claimed. Oil production in Libya almost ceased, which meant that approximately 1.7% of the global oil output disappeared. That loss was enough to cause the oil price to increase by 17%. Even students at the storefront university in Chicago from where I obtained my economics degree should be able to calculate and interpret the short-run elasticity of the oil price from those numbers, and, if they are hooked on nonsense about speculation, also realize that OPEC receives all the help it needs from large oil producers elsewhere who, surprisingly, prefer high to low oil prices, although they do not shout this preference from the highest mountains or skyscrapers in their countries.

They did not have anything to shout about to persons like Messrs. Anschutz, Davis, Icahn and Rainwater, however. These four obtained their 'smarts' from the kind of analyses found in this chapter, where one of my main intentions is to avoid the lies and misunderstandings that characterize a large part of the energy economics literature and that have found their way into the executive suites of many governments. I therefore close with a suggestion for the genuinely curious. *Do not waste your valuable time reviewing the work of amateurs, regardless of the prestige of the universities that pay their salaries.*

Energy and Economic Growth

This is a topic that I did not intend to include in this section or chapter. However, while going through my references, I encountered an article with the title "The Role of Energy in the Industrial Revolution and Modern Economic Growth". I consider this article 'crank', and I suggest that the reason it was published is because the referee or referees were either intoxicated or impressed by the large number of equations it contains, none of which make any sense in regard to what has taken place in Sweden over the past 60 or 70 years, even though Sweden is the country whose economic growth is being examined by the authors of this foolishness.

To begin, readers deserve a simple but correct explanation of what took place in Sweden in its climb from poverty to prosperity. Sweden was a poor country prior to the Second World War, and it was this war that gave Sweden its entry into the elite of industrial nations. This does not mean that Sweden would not have achieved its universally recognized industrial success without gunfire, but hardly as soon, because the war resulted in Social Democracy replacing incompetence as the theoretical basis of the Swedish economy. Social Democracy means many things, most of which I do not know nor need to know, but first and foremost it means education for everybody.

When Josef Goebbels, the German propaganda minister, declared 'Total War' on the Allied powers, the government of Sweden made it known that the country would remain strictly neutral, but implicitly took steps to practice Total Defence. Total Defence meant a partial (but efficient) mobilization of the available technical brainpower, together with — to my way of thinking — the establishment of a superior educational system: an educational system that was in rhythm with Swedish cultural values. Sweden eventually came into possession of industrial enterprises and engineering talent that ranked as shining stars in the international industrial realm, and somewhat later I became familiar with two modern Swedish economic miracles.

In a country of roughly 7.7 million people, approximately one million apartments were constructed in Sweden in 10 years, and later, 12 nuclear reactors were constructed in just under 14 years. Where the first of these two 'miracles' is concerned, one of my American students called the apartments "cubby holes", while the present Swedish Energy Minister believes that the nuclear reactors can easily be replaced by solar and wind-based equipment, and she was able to entertain the comatose delegates at what was labelled an 'energy summit' sponsored by the newspaper *Svenska Dagbladet* with a sub-intellectual résumé of her thinking on the subjects of oil, natural gas, nuclear energy, and coal.

In my early energy economics lectures in Sweden and Australia, I often referred to oil as the lifeblood of an economic system. This was because I was constantly running into that expression in the popular literature. However, the correct designation is not oil but energy. In a load of bunkum foisted on graduate students at Uppsala University, a professor from the University of Stockholm attempted to prove that motoring and groceries are in competition for the limited amount of oil-based motor fuel that will be available later in this century, where the assumption is that, e.g., corn-based ethanol can be processed into a viable and economic substitute for petroleum products.

I prefer not to take part in a discussion of this subject at the present time; however, I would like to stress that the main thing to worry about in the next few decades is not a reduction in economic growth, but unexpected negative spikes in per-capita real incomes that may be caused by unavoidable political events. The mathematics here is very different from that in the article mentioned above. What it mostly involves is adding and subtracting, although I would not be surprised if some secondary-school calculations of the type presented to beginner classes at Boston Public were useful. I also suggest that scholars who choose to deal with this topic should do so when totally awake, and not after being lectured to by badly-taught professors who think that irrelevant mathematics splattered across a white or black board can always replace common sense.

7. On the Sunny Side of the Nuclear Street

"Work on your charisma."
— *Dr. Albert Speer's advice to an American soldier*

This item in my countdown is a shortened version of a brilliant keynote address that I really and truly wanted to present at a major conference that took place in Stockholm at the Stockholm School of Economics. The subject I wanted to deal with was nuclear energy, which is more important than ever because of the tragic events that have taken place at the Fukushima nuclear complex in northern Japan. Of course, eventually it became clear that I was unlikely to be invited to present a keynote address on nuclear energy in Stockholm, nor allowed to give any sort of address on any subject in that marvellous city.

Should exclusion be my fate indefinitely, however, I doubt whether I will complain, because some of the following discussion has already been published in *Concordiam*, which is a journal sponsored by the George C. Marshall European Center of Security Studies, that in some ways might be described as an organization dedicated to keeping the spirit of the Cold War alive. Therefore, if any *pique* emerges because I elect to make my thoughts available to a wider audience, and should I return to the United States in the near future, I would like to express my willingness to give a charismatic lecture to a distinguished seminar at the Guantanamo Detainment Center or a more conventional educational facility. That is why I mentioned Dr. Speer above.

There is also a possibility that even if I had been given permission from the conference bosses to take part in a Stockholm gig, it is not certain that I would have been encouraged to discuss the subject of nuclear energy. This is because I would have insisted that in the medium to long run, nuclear is the optimal source of *base load* electric power, and in the very long run, technology might enable smaller nuclear equipment to handle *peak loads*. Not only is nuclear energy inherently the most reliable, but in the long run it could turn out to be the most flexible because, in the coming decade or two, it seems likely that technological progress can do more to economically upgrade or obtain a greater fuel efficiency from nuclear equipment than is possible for various other sources of electricity.

The basic argument here is simple, and it follows talks and comments that I have presented and heard in many parts of the world, and where I often began my contributions by explaining why a peaking of the world oil (plus oil-like liquids) supply is certain, and perhaps sooner than many experts think. In the course of those expositions I also reminded my audiences that even if a peak does take a long time to arrive, an unambiguous *flattening* of the global 'liquids' output curve will have an unpleasant psychological as well as economic significance, and the resulting macroeconomic stresses in all except a few very lucky countries could mean output and employment dislocations at least as great as those following the oil price upsurge of 2008. (Here readers should once again examine the work of James Hamilton (2009).)

The question now becomes, what does this have to do with nuclear energy, and as I enjoy mentioning in my lectures, the short answer is every-thing. In both a real and an abstract sense, oil provides a benchmark price for the world energy economy, and so in a rational world, immediate and perhaps dramatic steps would be taken to reduce the vulnerability of indus-trial countries to periodic increases in the price of oil. The claim here is that additional nuclear is the most economical source of the *extra* energy needed to supplement the non-nuclear energy diversity on which the future pros-perity of the country in which you are living must be built, even if no nuclear facilities presently exist in that country. For instance, decision makers in a non-nuclear country like Denmark should do everything possible to con-vince the decision makers in surrounding countries to increase their nuclear inventories, even though they themselves provide voters in their own country with unsound fantasies about how wind energy will enrich their future.

Diversity is a controversial concept, because it can mean radically dif-ferent things to different individuals. In a brilliant and easily read article, Richard Rhodes and Denis Beller (2000) say the following: "Because diversity and redundancy are important for safety and security, renewable energy sources ought to retain a place in the energy economy of the century to come." By itself, this statement is enough to warm the heart of every envi-ronmentalist between Stockholm and the Capetown Navy Yard. But Rhodes and Beller continue by insisting that "... nuclear power should be central. Nuclear power is environmentally safe, practical and affordable. It is not the problem — it is one of the solutions."

One of the solutions! I wonder what Mr. Axelsson, the director of the Swedish *Naturvårdverket* (which in the lingo of George Orwell resembles an indoor welfare scheme), would have to say about that assertion if he were to pay a visit to some lecture room in which I am giving one of my harangues on nuclear energy. (Note: *Naturvårdverket* = Environmental Protection Agency.)

Axelsson once published an article in a Stockholm morning paper dealing with replacing nuclear with wind turbines, and which, in an educational institute of high or low status, would have caused him to share the same fate as myself after my first year in engineering school, by which I mean expulsion. The strange thing is that he must know that the *cost* of electricity in Sweden — which is determined by nuclear and hydro — has been among the lowest in the world, while the cost (and price) of power in Denmark — perhaps the promised land of wind energy — is the highest in Europe. What he might or might not know is that while Swedish managers and engineers are sometimes capable of singing the praises of wind and solar in public, in private they have roughly the same opinion of those items as I do: a certain amount of renewables and alternatives might be justified in an optimal energy program for Sweden, but given the lack of a comprehensive energy plan, nobody can say exactly what these should be, nor how much. (And notice please: Swedish electricity prices increased because of the increased export of electricity.)

The attempt to prematurely introduce wind and solar on a large scale in many countries or regions of countries hardly deserves to be called irrational, although admittedly it might be good business as long as the subsidies keep coming. As things are fortunately turning out, the anti-nuclear booster club is unexpectedly losing its authority in a country that was one of its most important strongholds, by which I mean Holland. That one-time sworn enemy of nuclear energy now has a government in which there are persons who openly ridicule the subsidies required to keep windmills turning, and at the same time argue in favor of addressing the nuclear option.

Moreover, since the average capacity factor for wind in, e.g., Sweden is at most 25 percent, a back-up for wind will always have to be available for this and other highly industrialized countries. As far as I am aware, 'back-up' in small or large amounts are seldom or never mentioned in any country where I have worked or visited. As a result, I feel comfortable in stating that any calculation in Sweden showing wind as having any sort of a cost advantage over nuclear will eventually be characterized as preposterous. Note the word "eventually", because at the time this is being written, self-appointed Swedish energy experts refer to the tragedy in the Fukushima nuclear complex day and night, and although there would not have been a disaster without the tsunami, and Sweden has no tsunamis, some of them want the entire Swedish nuclear sector scrapped.

In a long and complicated article that was mailed to me several years ago, five important energy researchers present an argument for nuclear power as a hedge against uncertain fossil fuel and carbon prices. That article also contains some helpful information about the cost of nuclear power during the years 2005–2006, or perhaps slightly before. Some unexpected increases may have taken place since that time, where I am thinking in particular of *ex-ante* and large *ex-post* costs associated with the European Pressurized Reactor (EPR) that may be ready to go on line in Finland, and which in terms of capacity ($= 1,600$ megawatts) is the largest reactor in the world.

The trouble that has been experienced in Finland with this new reactor is really quite simple. It is a 'one-of-a-kind' — or 'custom built' — reactor. In a decade or so, reactors of that size (and larger) will almost certainly be standardized, largely put together (i.e., assembled) in factories, and should cost much less. The final cost of the present Finnish reactor might come to 8 billion dollars (mostly resulting from the construction/installation time being 8 years), and, as I was informed recently, the firm in charge of delivery/installation — Areva, of France — will have to "eat" 3 billion of that amount.

Madame Lauvergeon, the former director of Areva, was clearly not pleased with these developments, but although she has not commented in detail on any misconceptions that may have surfaced during the Finnish venture, she has expressed some apprehension about the reported ability of Chinese firms to put a 1,000-megawatt reactor in operation from 'ground break' to 'grid power' in less than 5 years (and its 'life' would be 60 years, or more in my opinion). According to estimates mentioned earlier in this book, the 'overnight cost' of these reactors could be extremely favorable (and a more detailed look at this issue can be found in Chapter 6 of this book). If true, this means that the 'workshop of the world' has found yet another *niche* which can help the Chinese government to realize its goal of obtaining and exploiting the most efficient and competitive economy in the world.

Something of particular interest to me in the article mentioned just above is the following statement: "The Finnish experience shows that if well-informed electricity-intensive end users with long time horizons are willing to sign long-term contracts, then nuclear new build can be a realistic option in liberalized markets." This is almost — but not quite completely — correct, and as a result I have exchanged opinions with one of the authors of that article, politely explaining to him in scholarly language that evidence seems to make it clear that most "liberalized" (i.e., restructured) electricity markets are 'offbeat' or 'goofy' or 'wacko' or 'off-the-wall', and without a foundation in conventional (or *non-voodoo*) economic logic.

In the interest of fairness, I should probably mention again something called the 'Jevons Paradox', so named after a 19th century British economist. The paradox here is that if energy production is made a great deal more efficient, considerably more energy might be used, and as a result there would be a larger demand for various exhaustible resources, including energy resources. The UK Energy Research Centre has apparently done a great deal of research on this subject, and has confirmed that energy efficiency may result in an increase in the consumption of energy and other resources as a result of what were called 'rebound effects' earlier in this chapter.

6. Possible Shale Fallacies

Many bystanders are practically in a tizzy when they begin to talk about natural gas. Perhaps this is perfectly natural, because what many influential partisans of this resource believe is that shale gas and oil will restore the energy supremacy of the U.S., and perhaps deal a 'knockout' blow to OPEC, which has been a supreme figure of hate since the first oil price escalation (in 1973), when the governments of various OPEC countries — particularly those in the Middle East — came to the conclusion that the rich oil deposits in their countries belonged to local citizens, and not to the large oil 'majors' in the U.S. and the UK.

Funny things have happened since the shale revolution began. For instance, one of the largest and most profitable energy companies in the world — ExxonMobil — bought shale gas properties for 31–41 billion dollars, and according to a recent admission, they are now "in a world of hurt" due to the abnormally low gas price. More significantly, the largest shale deposit in North America — the Marcellus deposit — has been downgraded by two-thirds by the United States Geological Survey. Perhaps the 'funniest', however, is the insistence by a large number of self-appointed experts — including the President of the United States and his advisors — that the U.S. is now the proud possessor of a verified 100 years' worth of natural gas reserves. (For the record, at the end of 2012 the U.S. had the largest gas reserves in the world, with 681 billion cubic meters ($= 681$ Gcm), while Russia was in second place with 592 Gcm.)

Let us deal with this claim of 100 years' worth of shale gas by taking a careful look at Figure 1. The diagram in Figure 1(a) is a logistic curve. What happens in that diagram is that the present amount of a resource (i.e., reserves) — for example, the reserves of oil or natural gas — is Q', and as production (q) takes place, we move up the vertical axis toward Q'. As is the situation with logistic curves, for mathematical convenience we never

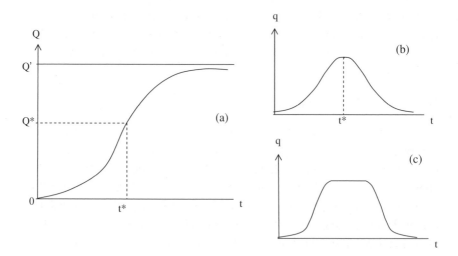

Figure 1. Cumulative Production and Typical Output Curves for Oil and Natural Gas

reach Q', but that is unimportant. Now, the slope of this curve (which is output per time period, or q) is shown in the normal-like curve in Figure 1(b), and at time t^* the output of the resource peaks. This peaking is discussed in the chapter on natural gas in this textbook, and it has taken place in a number of important oil-producing regions (such as the U.S. and the UK-Norwegian North Sea).

If we take the *entire world*, with Q' as the estimated reserves, then we are somewhere below Q^* and to the left of t^*. If we limit our consideration to the situation of natural gas in the U.S., and calculate the reserve-production ratio, it might well be 100, but this does not mean that gas will be available at the present or a higher output for 100 years. Instead, for geological and economic reasons, a peaking of the output should be considered, and for a reserve-production ratio of 100 years for U.S. gas, a good approximation for the time to peaking (t^*) is 50 years, because logistic curves work that way. Moreover, if q (which is annual output) is increasing, which in reality it usually is, then the approximate time to peaking is less than 50 years. Note: the important thing is peaking, and not $Q'/q = 100$ years, as might be the case today for the U.S.

But there is a problem. If we look at the output curves for actual oil or gas deposits, almost all of them do not look like the one in Figure 1(b). Instead, they look like the one in Figure 1(c), where instead of a unique peak we have a plateau. What this means is that the logistic curve in Figure 1(a) must be adjusted if we want a realistic output curve, which in turn means that there must be a 'flat' portion on both sides of Q^*. With luck, this should give us something like Figure 1(c). Conceptually this is not difficult to think about or arrange, but conventional mathematical equations that would move us from Figure 1(a) to Figure 1(c) might be difficult to formulate.

This humble person does not require any "conventional mathematical equations", or, for that matter, the other kind. If we are talking about real science instead of make-believe science, then it means that we do not say that there is 100 years' worth of natural gas available in the U.S., although if output is constant then there might be 50, and with increasing output perhaps much less.

When examining this topic, there are a number of important considerations that come into play. Some can be found in this textbook, or in my earlier textbooks, or in my book on natural gas (1987), and if this chapter were more comprehensive I might feel that environmental and production problems associated with hydraulic 'fracking' should be given a detailed treatment. Unfortunately I must leave that to others, but I suspect that the bottom line for shale gas could be a disappointment for many observers.

In the event that it isn't, an association called America's Energy Advantage (AEA) has come to the conclusion that the large-scale exporting of natural gas by the U.S. is a mistake, and the increased domestic production of oil and gas that may be possible should be a major component in what the brilliant *Financial Times* energy journalist Ed Crooks (2013) calls "a wider renaissance in the country's manufacturing through the availability of cheap energy".

I think that the correct expression here should be "cheaper energy", because as yet there is little sign of genuinely cheap *energy* in the U.S. other than in the natural gas sector, and nobody knows how long that will last. Instead, there is an outpouring of lies and misunderstandings of the type that led to 110 members of the U.S. Congress signing a letter calling for the approval of more export permits for U.S. gas.

My reaction here is the following: what do *they* know about the realities of the U.S. energy supply? America's Energy Advantage probably has some of the best geologists and engineers in the U.S. at its disposal, and, like me, they believe that the energy found in or near the U.S. could provide America with a substantial economic advantage if used correctly, which may or may not happen. Incidentally, the 'energy advantage' once possessed by the UK may have been frittered away by excessive exports.

At the time this was written, the 'spot' price of natural gas in Asia was four times the price in the U.S., which might explain the enthusiasm of those U.S. congressmen and congresswomen for exports. (The spot price is the price quoted for the immediate purchase or sale, and often delivery, in the short-term life of an asset.) Apparently, that Asian price was mostly due to the lack of 'spot cargos', which could be caused by the preference of many buyers for long-term arrangements. (Depending upon how contracts are written, some buyers in Asia are obtaining large amounts of gas for prices that are under the current spot price.)

Eventually, if present shipbuilding trends are maintained, enough LNG vessels will be constructed to provide Asian buyers with the gas they desire.

When (and if) that happens, there might be a partial convergence of global natural gas prices. Aside from that, if the amount of gas in the U.S. is actually much smaller than commonly thought, which is possible, then promoting large exports is equivalent to placing the incomes of energy company executives and stockholders above those of American voters. The same situation has prevailed in Sweden with electricity, where academic economists and other so-called experts convinced politicians to favor an export and deregulation policy that was destructive for both Swedish households and industries.

Unfortunately, one of the disadvantages associated with clarifying some of the aspects of this topic is that many of the persons attending seminars and 'workshops' on energy economics are totally ignorant of basic energy and theoretical economics, including issues with which they must often deal, such as the pricing of the natural gas and electricity they consume. Having taught economics in many universities, I generally overlook this shortcoming, except that some of these individuals not only attend seminars, but are invited speakers at others, where they have convinced somebody important that they are up-to-date on energy topics. One of the most pathetic meetings was at the Stockholm School of Economics, where natural gas was ostensibly discussed.

There is no mathematics in this section, because although mathematics (and economic theory) dealing with the *crucial* topic of the effect of steep (deposit) depletion/depreciation rates on the output of shale gas and oil are extremely important, a mere 'flash' of algebra would cause some readers to immediately discard this offering, and turn instead to the illogical misunderstandings about shale that are in full flower in virtually every part of the world. Put less dramatically, mathematics would only confuse the present issue, and intelligent seminar or course participants do not need it, even if some of them say or think that they do. The 'numbers' that will be given directly below tell the entire depreciation story.

Readers should also note that each fracking well uses millions of gallons of water, and a reputed main constraint on future production is the lack of water required for the production process. This could be why fracking has moved so slowly in China, despite reports of a huge amount of shale oil and gas in that country.

Executive Continuation

Arthur Berman, who is a very controversial figure in the great oil and gas world, says that exploiting shale resources is a "retirement party" rather than a revolution: "It is a last gasp." By way of contrast, Aubrey McClendon, founder and former CEO of Chesapeake Energy, has called Berman "a third-tier geologist", and perhaps the reason he feels so strongly about Mr. Berman is because, as indicated by Asjylyn Loder (2013) — who cites (official) Oklahoma State Records — a large Chesapeake well near Oklahoma City (U.S.) that initially pumped more than 1,200 barrels of oil a day (in 2009) now produces less than 100 barrels a day. That is hardly the kind of information that Mr. McClendon wants shouted to the high heavens.

Needless to say, this decline must have been a traumatic experience for Mr. McClendon, as well as for any investors he charmed into believing that sharing the financial burdens associated with the production of shale oil or gas would result in their also sharing some lovely profits. My advice to all readers with a serious interest in this topic is that they should not forget that unexpected and economically unwelcome fact of life: 100 barrels of oil realized instead of the 1,200 (or slightly less) expected is a disappointment that cannot be easily absorbed by rational investors! Nor can the fact that as recently as 2013, there was talk of Chesapeake's expenses being much larger than its revenues. Word was also going around that Chesapeake might abandon shale 'plays'.

In the immensely relevant article by Loder (2013), the *bottom line* is that "America needs 6,000 new wells a year, at an estimated cost of $35 billion, in order to maintain current oil production." I am also a specialist in formulating bottom lines, and on this occasion mine is as follows: given the amount of money that has been spent financing the wars of the last decade, 35 billion dollars a year for a *VERIFIED large amount* of a good as valuable as energy is a trifle for a country (America) with a Gross Domestic Product (GDP) of more than 16 trillion dollars. But admittedly, it would be extremely discouraging if, in the medium to long run, excessive drilling increased costs and greatly diminished supplies of gas and oil in the most desirable shale drilling sites, while money was available to pay for wells that — were they as productive as earlier commitments — might be capable of maintaining current output, particularly since the demand for natural gas is increasing all the time.

J. David Hughes (2013) calls this "The Red Queen Syndrome", *which means that the higher production goes, the more wells that must be drilled to offset the inevitable decline!* Moreover, as is quite clear from what is taking place in the U.S. at the present time, the *rate* at which wells are drilled influences the rate of decline. Probably the most important observation though is the one made by OPEC's Secretary-General, which is that "U.S. shale producers are running out of *sweet spots*" (i.e., places where it is certain that drilling will yield enough oil or gas to cover all production costs).

Shale Myths and Meaning

According to a recent article on the important site *321 Energy*, the shale revolution is just beginning — in other words, the good shale times are just starting to roll — and ostensibly this will be beautiful for the American economy. The opinion in this section is that on the basis of existing evidence, that tantalizing belief is doubtful, although it might be true that *shale oil and gas can be very important elements in the American energy panorama, even if they do not 'change any games' or incite any 'energy revolutions'!*

Oil shale, also known as kerogen shale, is an organic-rich fine-grained sedimentary rock containing kerogen (a mixture of organic chemical compounds), from which shale oil can be extracted. ("Organic" signifies the presence of carbon in the material under consideration.) Shale oil is a substitute for conventional crude oil, and a great deal of effort has gone into convincing ordinary citizens as well as decision makers that the U.S. possesses an enormous amount of exploitable shale. Whether this 'enormous amount' can always be profitably extracted is quite another matter, because, as with conventional oil, there is a substantial difference between *resources* and *reserves*, and this gap has hardly changed over the last 3 or 4 decades. Remember: reserves (or extractable oil) are a percentage of resources (or 'oil in place'). Perhaps I should emphasize that you should not believe everything you read or hear about shale *or* conventional oil and gas, since this is a topic with which charlatans seem to have established a special relationship.

But regardless of what you or the billionaire CEO of Chesapeake decide to believe, the economics of shale oil and shale natural gas are unlikely to become a new romance for this humble teacher of energy economics. I mentioned shale oil in my book on oil (Banks, 1980), and just a few months later I gave what I thought was a brilliant lecture on the subject of oil and gas at a workshop in Vienna. The same evening I was informed by an American business executive that if I were serious about shale, then I was a "fool", because in his part of the United States there was not sufficient water to 'fuel' the shale production process. (According to Jeffrey Michel (2013), an MIT graduate who is probably the leading energy economist in Germany, the same may be true of China, which could be the reason why the Chinese government and Chinese managers do not appear to be rushing to exploit their reputedly plentiful reserves of shale oil and shale gas.)

As a result, when I published my book on natural gas (1987), the word "shale" was not mentioned, although of course the basic technology employed to obtain shale natural gas (by some variant of *fracking*) was well-known. (Fracking involves pumping water, sand and chemicals into a deposit under high pressure, which — in conjunction with horizontal drilling — liberates gas or liquids and causes them to reach a well.)

I would like for everyone to understand that despite my disgust with some of the things I have heard or will hear about shale, I consider the best outcome of this shale 'hullabaloo' to be a consensus by highly qualified geologists that shale oil and shale gas deposits are intrinsically valuable assets that can be exploited over a long period of time. In addition, they will benefit more than just a few energy firms and so-called 'players', even if there is a possibility that shale deposits may not be worth singing about in karaokes where patrons are especially concerned with the 'national common good'.

Furthermore, *if expectations are moderated*, shale gas and oil can be valuable components of an *optimal national energy program that also involves nuclear and renewables*, even if shale oil and gas provide only limited and regional market opportunities. At the same time, there is no point in denying that things have happened on the shale oil and gas scene that were definitely not expected or desired, and will probably happen again. As an example, I can mention a brilliant firm like the energy giant Exxon reportedly taking a shale gas 'bath' for 41 billion U.S. dollars.

My explanation for the *faux pas* of Exxon is that management expected more from the new technology than the new technology was capable of delivering. Even experienced engineers were dazzled by the barrage of drivel about 'game changing' and/or 'a golden age of natural gas', much of which was launched by propagandists who could not distinguish between gas and a hole in the ground. Among these pretenders I can cite executives of the International Energy Agency (IEA), whose goofy predictions about the future of crude oil I discussed at great length with students during my tour at the Asian Institute of Technology (Bangkok). Surprisingly, their work was favorably referred to by MIT energy researchers like Henry D. Jacoby and Francis M. O'Sullivan (see Jacoby, O'Sullivan, and Paltsev, 2012), who perhaps believe that generous endorsements can sometimes pave the way for gratis plane tickets and visits to the IEA headquarters in the heart of magnificent Paris.

Some Essential Background

> *"Why not use your mentality, wake up to reality!"*
> — *Cole Porter ("I've Got You Under My Skin")*

A recent article on *321 Energy* that has caused me some unease was an interview carried out by the 'boss' of the site *OilPrice.Com*, Mr. James Stafford, who almost a year earlier strongly objected to my saying that one of his contributors had published an article on his site which indicated that the United States (U.S.) possessed at least as much oil in proved and hypothetical reserves as the rest of the world combined. The good Mr. Stafford labelled my opinions and research "garbage".

"Garbage" is a sensitive word for me, because after I was expelled from the Infantry Leadership School (at Fort Ord, California), I was assigned to work on a garbage truck for several months before eventually being assigned to the famous 38th Infantry Regiment. In any event, what Mr. Stafford did was to find a well-known economist, Professor Tyler Cowen, who has an excellent knowledge of the relationship between energy and the macroeconomy, although his belief that the shale gas "revolution" is on the high road and — *ceteris paribus* — can restore the dynamism in the American economy is without justification.

But let us be clear about one thing. Although I taught international financial economics and macroeconomics at Uppsala University (and in Prague and at several universities in Australia), at the present time I make a point of staying away from the details of that subject. What I do know however — and perfectly — is that it is very unlikely that enough natural gas and oil can be removed from the various shale deposits now available in the U.S. to return that country to a semblance of the macroeconomic stability (or sanity) that existed a few years prior to 2008, at which time the rise in the oil price accelerated. In case you have forgotten, it was in the middle of 2008 when the price of oil touched $147 a barrel, and the global economy went off the rails.

This last observation should never be forgotten: the oil price touched $147, and there were predictions by many experts that the oil price was on its way to $200 per barrel. Furthermore, it was the destructively high oil price, and *not* certain financial market irregularities, that put the skids under the global economy. The financial crisis that many observers talk about came *after* the recession began. More importantly, the near collapse in industrial production that took place in that dramatic year has not been completely reversed. For instance, industrial production apparently "peaked" (i.e., deviated from the *ex-ante* long-term trend) in the U.S. in 2007 or 2008, and according to Floyd Harris — a contributor to the *New York Times* — "stayed below the peak for 68 months". Returning to Professor Cowen, and perhaps Mr. Stafford, their hypothesis seems to be that since expensive energy created the problem, inexpensive energy (e.g., in the form of shale gas and exotic oil resources) might solve it.

I can accept that reasoning with a smile on my face. However, in the case of the U.S., it might be necessary to have access to a very large amount of comparatively inexpensive energy if a sustained upturn in the macroeconomy is desired, and I am sure that some doubt exists in the minds of many advisors to important politicians and bureaucrats about the availability of large amounts of comparatively inexpensive energy derived from shale, and furthermore, obtainable *over a considerable number of years!* By "comparatively inexpensive" I mean, e.g., natural gas at or fairly close to the present price, and available for a long enough period to fuel increasing amounts of expensive vehicles and industrial machinery, without experiencing an increase in price that takes it to the level that natural gas sells for in Europe or Asia.

Better Behavior and Its Ecstasies, and a Conclusion

"The U.S. energy revolution will end the old OPEC regime."

— Ed Morse

You aren't trying to fool us, are you Ed, because if you are and you decide to attend the superb courses in energy economics or game theory that I hope to be invited to teach at Uppsala University soon, I would have to ask you to leave the classroom until you learned the meaning of the word "revolution"! In any event, I suspect that there will not be an energy revolution in the U.S. until the nuclear reactor that multi-billionaire Bill Gates is financing is up and running, and unfortunately, maybe not even then.

In a forum I contributed to before being brusquely informed that my modest talents were no longer desired, I often complained about people reading newspapers instead of books in order to obtain the knowledge they need to understand applied energy economics. In my opinion, the problem was the sub-optimal backgrounds of the editors of some of those publications, most of whom have never studied engineering or economics. For example, read below what an editor of the *Wall Street Journal* (June 22, 2011) has to say about the way that global energy has and should develop:

> "Yet, beyond our merits, the Lord has recently smiled on us in the form of shale gas. . . . Don't bet on Mr. Medvedev. Bet on the crude logic of Russia's declining energy power, which Western policy should do everything possible to exploit, to deliver better behavior in Moscow."

This foolishness about "better behavior in Moscow" is intended to certify the credibility and/or diligence of a certain *editor* of a famous U.S. newspaper. Note again the expression "Russia's declining energy power", when in point of truth Russia's "energy power" is indeed powerful, and can only increase. One of the reasons why it will increase is that Russia is the largest country in the world (by landmass), and a sizable fraction of its geology is similar to that of North America, and thus Russia can also expect a shale oil and gas bonanza of sorts. Another reason is the availability of the dynamic Chinese market. Today China has 100 million cars as compared to 250 million in the U.S., but expectations are that by 2030 there will be 300 million cars in China. If you are concerned with the future price of motor fuel, and this estimate of Chinese automobile ownership is correct, then I suggest you hope that the shale gas revolution that Professor Cowen expects to take place in the United States also takes place in China and Russia. And let me note that Bloomberg journalists have started to examine the issue of an intensified Russia–China energy relationship.

Almost certainly there are plenty of pseudo-scholars in North America and Europe who will tell you that the Russians are inherently incapable of exploiting their geological wealth (for example, their massive 'shale power'), but it so happens that for those of us who know the story of the main battle tanks of Russia and the U.S. during WWII, betting on Mr. Medvedev and his colleague Mr. Putin makes a lot of sense. The nonsense that we still occasionally hear about the engineering and scientific frailty of Russia, which implies an inability to shove a pipe into the ground and obtain oil or gas if they are actually present, is about the same as the wishful thinking that I once heard about OPEC and the reputed inability of its technicians and managers to make the most of its oil reserves, which means OPEC countries cooperating in an oligopolistic manner that results in a maximization of individual incomes.

Professor Cowen recognizes that what he calls *The Great Stagnation* "first shows up in the data in 1973, when income growth slows and pro-ductivity growth falters" (Cowen, 2011). I pointed this out in my oil book (1980) and in my book *Scarcity, Energy, and Economic Progress* (1977), and the reason I took note of it is that I was familiar with the research of Professors Dale Jorgenson and Edward Hudson, although it was barely noticed by most economists at the time, and the same is true now. Cowen (2011) also says, "It's hard to avoid the conclusion that this has something to do with the end of the age of cheap energy."

In the interview Professor Cowen gave to Mr. Stafford, he suggests — as does Mr. Morse — that the "shale gas boom" will bring an end to the present era of expensive energy. This is probably, though not certainly, wrong. Right or wrong, I do not mind hearing this from Mr. Morse, who is managing director and head of commodities research at Citigroup, because I cannot imagine my intelligence degenerating to a point where I require help from the commodities research people at one of the large financial institutions. What I do hope, however, is that since Professor Cowen undoubtedly knows a great deal about how the global economy works, he will avoid giving inter-views to people who have certain 'agendas', as for instance the gentleman who was kind enough to make him the subject of an interview dealing with shale. Unless I am mistaken, the agenda here is desiring to be thought well of by movers-and-shakers who want the world to believe that talk or thought about energy limitations is nonsense. As it happens, I am also optimistic about energy, but often sceptical about a lot of the things I am told regarding natural gas and oil.

In any event, the lies being told about oil are almost too grotesque for scholars like myself to accommodate. According to Deborah Rogers, shale oil wells are declining even faster than shale gas. She supports this opinion by saying that according to well production data filed with North Dakota authorities, daily oil production per well in the Bakken Shale has already peaked, and older wells are declining so rapidly that new wells cannot compensate for this decline.

The views of Ms. Rogers are apparently controversial, but when observers refer to production data that are filed with state authorities — as was also the case with the Bloomberg journalist Asjylyn Loder — then it is time for dedicated researchers to stop playing games and to get down to real business. For instance, they can confirm or deny the statistics provided by Deborah Rogers, and also tell curious persons like myself why virtually all the production of shale gas is taking place in North America instead of at least some of the other 12 regions that have registered the presence of shale deposits that probably can be economically extracted. (A very small amount takes place in China.)

Moreover, many persons who want us to accept 'crank' evaluations of energy issues are not interested in what is known as the common good, but instead are desperate to join the ranks of the rich or super-rich. While I am willing to help them achieve their goals if some money changes hands, I have no use for their wisdom where items like 'external' costs are concerned. In the case of North Dakota, these costs result from taxes on energy companies not compensating for various kinds of environmental (or 'collateral') damage.

I conclude this exposition by reminding readers that an intense discussion about fracking is taking place in France. The U.S. Energy Information Agency (EIA) believes that France possesses 137 trillion cubic feet of technically recoverable natural gas. Ostensibly the reason for not exploiting this bonanza is environmental, which is an opinion I accept because before publishing my oil book, I was told of an executive in a large energy company describing the air and water pollution associated with obtaining oil from shale as highly injurious. That executive was Donald E. Carr, and he made this clear in a brilliant and non-technical book (Carr, 1976) that should be read by everyone interested in energy.

Carr's book was published in 1976, and it might be true that production techniques for obtaining shale oil and gas have been greatly improved since then, such that deleterious emissions have been substantially reduced. This is something the current French president is unlikely to dwell on, however, since he has put his ambitious shoulder to what he thinks is a semi-populist wheel, and presumably requires the votes of part-time as well as committed environmentalists in order to obtain another term in office. Thus, in France, as in Germany, it *could happen* that we are being treated to the spectacle of hypocritical politicians jeopardizing the futures of their foot soldiers in order to enjoy more years of well-paid notoriety.

It does no harm to note that politicians in many industrial countries are reluctant to commit themselves on energy topics, even though they have access to the best opinions on these topics, and moreover have access to persons who can present these opinions to the general public in the kind of language that voters understand and appreciate — the language of money. In the UK it is now clear that shale investments may not be as lovely as once thought, and so 'players' are looking for equity partners to share what might become an odious burden. That is to say, equity partners (= investors) who are impressed by what they *think* is taking place in the U.S., and are frantic to gain some of the millions they believe will be up for grabs when (or if) shale gas and oil starts flowing like they hear every day that it is destined to flow.

5. Energy and the Best Brain of the 20th Century

> *"If you meet someone at a party who says that he is Napoleon, you don't start discussing cavalry tactics at Waterloo."*
> — *Professor Robert Solow (MIT)*

Well that depends, Robert. If he is the gentleman who gave the party, and you would like to receive another invitation from him some day, you might feel it wise to suggest that if Napoleon's troops had been riding elephants or dinosaurs instead of horses, he might have enjoyed another few years in gorgeous Paris instead of being turned over to that nasty Sir Hudson Lowe on St. Helena.

Until about 2008 it was the oil optimists who gave most of the parties, or at least supplied the music. It is highly significant — and enjoyable — that we only encounter a few of those people at the present time, although it continues to be annoying when we suddenly find ourselves confronted with humorless pundits who reject mainstream economics, geology and statistics, and denounce the oil market realism that is occasionally showcased by our sterling media.

Notice the term "realism", because you might still be told that even if the *discovery* of oil were *passé* (i.e., terminated), there would still be 40 years of comparatively inexpensive oil in our future. This is because 40 years or thereabouts was (or is) the global reserve-production (R/q) ratio. However, for reasons described in an earlier textbook (Banks, 2007), I conclude that the R/q ratio is mostly irrelevant, and the same is true of the anticipated year in which the global output of oil will peak. *When the price of oil can touch $147 a barrel (= $147/b), as it did in the summer of 2008, and during this period distinguished commentators were talking about it moving to $200/b or higher, then it is clear that some kind of geopolitical peak has already been reached.*

The explanation for *that* geopolitical summit was the apparent peaking (or 'flattening') of conventional non-OPEC oil production several years earlier, as well as the kind of sophistication that I expected OPEC countries to show when I published my oil book many years ago (1980). Fortunately, I was about 20 years off-target, because at one time a shortage of oil could have resulted in a very ugly political and economic scene, particularly if large oil-consuming countries elected to compete for a piece of the remaining supplies with the aid of military assets.

The question can thus be asked how John von Neumann — often called the best brain of the 20th century — might have approached this issue. This is not the place to elaborate on game theory, but I happen to believe that if von Neumann had thought that the later contributions of, e.g., game theorist and Nobel laureate John Nash were of great import, he might have devoted a few minutes of his time to deriving them for the book he co-wrote with Oskar Morgenstern (von Neumann and Morgenstern, 1944). (Nash's life and work were turned into a mindless burlesque in the film called *A Beautiful Mind*, to which he apparently gave his approval.)

Accordingly, I can picture von Neumann saying that the present-day oil market game is something where a transfiguration of his famous *maximin* theorem might be applicable. A misinformed student once grandly informed me that the maximin theorem strictly applied to two-person conflict-like situations in which the interests of players were in strict opposition; but it happens to be true that von Neumann intended any two-person conflicts in his book (or the books of other game theorists) to be the cornerstone of a scheme in which there are many players, and in addition there can be a certain amount of cooperation. Some of this thinking can be found in the latter part of his book.

More importantly, as William Poundstone (1993) brilliantly noted, von Neumann's game theory was only "tangentially about games in the usual sense". Simple games featuring a well-defined form of computation are what you might confront in your Economics 201 textbook — or at least in its first half — but according to Poundstone, von Neumann once said that "Real life consists of bluffing, of little tactics of deception, of asking yourself what is the other man going to think I mean to do. And that is what games are about in my theory." The other man or *woman* is probably the correct usage here.

In the summer of 2008, as I was presenting a lecture on oil at the Ecole Normale Superieure (Paris), the oil price appeared to be on the verge of moving off the Richter Scale. Neo-classical or orthodox explanations of oil price formation had been provided some years earlier by the Chicago Nobel laureates Milton Friedman and Gary Becker, but as usual they completely misunderstood the economic and historical forces that were at work and would soon coalesce. First and foremost, the Middle East producers of oil intended to extend the 'life' of oil reserves, which logically meant restricting the amount removed as the oil price escalated. Instead, as much oil as possible would be left in the ground, and when eventually extracted, a large fraction would be used to produce oil products and petrochemicals. As simple as this sounds, it was not widely understood by governments of the oil-importing countries, nor was it made clear to these governments by the large oil-producing firms.

Like von Neumann, the Middle East producers dismissed (or ignored) the short-run supply-demand *equilibria* familiar to beginning and advanced students of mainstream economics, and instead formulated and eventually began to follow an elaborate strategy for economic development. The key word here is "strategy", and to paraphrase Antoine de Exupery, "a goal without a strategy is a dream". In the first part of your economics textbook, 'strategy' typically consists of automatically reacting to existing prices. While the strategy which Middle East producers have now apparently adopted features some *reacting*, it is primarily concerned with doing what is necessary to keep the oil price rising. For instance, during the Libyan drama, the Middle East producers' talk about replacing Libyan production — but not doing so — was precisely what John von Neumann would have advised: *just talk!*

Spin-offs of von Neumann's work are too extensive to take up in this short discussion, but a careful reading of these materials makes it clear that the only sensible thing for producers of oil to do is to collude, assuming that it is legally possible and that the rewards of collusion are *coalitionally* rational: i.e., rational in the sense that in the long run, all coalition members receive a payoff that is roughly commensurate with their contribution to the 'price-fixing' exercise. Coalitionally rational quotas of the kind theoretically practiced by OPEC constitute what is known as the *core* of a game, and in theory imply stability. This is why, incidentally, it is not certain that Iraq will live up to some of the curious predictions of its eventual production that are now being liberally circulated by *ad hoc* oil-market connoisseurs.

Regarding consumer strategies, Tsuyoshi Inajima and Yuji Okada (2013) have reported a plan by Japanese buyers to form a consumers' cartel (an oligopsony) to buy LNG (liquefied natural gas), presumably since Japan is the largest importer of that commodity in the world. I have probably taught oligopsony theory to several hundred students, but this is the first time I read or heard anything about the likely application of that theory. With luck, some readers of this book should be able to tell me more about it in a few years.

Finally, since the subject matter above has to do with the alleged best brain of a century that was crucial for many readers of this book, whether they were born or not, it might be useful for me to say what I think would be the most important subject for that brain to consider if he were with us today. By that I mean population, and here I remember some conversations with former colleagues at the University of New South Wales (Sydney,

Australia), especially Tom Moder Mozina, who is a dedicated student of Asian economics, including Japan.

I know a little about Japan, having lectured at conferences there, and also served in the U.S. Army at Camp Majestic (Gifu), the wonderful port city Kobe, and the live firing ranges close to the base of Mount Fuji. According to *Bloomberg Business Week*, Japan is in deep trouble because that country is growing older too fast. A diagram in a recent edition of that publication shows India, Egypt, Colombia and Mexico as the four countries with the smallest fraction of their population over 65 years of age, while in the (Bloomberg) 'jumbo' position the diagram shows Japan, Germany, Italy and France (where the latter is tied with Spain). In other words, implicitly, because of a shortage of nimble brains and hands, the last four countries are supposed to be going nowhere in a hurry.

I would not blame my students for believing that this was true if it were not for an explicit outburst of misapprehension in the discussion accompanying that diagram. To quote the Bloomberg expert masquerading as an expert on Japan, "to offset labor shortages, Japan has begun easing immigration requirements for highly skilled workers. So far the program has fallen short of its modest target: under a quarter of the 2,000 professionals it sought have come to work in Japan."

The thing for readers of this book to understand is that an overwhelming majority of the Japanese do not want foreign 'workers' in their country, highly skilled or not. What they want is for their political masters to stop playing the fool, and to reproduce the economic performances that were in line with the mentality of John von Neumann, and which I repeatedly told my international finance students about before I decided to concentrate on energy economics. In a short but brilliant lecture that I attended a few weeks after starting my three-year 'tour' at the Palais des Nations (in Geneva, Switzerland), my colleagues and myself were informed that Japan's development plans were generally regarded by economists in that noble structure as a role model for industrialization and economic progress.

Moreover, unless growth in the global economy accelerates at a higher rate in the near future, there will be thousands of brainy engineers roaming the streets of India, Egypt, Colombia and Mexico begging for work before their valuable analytical skills are dissipated by idleness. Instead, imagine being a qualified engineer or technician and, after arriving at Kobe's airport, proceeding to your apartment on or near one of the sensual hills in that exotic city.

I would also like to mention something that the editors and journalists of the *Bloomberg Business Week* are probably incapable of understanding, but John von Neumann might after a short talk with me. Technology, economics and geography are working in Japan's favor, assuming of course that their navy does not pay another early-morning visit to Pearl Harbor. The message that needs to be understood and appreciated here is that for many countries, though not Japan, the problem is not going to be too few people, but too many. As far as I can remember, almost every taxi driver I talked to in Germany and Australia was clear on this point, but only a few academics are.

4. Some Unfriendly Economic Comments on Another Green Fantasy: Roadmap 2050

According to Hughes Belin (2010), *"Tout Brussels"* once gathered in Brussels (Belgium) on a wonderful spring day for the presentation of 'Roadmap 2050', by which he meant the flamboyant occasion on which the European Climate Foundation (ECF) — a so-called think tank — unveiled yet another green fantasy about how Europe could be decarbonized for little more than lunch money. As will be noted later, what we are actually talking about here is a project that could necessitate deploying trillions of U.S. dollars; however, once the investments proposed in the Roadmap are carried out, and Europe's existing energy infrastructure is replaced with certain low-carbon alternatives, electricity prices in the long term will supposedly be constant, and dangerous levels of greenhouse gas emissions can be avoided. Put somewhat more technically, (discounted) short-term costs, though high, will be at least compensated for by (discounted) long-term benefits.

In case you have forgotten your theoretical welfare economics, the mathematical and economic details of this arrangement are straightforward, albeit tedious, and once we leave the classroom for the real world, we might encounter some very disturbing prospects. However, the environmental correspondent of the influential *Financial Times* (UK), Fiona Harvey, does not seem to be bothered by unanticipated and/or troubling occurrences, and apparently believes that not only will fossil fuel power stations be banished from the face of the earth, but nuclear facilities will also be eventually liquidated.

As I have attempted to explain to Ms. Harvey and her colleagues for many years in a dozen or so of my articles, the optimal power generation strategy will feature a host of renewables and alternatives — and not just those mentioned by the Roadmap theorists — but also (probably) a moderate and sustainable increase in nuclear energy, and in particular — though not exclusively — energy from the next generation of nuclear equipment (Gen 4) when it becomes available in a decade or so.

According to Ms. Harvey, Mr. Matthew Phillips of the ECF said: "When the Roadmap 2050 project began it was assumed that highly-renewable energy scenarios would be too unstable to provide sufficient reliability, that highly-renewable scenarios would be uneconomic and more costly, and that technology breakthroughs would be required to move Europe to a zero-carbon power sector. Roadmap 2050 has found all of these assertions to be untrue."

Well, Matt, with all due respect, as well as a profound understanding on my part that you could hardly do less than to praise Roadmap 2050 to the high heavens, I would like to emphasize that regardless of whether or not you and yours believe those initial *assumptions* you quoted, they are not only true, but much truer than you, your friends and neighbors, and the ladies and gentlemen at the Roadmap launching site could possibly realize. Furthermore, the day after the Brussels extravaganza, I attended a seminar on coal in Stockholm, where among other things, during a lull in the proceedings, one of the speakers repeated to me some of the bunkum from the Brussels gig. In addition to waffle about Carbon Capture and Storage (CCS), he mentioned the wind farms that could be placed in the North Sea, and solar farms constructed in, e.g., Spain, with all of these connected by a super-grid, and as a result, providing the kind of reliable power that conventionally is not expected from items like isolated wind and solar, where *capacity factors* can be extremely low.

I of course had no choice but to inform that gentleman, and some others, that Roadmap 2050 was one of the most grotesque misconceptions ever presented to audiences of sober engineers, researchers and executives. What I did not inform him and the other speakers was that I understood the reasons why Roadmap creators and disciples supported such obvious nonsense, foremost among which was the fact that it would provide many jet-setters with an abundance of interesting — though socially unproductive — work that might last until they began drawing survivors' benefits.

The Roadmap was also taken up by James Kanter (2010) of the *New York Times*, in a short but informative article. That article deserves considerable attention, because according to Mr. Kanter, the cost of constructing the super-grid and providing facilities for reducing emissions by 80 percent is about 9.5 trillion dollars. Quite naturally, this is an informal estimate, and perhaps an exaggeration, but even so I suspect that in a seminar room or conference, it would be difficult for a lecturer to get the words "9.5 trillion dollars" out of his mouth, because it would immediately amount to a condemnation of the Roadmap and anyone dense enough to believe in it or anything like it. Here I want to mention that the Roadmap is very unlikely to be supported by engineers and economists who have nothing to gain financially and/or career-wise, and this is particularly true in France, where cost-benefit calculations are more or less routine. Needless to say, in Sweden, engineers would give this bizarre 'Roadmap' deception an exaggerated amount of respect, because it would be a social and also an economic mistake to compromise their green credentials. Another organization that has indicated support for the Roadmap is the Stockholm Environmental Institute.

As far as I am concerned, the principal interest of those endorsing the Roadmap is in making money, which also includes confusing the general public in Sweden and elsewhere about the ability to reduce CO_2 emissions with 'Carbon Capture and Storage' techniques and cap-and-trade schemes. I regard cap-and-trade as an extravagant scam, and I think it necessary that everyone who reads this unfriendly discussion should know that it has been rejected by the economists who first proposed it, Professors Thomas Crocker and the late John Dales, and also the leading climate scientist and believer in (*anthropogenic*) global warming, Dr. James Hansen.

When Roadmap 2050 was introduced in Brussels, playing in the background was Shirley Bassey's melodic rendition of "History Repeating". Considering the high costs and low benefits that the Roadmap entails, having Ms. Bassey bellow her famous "Goldfinger" would have been more appropriate, especially if she had been accompanied by a 'pick-up' rap chorus recruited from the lucky men and women enjoying tax-free employment in that up-market part of Europe.

3. Not So Nice about Oil

Recently I gave a lecture in Paris at the Université de Paris (Dauphine), in which I did my usual best to explain to a group of energy economics students that OPEC has now taken command of the oil price. It is possible that some of those students had no problem with my contention, because as many observers now understand, when the present macroeconomic *malaise* blows over, OPEC will likely move to increase its already impressive income by occasionally raising the price of a barrel of oil by a few more dollars. As it happens, the OPEC "hawks" constantly express dissatisfaction with the present price, regardless of what it is. One thing, however, should please them: according to forecasts by the International Energy Agency, OPEC income topped one trillion dollars in 2011, and according to my limited research, the same was definitely true in 2012 and 2013.

I can remember a time when anyone even thinking that OPEC would someday be on the receiving end of that kind of money would keep his thoughts to himself. But now, of course, the sky is the limit. Chinese demand continues to move upward, and according to McGraw-Hill's information arm, the Chinese demand reached 8.45 mb/d in April 2013, which was almost 13 percent above the demand of that country a year earlier.

What will happen to price if this kind of demand growth continues is difficult to say, although according to my way of viewing the issue, it will not be anything nice for those of us on the buy side of the global oil market. The next oil price escalation could begin at more than $100/b (for some average of WTI and Brent oil), which is something that would be wonderful to avoid.

Not everyone believes that the situation is as serious as I like to claim, however. For example, not too long ago Sandrine Torstad, who is in charge of market analysis at (Norway's) Statoil, claimed that the present large inventories of oil will have to be decreased before the price goes on a rampage. I put my usual diagram and mathematics on the whiteboard at Dauphine to explain why normally this was good reasoning, but I made it clear that I am no longer certain that my logic makes as much sense today as it did earlier. OPEC's position has simply become too strong.

At the end of 2008, with the oil price rapidly moving south, OPEC began a production-decrease program of 4.22 mb/d that resulted in certain highly paid oil experts once again seeing their predictions turn sour. What happened was that the oil price began a climb that — on the basis of what was happening in the world macroeconomy and financial markets — should not have taken place. That steady climb culminated with the (WTI) oil price close to $90/b, which was boosted to over $112/b by the military venture in Libya.

The Secretary-General of OPEC, Abdallah Salem El-Badri, recently said that "the emergence of oil as a financial asset, traded through a diversity of instruments in futures exchanges and over-the-counter markets, may have helped fuel excessive speculation to drive price movements and stir up volatility". I had been lecturing in Paris for almost two hours by the time I got around to that subject, and had a slight problem remembering what I had spelled out in detail in the long papers that I recently published on oil, but even so I had enough steam left to insist that the oil futures markets *follow* rather than 'lead the way'.

Of course, to understand this issue, no economics is necessary — just common sense. There were three or four pronounced oil *spikes* after 1973, which did not lead to a sustained rise in the oil price. However, when the oil price fell to $10 a barrel — and perhaps lower — at the end of the 20th century, the higher OPEC councils decided that they were no longer going to sell their precious oil at bargain basement prices, nor was it necessary, given the oil reserves situation referred to on several occasions in this book. The oil price began moving up, and when the oil supply situation became completely clear to OPEC managers, they made their move. Anyone who looks at a plot of the oil price from about 2004 onwards should get the message, and if they still do not, they should get it on the basis of what happened in the beginning of 2009, when the demand for oil was falling because of bad news in the world macroeconomy. OPEC quite simply reduced its supply of oil, and because there were no other supplies that could be brought to the market in the short run, or maybe in the long run, the price of oil quickly recovered. In fact, it increased by about $40 in a very short time, which is something that would have been unthinkable a decade earlier, given the state of demand.

At the same time I made it clear to those young students that if I were in the place of Mr. El-Badri, I would have said exactly the same thing. After all, that gentleman does not work for the governments of North America, nor the careerists directing the European Union, nor the Association of

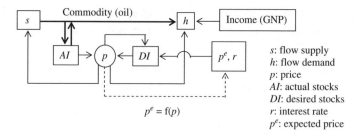

Figure 2.   'Stock' and 'Flow' Supply and Demand in an Oil Market

Good Housekeepers that I used to occasionally hear about when I was a boy in the United States. If the television audiences want to believe that OPEC countries prefer less money to the long green now pouring in, and therefore — as we sometimes hear — will join the oil-importing countries in keeping the oil price from 'breaking' $150/b again, then as far as I am concerned, they deserve what they will get. Once again I can refer to the formula that I constantly touch on in this book: we need a big package of the right renewables and alternatives, and we need it soon, and we will likely also need more nuclear energy to make that package optimal.

Before leaving this topic, I wish to draw my readers' attention to Figure 2, which is also found in Chapter 4 of this book. Since some persons may concentrate on this chapter only, that diagram deserves to be seen again here. The first step is to appreciate that Figure 2 is both simple and logical! For example, it is NOT the kind of flow diagram that we see in our favorite electrical engineering textbooks.

The focus of Figure 2 is on actual stocks (*AI*) and desired stocks (*DI*), and not the flow supply and flow demand so familiar from Economics 101. It is a stock-flow rather than a flow model, and for a price equilibrium, *AI* must equal *DI*. If not, flows take place until the equality signifying equilibrium comes about. If, for example, *DI* > *AI* because it is expected that price will increase, then price will increase as an attempt is made (via an increase in flow demand) to increase stocks (i.e., *AI*). But at all times the key items in this short-term price formation model are stocks (i.e., inventories), which conceptually — in the short run — are more important than the supply (*s*) and demand (*h*) *flows*! The mechanics of this market (and also futures and options markets) are explained in detail in Chapter 4 of this textbook, where it is emphasized that equilibrium means $AI = DI$, and not $s = h$. My students must understand this perfectly, and it — and Figure 2 — must be presented and explained on every examination.

One more example might be useful. If $AI > DI$, and thus is a non-equilibrium situation, then price falls in order to make the excessive inventories more attractive to consumers. The sequence is $AI > DI \rightarrow p \downarrow \rightarrow h > s \rightarrow AI \downarrow \rightarrow AI = DI$. What about when $DI > AI$? Then the adjustment procedure is equally simple: $DI > AI \rightarrow p \uparrow \rightarrow s > h \rightarrow AI \uparrow \rightarrow DI = AI$. Pedagogically, dealing with this issue as is done here makes more sense than writing a differential equation in which the degree of the equation is arbitrary. (Readers can now explain why we do NOT have an equilibrium if we have $s = h$, but $AI \neq DI$.)

2. Remembering the Stern Review of Climate Change: An Unfriendly Note

"It's absolutely unbelievable what's going on. We're living in just about the most dishonest time in the history of man."

— Stephen Jarislowsky

In preparing this textbook, several widely publicized topics were deliberately kept to a minimum. Perhaps the most notable was the *Stern Review on the Economics of Climate Change*. On several occasions I heard this document mentioned when I was visiting professor at the Asian Institute of Technology (Bangkok), and since I had encountered some information about it earlier, I took the liberty of informing my students that it was not to be discussed in my classroom in the course of what might be described as normal business, and under no circumstances was I prepared to regard it as a suitable topic of conversation during my precious leisure hours.

Several years ago Lord (Nicholas) Stern (of Brentford) appeared in Stockholm in order to clarify for a large audience at the Royal Institute of Technology the conclusions reached in his widely advertised analysis of the cost of preventing various environmental calamities. I was not invited to attend this 'gig', because it was probably believed by the organizers that had I been present, there might have been what is sometimes called an 'incident'. This was definitely not certain, because I would have been

willing to exercise a maximum of self-restraint to avoid informing Lord
Stern that many leading economic theorists regard his work on this topic as
scientifically meaningless.

"Meaningless", and in the light of the finished product known as the
'Stern Review', pedagogically insignificant. To my way of thinking, that
document — or at least the very small portion that I forced myself to
examine — is an insult to economics teachers like myself, and also to students who require our guidance where their reading materials are concerned,
but who unfortunately may have been informed by advisors, supervisors,
mentors and itinerant charlatans that the Stern Review is state-of-the-art
knowledge.

There is also this matter of innocent bystanders who pay taxes in order
for — among other things — systematic and professional attempts to be
made to ascertain the extent and mechanics of e.g., atmospheric pollution,
and if necessary, to suggest or devise efficient programs for reducing the
dangers it may pose to their health and/or their standard of living. I also
want to emphasize that the few analytical fragments I perused of the Stern
Review reminded me of the esoteric contributions that once filled a journal
called *The Review of Economic Studies*, which played an important part in
my graduate education at the University of Stockholm, and which contained
articles that later appeared on my reading list when I taught mathematical
economics in Uppsala and Australia. Eventually I smartened up, however,
and I can proudly state that I have not read nor taught anything specifically
from that or similar journals for more than 20 years, regardless of what
close or distant colleagues feel is pedagogically appropriate.

Something else should be pointed out before we proceed further in this
section. The team working on the Stern Review came to 23 men and women,
as well as many consultants. In other words, it may have been true that
anyone who had anything positive to say about Lord Stern's approach to
this topic — regardless of how inappropriate or loony-tune it might have
been — was welcome to contribute.

Graded "F" for Failure

After scrutinizing a few pages of the Stern Review, the first conclusion
I formed was that if it is true that we are now living in the most dishonest

period in modern times, as a Canadian billionaire once claimed, then the so-called research done by Stern and his team might best be described as an outrageous waste of money — money that could have been used to provide young economics students with a portion of the education they both need and deserve. This is perhaps a reason for Professor Richard Tol saying that if the Stern Review had been presented to him as a Master's thesis, he would have graded it "F" (for failure).

But please take note that I am not criticizing nor denigrating climate scientists, economists, journalists, or anybody else who takes the position that global warming may be a clear and present, or nearly-present, danger. For instance, I am completely uninterested in the fact that in a recent poll conducted by the Pew Research Center in the United States, climate warming turned out to be in 20th place among 20 anxieties that persons who responded to the poll were asked to rank. As far as I am concerned, more can and should be done by the governments of many countries to improve the environment. And why not? When President Franklin D. Roosevelt was mobilizing the American people for the most brilliant war effort in human history, he made it clear that everyone had to participate: women, men, and even some children. The same is true here, and more importantly, invest-ments in environmental improvements that are financed by taxpayers, with or without their approval, make more economic sense than investing in stupid wars on the other side of the world.

There are no surprises in the Stern Review, at least for my good self. Given the limitations of economic theory, as well as the publications and other career achievements of Professor Stern, there was only one logical approach to this new project bearing his name, which was to assemble or borrow from an analytical model that would permit an overblown and ram-bling discussion of cost-benefit analysis to take place over 700 pages. The implicit or explicit model apparently chosen by Professor Stern, and which might have been chosen by myself, or for that matter by party animals pos-turing themselves and courting attention at the bar of an Uppsala University student club, could only be the one published by Frank Ramsey (1928) in his famous article "A Mathematical Theory of Saving".

What is the reasoning behind this choice? I do not know exactly how much of the Ramsey model Lord Stern has borrowed, nor do I want to know, but I am aware that a key result from Ramsey's (1928) article appears in the Stern Review. Moreover, the basic focus of such a model can be easily explained to a majority of economics undergraduates, and especially those pursuing their education in the kind of university or institution where environmental studies have almost the same status as theology.

Equally as important, a Ramsey-type model can be *summarily* enriched in case someone visiting a lecture on global warming decides to play ego games with the lecturer by impugning the logic of Ramsey's construction. For instance, a skilful economics lecturer with a modest amount of integral calculus at his or her disposal should be able to convince a drowsy audience of environmental 'tag-alongs' that this approach provides a useful insight into whether benefits from investments undertaken to avoid future environmental deterioration outweigh the present costs of these investments.

This kind of model often reduces to determining the discount rate (or rates) that should be used to evaluate future benefits. A discount rate says something about *substitution* between present and future benefits and costs. For instance, Professor (and Nobel Laureate) Kenneth Arrow and his collaborators have recommended that consideration should be given to "whether the benefits of climate policies, which can last for centuries, outweigh the costs, many of which are borne today". Arrow and his associates stress that dealing with this matter involves the rate at which future benefits are discounted, and the thing that caught my attention was his mentioning that an analysis of this sort might be useful for considering the disposal of nuclear waste.

I had a short talk with Professor Arrow when he was in Sweden to receive his Nobel Prize in Economics, and he gave me some invaluable advice on the book I should use to teach mathematical economics. But even so, I feel it necessary to say that *marginal adjustments of discount rates of the kind discussed in the Arrow article have little or no relevance when dealing with any major issue having to do with nuclear energy!*

I still remember teaching — without much enthusiasm I might add — the theory and use of discount rates, but where nuclear energy is concerned, the important answers are in the nuclear past and present, and not in fantasies and estimates about the future. They are in, e.g., the economic and political history of nuclear and other energy investments in countries like Sweden, and what has happened and will happen in China. Evaluating benefits from nuclear energy primarily involves a thorough understanding of the likely evolution of nuclear technology and the availability of nuclear fuel, although it seems that obtaining this understanding is as difficult as making sense of the Stern project and the strange belief that it was worth launching.

The Swedish Energy Minister has proposed altering Sweden's energy architecture in such a way as to obtain enough renewables (i.e., wind and solar) to replace all of Sweden's nuclear assets. Fortunately we know something about how this absurd strategy would play out without bringing discount rates into the discussion, because the cost of electricity in Denmark — which is generally considered the 'promised land' of wind-based energy — is the highest in Europe, while the cost of electricity in Sweden — which has a large nuclear commitment — was among the lowest, and would be *the* lowest if Swedish governments avoided the advice of quacks. It is also very likely that the nuclear waste problem will be solved in a decade or less by the design and production of integral fast reactors that reprocess all waste within the nuclear cycle. The multi-billionaire Bill Gates appears to be financing the construction of one of these devices.

Occasionally I find it necessary to engage in impromptu debates about nuclear energy, which in order to dominate, I find myself saying that nuclear energy might be capable of playing an important role in any program designed to reduce emissions of carbon dioxide (CO_2). In addition, because of its reliability, I like to claim that the use of nuclear energy increases the value of renewable facilities.

My argument in this matter usually turns on my knowledge of the situation with greenhouse gases in Sweden and France, which unfortunately means that this hypothesis may not be of interest to energy experts in many other countries. Frankly I do not care, because the teaching of energy economics is so pathetic in most countries that I lack the patience to provide the necessary explanations. I do however make a point of repeating as often as possible that when Sweden rushed into a nuclear commitment, it was possible to assemble in just under 14 years an inventory of 12 reactors. Furthermore, this was done by a workforce with only a modest background in nuclear matters. It is the presence of these reactors and the large hydro-electric sector that explains why — until the curse of electric deregulation came about — Sweden had one of the lowest electricity prices in the world where the supply to households was concerned, and probably the lowest price in the industrial sector.

As to be expected, it is always possible to find ladies and gentlemen who are thrilled to the marrow in their bones with the waffle presented in the Stern Review. Take for example Joan Ruddock, former Parliamentary Under-Secretary of State for Energy and Climate Change in the UK. She dismissed the criticisms of Stern by distinguished economists like Professors Richard Tol, Partha Dasgupta and Martin Weitzman because — according to her — these economists suffer from "a fundamental misunderstanding of the role of formal, highly aggregated economic modelling in evaluating a policy issue."

I am afraid that I must reject that point of view, my excellent Ms. Ruddock: they do not misunderstand! You and I are guilty of that shortcoming! You because your education in economic theory is inadequate, and me because although I am a humble teacher and academic energy economist, the only interest I have or need in what you mistakenly call "aggregated economic modelling" is the slender amount necessary for me to write and publish items like this book and any other energy economics textbooks that I may decide to undertake in the future.

1. A Disobliging Comment on Electric Deregulation

In the summer of 2001, a few months before the 9/11 attacks on the World Trade Center and the Pentagon, I was invited to Hong Kong as a visiting professor and university fellow for the purpose of lecturing on electric regulation and deregulation. My visit was at least partially sponsored by one of the foremost (electric) power companies in Hong Kong, and what they wanted me to do was to inform university teachers, journalists, students, and anybody else I came into contact with, that electric deregulation was a crazy and unworkable concept that would bring misery into the lives of many consumers of electricity.

This needs some explanation, by which I mean why would a large power company want me to travel halfway across the world to ridicule electric deregulation? The answer is that the directors of that company knew that electric deregulation was a lost cause, or to quote Jean-Paul Sartre, "a fire without a tomorrow". In Sweden though, as in many other countries, it did not make any difference to the directors of power companies what it was, because they were primarily concerned with putting themselves in a position where they could take the money and run. For instance, deregulation made it possible for one Swedish power company to shift much of its attention to Germany, where it specializes in making absurd claims about its program for a "green" future.

Things are different in China. A deregulation failure in Hong Kong means something very different from a failure in California or Sweden. In Sweden and many other countries, deregulation miscarriages result in disparaging articles in newspapers or business magazines, but the poor consumers are left to gnash their teeth and curse. On the other hand, in Hong Kong, somebody important might confront the executives responsible for the misfortune, demand an explanation, talk to them in a manner that sergeants in the American Army once talked to recruits, and perhaps ask to examine some bookkeeping and other paperwork. I do not think it necessary to tell you how this could turn out, because the Chinese government does not make a practice of applauding incompetence.

In a privatization conference held at Södertörn University in Stockholm, where I presented a more dramatic version of this brief section, a very intelligent young Finnish woman asked me what I meant by deregulation failure, to which I replied it was obvious. Deregulation failure is when consumers and firms are promised lower prices, but instead prices increase. This can be put another way: electric deregulation may succeed in seminar rooms and conferences, but if its record means anything, it has failed dismally in the real world. An extensive analysis of this topic can be found in the book by Professor Lev S. Belyaev (2011).

In case you have forgotten, in Southern California electric deregulation led to the wholesale electricity price increasing by 533% in about 8 months. One of the high points of that fiasco was the California state government paying billions of dollars to firms generating electricity, with some of these firms called "out-of-state criminals" by California Governor Gray Davis, because they gamed the system by pretending that they could not supply more electricity.

Electric deregulation failed in South Australia. It failed in almost every state in the United States of America where it was attempted, and in my former home state, Illinois, a state official — Kimery Vories — reported that deregulation resulted in the price of electricity increasing by 40%, all at once. As already noted, it failed here in Sweden too, and as I told colleagues and students in Bangkok a few years ago, electric deregulation in Sweden seems to mean that the largest power company in Scandinavia has been awarded a gold-plated licence to make fools of the consumers of electricity.

Electric deregulation very definitely failed in Norway, although in that country some comedy was added to the process. Consider the following exchange of comments:

> *"What we've seen this winter is that the deregulated market works for (electric) companies, but not for consumers."*
> — *Hallgeir Langeland (Norwegian Parliament, 2003)*

> *"I have to expect to be a scapegoat (for deregulation failure), but people know they can't blame me for the policy."*
> — *Einar Steensnaes (Norwegian Energy Minister, 2003)*

My comment on this *repartee* is that nobody with his or her head screwed on right would blame you, Einar. The people to be blamed are the outraged electricity consumers who sent you letters filled with insults and undiplomatic language, and who questioned your intelligence and your qualifications for the position you held when the 'spot' price for electricity in your country escalated by a factor of almost 30 during one brief period.

I remember giving one of my sermons against deregulation in Lima (Peru), and fortunately I got out of that country just in time, because when they initiated that goofy experiment some shots were fired, as was the case in the Dominican Republic. Deregulation failed in Brazil too, where Lutz Trevesso, CEO of a large power company, said that deregulation would create more problems than it solved.

You have heard what *I* think of deregulation, so now let us turn to some other opinions. U.S. Senator Ernest Hollings brusquely abandoned the deregulation sinners who had seduced him into the ways of 'liberalization', and began to call himself a "born-again regulator". Another U.S. Senator, Byron Dorgan, was more explicit. He put it this way: "I've had a belly full of being restructured and deregulated, only to find out that everybody else gets rich and the rest of the people lose their shirts!" (*Financial Times*, April 22, 2003). A headline in the *New York Times* (July 15, 1998) read as follows: "Deregulation fosters turmoil in power markets!" Personally I am very fond of Governor Gray Davis' judgment: "At the mercy of forces that show no mercy." Governor Gary Locke of Washington (State) offered an important thought on the bad news resulting from the deregulation travesty, concluding that since the government caused the suffering, it was up to them to cure it. And last but not least, U.S. Congressman Peter DeFazio put it this way: "Why do we need to go through such a radical, risk-taking experiment?" Congressman DeFazio answered his own question by saying "it's because there are people who are going to make millions or billions!"

And finally, when I first began to study regulation and deregulation, the leading scholar in the field was Professor Alfred Kahn. Once the electric deregulation failures began, he made the following statement: "I am worried about the uniqueness of electricity markets. I've always been uncertain about eliminating vertical integration. It may be one industry in which it works reasonably well."

I am not worried at all, ladies and gentlemen, because my interest in this topic is not vertical integration. It is the supreme importance of electricity as compared to certain other energy options. For example, there may be passable substitutes for natural gas, but — everything considered — there are no substitutes for a large supply of inexpensive and reliable electricity, especially if we are considering modern and civilized countries whose citizens and/or voters are concerned about their futures.

I asked the young Finnish economist mentioned above what she prefers: the *choice* that is supposedly one of the rewards of electric deregulation, and which a self-proclaimed expert on this topic like Professor David Newbery thinks is especially important, OR *sustainably low electricity prices*? Unfortunately, she was unable to provide a straightforward answer, informing me instead that she did not know what I was talking about. If I had been carrying aspirin, I would have reached for it, because once again I was forced to realize that I still have a great deal to do where clarifying this issue is concerned.

Finishing Touches

> *"Energy is the major factor of social well-being."*
>
> — *Professor Earl Cook*

So the countdown is finally over, and I hope that readers are able to congratulate themselves for being able to read this chapter without consulting their notes or asking for help with the mathematics, because there were no mathematics. But there are still quite a few details that serious readers should digest before they inform friends and neighbors that they are on the highway to energy economics stardom. I want you to reach that objective, and so I will take the liberty of advising readers to keep rereading this chapter until they learn it perfectly, just as I intend to do.

Frankly I am amazed at how many mistakes are made by persons attempting to convince colleagues and competitors about various 'rights and wrongs' in energy economics, and in the forums to which I contribute, I am also amazed at how easy it is for persons with a rudimentary knowledge of this subject to deal with individuals who are without that background, regardless of the education or intelligence of the latter.

Please remember that there is a rhythm to energy economics, and getting that rhythm — that swing — puts you ahead of the game. As for the mechanics of obtaining that wonderful rhythm, the following advice might help. There are many pages in this book, and I suggest that you examine some of them daily until you are *absolutely* fluent where they are concerned, and just as important, you are prepared to demonstrate your fluency at the slightest provocation. If you want to demonstrate it in class or seminar rooms, or at conferences or meetings, and you are not standing in front of the room, please remember that Albert Einstein always sat in the first row at the seminars he attended.

Where mistakes are concerned, please believe that I have made at least my share, but at the same time I enjoy remembering and mentioning that electricity market 'reform' (i.e., restructuring or deregulating) is a subject in which I am largely 'mistake-free'. In addition, a mistake that I do NOT want to make is to finish this discussion without indicating the components of world energy production at around the end of 2013. These are measured in British Thermal Units (Btu), which means that we avoid comparing barrels of oil with, e.g., tonnes of coal (see Table 1).

The global consumption of fossil fuels (oil, natural gas, coal) increased from 26,200 million barrels of oil equivalent (mboe) in 1965 to 80,000 mboe in 2012, and according to the Energy Information Agency (EIA) of the U.S. Department of Energy, by 2035 oil demand is projected to increase by over

Table 1. World and U.S. Energy Production

	World Production (2013)	U.S. Production (2013)
Crude oil	140,000,000 Btu	30,000,000 Btu
Coal	133,000,000 Btu	18,000,000 Btu
Natural gas	103,000,000 Btu	27,000,000 Btu
Hydroelectricity	28,000,000 Btu	2,000,000 Btu
Nuclear	22,000,000 Btu	7,000,000 Btu
Wind	5,000,000 Btu	1,000,000 Btu
Biofuels	2,000,000 Btu	0.900,000 Btu
Solar	1,000,000 Btu	0.015,000 Btu
Geothermal	0.290,000 Btu	0.081,000 Btu

30%, natural gas by 53% and coal by 50%. I would not be surprised if these forecasts are wide of the mark, but that does not make a difference. They are most likely in the 'ball park', and they are definitely better than a lot of the forecasts that we have to entertain.

We can also take a quick look at what is taking place in the nuclear world at the present time. The Chinese have 30 nuclear plants of 1,000 MW under construction, and over 100 are planned. Russia has 10 plants under construction, and 21 more are planned. According to the nuclear engineer and executive Malcolm Rawlingson, concrete is being or has been poured for 4 new reactors in the U.S., but there would be many more if there was no belief in that country that shale natural gas could solve their energy problems — "could", not *will*.

Apparently there are about 67 large power reactors being constructed around the world at the present time, and according to my research the average length of life of this new equipment should be at least 55 years, although it could easily be more if the people operating them do not fall asleep. But regardless of what it turns out to be, when these new plants are eventually consigned to the scrap heap — probably at some point in the closing decades of this century — it could happen that fossil fuels will be on their last legs, and in these circumstances I sincerely doubt whether there will be many engineers or managers who inform voters and politicians that windmills and solar panels alone can provide the base load power now being provided by coal and perhaps gas. Instead, by then it should be clear to voters and politicians alike that nuclear + renewables is an arrangement that is probably optimal.

There are several other things in Table 1 that deserve mention. In the U.S., natural gas has surprised many of us who remember the pessimistic predictions of Alan Greenspan, perhaps the best known of the U.S. Central Bank (= Federal Reserve) directors in recent decades. That gentleman once suggested that natural gas was a lost cause in the U.S. unless drastic measures were taken to obtain foreign supplies. Of course, the spectacular arrival of improved shale gas technology appears to have saved the day, although it is too early to attempt to forecast the 'denouement' of what is often called 'the shale revolution'.

The renewables in Table 1 are wind, solar and biofuels. I usually make an attempt to stay away from these, but according to the latest World Energy Outlook of the IEA, they will account for about half the increase in global power generation by 2035, with the result that China will be generating more than the U.S., Japan and the EU combined. The IEA also claims that renewables will account for over 30% of the global power mix by 2036, with wind and solar making up 45% of the expected increase in renewables.

The IEA has some other optimistic things to say about the growth of renewables in the global power scene, but I do not expect much of them to come true, although I want to repeat that a combination of nuclear and renewables might provide many benefits, assuming that governments do the right thing where taxes and subsidies are concerned. One of the things that must be remembered is that the IEA wants to stay on the right side of what they consider to be global opinion, which in turn means supporting renewables and, for the most part, ignoring nuclear power. IEA experts also predict that global energy demand will tend to shift to Asia, with India and Southeast Asian countries leading energy consumption. The word "leading" may be out of place here.

It is not easy to obtain a starring role when the subject is energy economics, and if you do obtain one it may not be easy to continue to live up to the expectations of your audiences, but there are usually things that they will know that you do not know, regardless of whom 'they' happen to be. For instance, despite my textbooks and several hundred articles, I do not know as much about oil pipelines as I would like to know, and it is only recently that I learned about the pattern of oil prices in Canada.

In this matter of oil pipelines, I was compelled to express surprise when I was told that plans are being made to transport Canadian crude thousands of miles from Alberta to, e.g., the U.S. Gulf Coast so that it has access to the international market, where its price is higher than in Canada. In my work I have always mentioned the West Texas Intermediate (WTI) and the Brent prices of oil, but while I had heard of other 'benchmarks', I generally ignored them. It is only recently that I heard of *Western Canada Select*, which is the regional benchmark for low-quality heavy oil, and was selling for \$63/b as compared to \$95/b for WTI, and sour crude from Mexico that was selling for \$86/b. More interestingly, I have just found out that exports of U.S.-produced crude are largely banned, which as an American citizen I think is marvelous, and recently there has been some discussion about forbidding the export from U.S. ports of Canadian oil that flows through pipelines in the U.S. I call this kind of behavior self-preservation, which is often called the first law of nature.

Another troubling issue attracting considerable attention just now has to do with utilizing solar radiation to obtain electricity. Generally, I have been skeptical about photovoltaics (PV), and therefore was not surprised to hear about a few spectacular bankruptcies of firms manufacturing solar cells and photovoltaic arrays. (What I did not know was that these bankruptcies were due to American firms not being able to compete with Chinese firms.)

My principal interest here, though, is the absurd intention of the Merkel government to replace nuclear energy with things like solar and windmills, but it is impossible to deny that there should be many places in the U.S. and elsewhere where the utilization of solar and wind makes economic sense. Furthermore, since Professor David Goodstein speaks well of this resource, I intend to be as positive as possible in the future, though not in the case of, e.g., Sweden. What I eventually expect to find is that both solar and wind have a great deal to offer, although not as much as people like Germany's chancellor and her foot soldiers ostensibly believe.

There is something refreshingly simple about PV, in that the electrical energy from cells/panels on the roof of a house provide direct (and not the usual alternating) current for appliances like the kitchen stove, washing machine, clothes dryer and some light fixtures, and with the aid of small converters that can turn direct current into alternating current, electricity could be supplied for the refrigerator. A recent governor of California thought that everyone should be in the market for these solar cells, since California is famous for its sunshine.

But what about solar power stations, or 'concentrated power' as they are called? Hopefully, in a few months, a solar power facility in the Mojave Desert (45 miles south of Las Vegas) named *Ivanpah* will become operational, and will sell enough power to California utilities to provide electricity to 140,000 homes. The story here is that the hot rays of the Nevada sun will boil water, which creates steam that will run turbines. The cost of that facility is reportedly 2.2 billion dollars, and according to my mathematics, subsidies of one sort or another must have been involved, because 2.2 billion dollars for about 400 megawatts of capacity may not be an optimal investment at this point in time. Rumor has it that concentrating solar power in installations like *Ivanpah* has not worked as well as expected, and PV is a better option.

According to Professor Mark Jaccard, who apparently has consulted for governments in many countries on this issue, California was a tragedy on the road to successful reform. The successful reform — if in fact it did arrive — must have made an appearance after I lost interest in this topic because, as emphasized in my lectures on this subject in Hong Kong, electric deregulation (restructuring) was an unadulterated curse. On the other hand, what Jaccard *specifically* indicates as success for the deregulation 'experiment', as it was often labelled, I unambiguously call failure, by which I mean the outcomes in England and Norway and at least two 'regions' in Australia. In fact, as I point out whenever I get the opportunity, the failure in Norway had both tragic and comical sides, where by the former I meant the reliance on Nordpool — the Nordic Electricity Exchange — and by the latter, the abusive mail sent by some citizens in that rich country to the ladies and gentlemen they thought responsible for pretending that something that was not broken needed to be fixed.

I am also completely *certain* about the development of nuclear energy. Nuclear energy will be accepted by every country whose citizens prefer low-cost to high-cost electricity. I also strongly believe that OPEC is the most important factor influencing the price of oil. I do not know, however, whether shale gas is going to be as important as certain opinion makers (or opinion manipulators) say, although I can confess that I would like to know, and perhaps some day I will find out. The problem for me at this point in time is understanding its success in North America, and at the same time its lack of success elsewhere.

Before continuing on the above line, I want to present more numbers. They are of course about natural gas, because that is currently the most discussed topic in energy economics. Total global gas reserves are 187,300 billion cubic meters = 187.3×10^{12} cubic meters. As for the countries holding these reserves, we have Iran (18%), Russia (18%), Qatar (13%), Turkmenistan (9.3%) and U.S. (4.5%). Thus, the U.S. is not the global powerhouse that it is sometimes claimed to be, at least where reserves are concerned. The strength of the U.S. is in production, and here we have U.S. (20%), Russia (18%), Iran (5%), Qatar (5%) and Canada (5%). Finally, for consumption we have U.S. (20%), Russia (13%), Iran (5%), China (4%) and Japan (3%).

Something that I once wanted to know is why aren't those famous laws of arbitrage providing the kind of outcomes alluded to in your Economics 101 textbook, since the Brent oil price at the present time is much higher than the West Texas Intermediate (WTI) oil price. (By arbitrage I mean buying in a 'cheap' market and selling in a 'dear' market, and as a result bringing the two prices closer together.) There would not be much of a problem here if instead of Texas and Europe we were talking about New York and Philadelphia, and instead of oil the item under discussion was designer clothes. In that situation, price differences might be quickly eradicated.

A gentleman whose fear of paparazzi has led to him calling himself "Old Trader" in the forum *Seeking Alpha*, provided a useful introduction to this issue, but after examining what he had to say, I realized that arbitrage (or its absence) is not as important as knowing what percentage of oil buyers are now buying at the Brent price. Personally, I am willing to believe that enough oil importers are paying the Brent price to provide OPEC with an aggregate income of a trillion dollars a year.

As alluded to just above, *ceteris paribus*, geography can explain the absence of arbitrage, at least in the short run. Moreover, as I never get tired of saying, if a large percentage of buyers are paying the Brent price, then the decision to intervene militarily in the Libyan *impasse* instead of with "diplomacy and dialogue" — as Pope Benedict suggested — must be one of the most irrational actions since December 11, 1941, when Adolf Hitler declared war on the United States of America.

The WTI (West Texas Intermediate) and Brent oil prices are still where many of us who study this price believed — and hoped — that it would not be until the global macroeconomy is fully recovered. Please allow me to admit that I have some problems understanding why, in light of all the good news about oil and gas that we receive from 'experts' in the U.S., the price of WTI this morning was $94/b, while Brent oil was $108/b. I hope that the people in oil-importing countries who were responsible for the *faux pas* in Libya and elsewhere now realize that the present oil price is a very bad piece of news from a macroeconomic point of view, because these wars have taught the very smart persons who manage OPEC how to utilize the strength of that organization.

There are a few equations in this book, and in preparing themselves to deal with them readers should think about studying and learning as much as they can of intermediate microeconomic theory, which in most coun- tries is still the course that comes after the introductory course or courses in economics. I would like to testify that there are marvellous intermediate books in economics now available, but readers should be aware that some are better than others. The same is true for energy economics. But no matter which energy economics book you have on your 'night stand' or under your pillow, the book *Out of Gas: The End of the Age of Oil* by David L. Goodstein is a must, though perhaps not for obtaining the precise date and hour when the Age of Oil will conclude. (As for an advanced microeco- nomics book, *Microeconomic Theory* by James M. Henderson and Richard E. Quandt (1980) is the book I used in Sweden, Senegal and Australia, and it is unbeatable. Strangely enough, I recommended this book to two graduate students at my university, and they looked at me as if I had lost the plot — which of course I had, because many students and teachers are no longer interested in the best books.)

As I have tried to make clear, this is the third textbook on energy eco- nomics that I have written, although I have written other books that contain energy economics. I have also lectured on energy economics in many coun- tries and many parts of the world. There is no doubt in my mind that this book should contain comprehensive discussions of renewables and alterna- tives, but then it would be too long. More importantly, I feel that I am not qualified to present this topic the way that it should be presented. If you, the reader, feel that you are, please begin as soon as possible, and keep working on it until you obtain what you and I need.

What I have attempted to provide here is a book that can be absorbed in one semester. If it is still too long, then the reader should concentrate on the first three chapters — the short course — and then turn to this chapter and perhaps one of the others, which should be learned perfectly. Ten years ago I would have suggested the chapter on oil, but today I favor the one on nuclear energy. Having completed this book, I intend to devote most of my time now to nuclear energy and its future. By that I mean the economics (and not the physics) of nuclear. True, I failed elementary physics (and elementary mathematics) twice during my first year in engineering school, but fortunately I think that at last I understand nuclear economics. If I ever encounter the Dean of Engineering at Illinois Institute of Technology in Chicago, I would thank him from the bottom of my heart, because his decision to expel me meant that I had to stop playing and take care of business.

A question now should be asked: what topic not treated in this book is going to be very important for energy economics in the future — or perhaps I should just say economics? I tried to answer that question more than 30 years ago, and the topic I gave was population. That is still the answer. As compared to the future, the answer for the present is education, especially secondary school and primary school education. I hope that our political masters remember this the next time they consider spending billions of dollars in order to participate in a stupid war in some remote part of the world.

A world population of 6 or 7 billion people is a completely different proposition from one of 10 or 12 billion. These numbers require some explanation. United Nations demographers somehow came to the conclusion that world population at the end of the 20th century would be 6 or 7 billion people, and would stabilize at that level. The number is already at 7 billion now, and current UN estimates are 9 or 9.5 billion by mid-century. Readers are invited to give this some thought and offer their own estimates, along with some comments about a future stabilization.

While doing research for my book *Scarcity, Energy, and Economic Progress* (1977), I paid some attention to a number of estimates of the future population. One very well-known Swedish professor stated that 20 billion people might fit comfortably on planet earth. I did not question this at the time, because although that estimate seemed absurd, it would not have benefited my career to have suggested that he was probably in error. For a while though, I was prone to quote Dr. Joseph Spengler, who calculated that in the very long run we should expect a population of 15 billion — if we were lucky.

A well-known scientific celebrity, Herman Kahn, estimated that between 20 and 30 billion people might find a place on this earth, with some of them enjoying a fairly high standard of bourgeois comfort. In case you do not know, numbers were not particularly important for Dr. Kahn, because he also believed that nuclear war was quite thinkable, although he apparently had difficulty convincing the likely victims of a nuclear clash that he knew what he was talking about. It has also been claimed that Dr. Kahn was the model for Dr. Strangelove in the film *Dr. Strangelove or: How I Learned to Stop Worrying and Love the Bomb*.

Other estimates could be cited, some of them even more grotesque, but there is a lesson here. In an ideal world, especially the world that has come to be known as Economics 101, both decision makers and readers of this book would start thinking about how the world in general, and their countries in particular, should look at mid-century, or even later. Personally, I would love to join these decision makers and readers, but I have to confess that where matters of this sort are concerned, I hardly know where to begin, except to say that energy and education might be more important than ever.

I can, however, mention Harold E. Goeller and Alvin M. Weinberg, who, when working at the Oak Ridge Laboratory, published some articles on what they called an 'Age of Substitutability', which was largely based on biological resources which they thought of as inexhaustible (which may or may not be true), and also a few plentiful minerals. Interestingly enough, one of my students came to the same conclusion and casually mentioned it in one of my lectures, but I brushed her off, because at that time I disliked thinking about any aspects of any part of economics that did not have something to do with mathematics or econometrics or some other kind of abstract presentation that gave the impression that economics was a subdivision of physics. If she happens to pick up this book now or at some point in the future, I would like to apologize, and also to inform her and other students of economics that the very great power of economic theory is only marginally the result of being able to write a few equations. Those equations are perhaps necessary, but definitely not sufficient.

I close with what I think of as a few words for the wise. At some point during this century, in a world in which there will be an excess of human beings but possibly a shortage of natural resources, it is very likely that only the strong or smart or both — or those who are protected by the strong or smart or both — are going to feel really comfortable, so it might be a good idea to do some serious thinking about the topics taken up in this book.

Bibliography

Adelman, Morris A. and Martin B. Zimmerman (1974). "Prices and profits in petrochemicals: an appraisal of investment by less developed countries." *Journal of Industrial Economics*, 22(4): 245–254.

Angelier, Jean-Pierre (1994). *Le Gaz Naturel*. Paris: Economica.

Baltscheffsky, S. (1997). "Världen samlas för att kyla klotet." *Svenska-Dagbladet*.

Banks, Ferdinand E. (2008). "The sum of many hopes and fears about the energy resources of the Middle East." Lecture given at the Ecole Normale Superieure (Paris), May 20, 2008.

——— (2007). *The Political Economy of World Energy: An Introductory Textbook*. Singapore and New York: World Scientific.

——— (2004). "A faith-based approach to global warming." *Energy and Environment*, 15(5): 837–852.

——— (2001). *Global Finance and Financial Markets*. London, New York and Singapore: World Scientific.

——— (2000). *Energy Economics: A Modern Introduction*. Boston, Dordrecht and London: Kluwer Academic Publishers.

——— (2000). "The Kyoto negotiations on climate change: an economic perspective." *Energy Sources*, 22(6): 481–496.

——— (1987). *The Political Economy of Natural Gas*. London, Sydney and New York: Croom Helm.

——— (1985). *The Political Economy of Coal*. Boston: Lexington Books.

——— (1980). *The Political Economy of Oil*. Lexington and Toronto: D.C. Heath.

——— (1977). *Scarcity, Energy, and Economic Progress*. Lexington (Massachusetts) and Toronto: Lexington Books.

Basosi, Riccardo (2009). "Energy growth, efficiency and complexity." Conference Paper; Center for the Study of Complex Systems, University of Siena (Italy).

Bauquis, Pierre René (2003). "Reappraisal of energy supply-demand in 2030 shows big role for fossil fuels, nuclear but not for non-nuclear renewables." *Oil and Gas Journal* (February 17).

Belin, Hughes (2010). "A heavy burden on EU policymakers." *European Energy Review* (April 14).

Bellman, Richard (1957). *Dynamic Programming.* Princeton: Princeton University Press.

Belyaev, Lev S. (2011). *Electricity Market Reforms: Economics and Policy Challenges.* New York and Heidelberg: Springer.

Bloomberg (2012). "U.S. cuts estimate for Marcellus Shale Gas Reserves by 66%." *Bloomberg.Com News* (January 24).

Carr, Donald E. (1976). *Energy & the Earth Machine.* London: Abacus.

Casazza, Jack A. (2001). "Pick your poison." *Public Utilities Fortnightly* (March 1).

Chenery, Hollis and Paul Clarke (1962). *Inter-industry Economics.* New York: Wiley.

Chevalier, Jean-Marie (2009). *The New Energy Crisis: Climate, Economics and Geopolitics.* London: Palgrave MacMillan (with CGEMP, Paris-Dauphine).

Clarren, Rebecca (2013). "Fracking is a feminist issue." *MS* (Spring).

Clemente, Jude (2012). "Shale gas in Europe: challenges and opportunities" (Stencil).

———— (2011). "Slaying the dragon? Three ways high oil prices impact China — and the world." Working paper, San Diego State University.

Cobb, Kurt (2011). "Can we believe everything we've heard about shale gas?" *OilPrice.Com* (June 17).

Cowen, Tyler (2011). *The Great Stagnation: How America Ate All the Low-Hanging Fruit of Modern History, Got Sick, and Will (Eventually) Feel Better.* Dutton.

Crooks, Ed (2013). "Gas export opponents ignite US shale debate." *Financial Times* (March 26).

Dieckhoener, Caroline (2012). "Stimulating security of supply effects of the Nabucco and South Stream projects for the European natural gas market." *The Energy Journal.*

Dittmar, Michael (2011). "The future of nuclear energy." Conference Paper; Institute of Particle Physics, Zurich, Switzerland.

Dorigoni, Susanna and Sergio Portatadino (2009). "When competition does not help the market." *Utilities Policy*, 17(3/4): 245–257.

Frank, Robert H. (2007). *Microeconomics and Behavior*. New York: McGraw-Hill.

Goodstein, David L. (2004). *Out of Gas: The End of the Age of Oil*. New York: Norton.

Hamilton, James (2009). "Causes and consequences of the oil shock of 2007–2008." *Brookings Papers on Economic Activity* (Spring): 215–261.

Harlinger, Hildegard (1975). "Neue modelle für die zukunft der menshheit." IFO Institut für Wirtschaftsforschung (Munich).

Henderson, James M. and Richard E. Quandt (1980). *Microeconomic Theory*. New York: McGraw Hill.

Holmes, Bob and Nicola Jones (2003). "Brace yourself for the end of cheap oil." *New Scientist* (August 2).

Hughes, J. David (2013). *Drill, Baby, Drill*. Post Carbon Institute.

Hunt, Tam (2011). "The peak oil catastrophe in waiting." *EnergyPulse* (March 10).

Höök, Mikael (2010). *Coal and Oil: The Dark Monarchs of Global Energy*. PhD Thesis, Uppsala: Global Energy Systems, Uppsala University.

Inajima, Tsuyoshi and Yuji Okada (2013). "Japan planning LNG to help out the price of imports." *Bloomberg News* (September 10).

Jacoby, Henry D., Frankie O'Sullivan, and Sergey Paltsev (2012). "The influence of shale gas on U.S. energy and environmental policy." *Economics of Energy & Environmental Policy*, 1(1): 37–51.

Kanter, James (2010). "Europe urged to share power across continent." *New York Times* (April 12).

Keane, Eamon (2010). "Shale gas hype" (Stencil).

Keller, Michael (2011). "Comment on article by Banks: 'A short but disobliging version of my recent lecture on electric deregulation'." *EnergyPulse* (September 10).

Kreps, David M. (1990). *Game Theory and Economic Modelling*. Oxford: Oxford University Press.

Kubursi, A.A. (1984). "Industrialisation: a Ruhr without water." In *Prospects for the World Oil Industry*, edited by Tim Niblock and Richard Lawless. London and Sydney: Croom Helm.

Loder, Asjylyn (2013). "Runs out fast." *Bloomberg Business Week* (October 20).

Lorec, Phillipe and Fabrice Noilhan (2006). "La stratégie gasiére de las Russie et L'Union Européenne." *Géoconomie* (No. 38).

Martin, J.-M. (1992). *Economie et Politique de L'energie*. Paris: Armand Colin.

Mazur, Karol (2012). "Economics of shale gas." *EnergyPulse* (October 3).

Michel, Jeffrey H. (2013). "Fracking und die Europäische Energieversorgung" (Stencil).

Ollevik, Nils-Olof (2007). "Och det som började så bra." *Svenska Dagbladet* (May 21).

Overbye, Thomas J. (2000). "Reengineering the electric grid." *American Scientist*, 88(3): 220–229.

Owen, Anthony David (1985). *The Economics of Uranium*. New York: Praeger.

Parfit, Michael (2005). "Future power." *National Geographic* (August).

Petersen, John (2013). "The great climate change fraud." *John Petersen's Instablog* (August 26).

Poundstone, William (1993). *Prisoner's Dilemma*. Oxford: Oxford University Press.

Ramsey, Frank (1928). "A mathematical theory of saving." *Economic Journal*, 38(152): 543–559.

Rhodes, Richard and Denis Beller (2000). "The need for nuclear power." *Foreign Affairs* (January–February).

Silverstein, Ken (2013). "High-profile scientists and environmentalists rally around nuclear energy." *EnergyBiz* (November 5).

Stackl, Erhard (2011). "Lessons from a dictator." Published in the *New York Times*, and reproduced in *The Observer (UK)*.

Stevens, Paul (2010). "The Shale Gas Revolution: Hype and Reality." A Chatham House Report (September).

Teece, David J. (1990). "Structure and organization in the natural gas industry." *The Energy Journal*, 11(3): 1–35.

Teitelbaum, Richard S. (1995). "Your last big play in oil." *Fortune* (October 30).

von Neumann, John and Oskar Morgenstern (1944). *Theory of Games and Economic Behavior*. Princeton: Princeton University Press.

Wallace, Charles P. (2003). "Power of the market." *Time* (March 3).

Watson, Jerry (2011). "Comment on article by Banks: 'A short but disobliging version of my recent lecture on electric deregulation'." *EnergyPulse* (September 10).

Watts, Price C. (2001). "The case against electricity deregulation." *Electricity Journal*, 14(4): 19–24.

Index